T0336036

A Selection of Image Understanding Techniques

This book offers a comprehensive introduction to seven commonly used image understanding techniques in modern information technology. Readers of various levels can find suitable techniques to solve their practical problems and discover the latest developments in these specific domains.

The techniques covered include camera model and calibration, stereo vision, generalized matching, scene analysis and semantic interpretation, multi-sensor image information fusion, content-based visual information retrieval, and understanding spatial-temporal behavior. The basic principles of each technique are reviewed, and current methods and practical examples are described. Research trends and recent results are also discussed, together with new technical developments.

This is an excellent read for those who do not have a subject background in image technology but need to use these techniques to complete specific tasks, and will prove useful for those pursuing further study in the relevant fields.

Yu-Jin Zhang is a tenured professor of image engineering at Tsinghua University, Beijing, China. He earned his PhD in applied science from the State University of Liège, Liège, Belgium. He was a post-doc fellow of Delft University of Technology, Delft, the Netherlands. He is also a CSIG and SPIE fellow. Dr. Zhang has published 56 books and more than 500 research papers.

A Selection of Image Understanding Techniques

From Fundamentals to Research Front

Yu-Jin Zhang

CRC Press
Taylor & Francis Group
Boca Raton London New York

CRC Press is an imprint of the
Taylor & Francis Group, an **informa** business

First edition published 2023
by CRC Press
6000 Broken Sound Parkway NW, Suite 300, Boca Raton, FL 33487-2742

and by CRC Press
4 Park Square, Milton Park, Abingdon, Oxon, OX14 4RN

CRC Press is an imprint of Taylor & Francis Group, LLC

ISBN: 978-1-032-42345-6 (hbk)
ISBN: 978-1-032-42350-0 (pbk)
ISBN: 978-1-003-36238-8 (ebk)

DOI: 10.1201/9781003362388

Typeset in Minion
by SPi Technologies India Pvt Ltd (Straive)

Contents

Preface

Image understanding technology is widely used in many applications today, and the discipline of image understanding has attracted much attention in the information community. This book introduces image understanding technology in a new way, by combining the contents, characteristics and styles of both textbook and monograph.

The book differs from pure textbooks on image understanding; it is also not a theoretical monograph on the technology. It offers comprehensive coverage of several commonly used image understanding techniques, from the essential concepts and basic principles through typical specific methods and practical techniques to research frontiers and the latest developments.

The book is suitable for readers who do not yet have a comprehensive grasp of image understanding, but need to use image understanding techniques to solve specific tasks. Readers should be able to quickly grasp the elementary knowledge to enable further study, find a suitable technique for solving a practical problem, and learn the latest development in their specific application domain.

This book does not attempt to cover all branches of image understanding technology, rather, it offers an in-depth discussion of seven selected image understanding techniques (based on image processing and analysis). These are: camera model and calibration, stereo vision, generalized matching, scene analysis and semantic interpretation, multi-sensor image information fusion, content-based visual information retrieval, and understanding spatial-temporal behavior. The presentations and discussions on each technique are self-contained.

The materials in this book are arranged in eight chapters with 47 sections and 120 subsections, with 164 figures, 41 tables, and 366 numbered equations. Over 200 key references are introduced and provided at the end of the book for further study.

Special thanks go to Taylor & Francis Group for the kind and professional assistance of their staff.

Last but not least, I am deeply indebted to my wife and my daughter for their encouragement, patience, support, tolerance, and understanding during the writing of this book.

Yu-Jin ZHANG
Department of Electronic Engineering
Tsinghua University, Beijing
The People's Republic of China
Homepage: http://oa.ee.tsinghua.edu.cn/zhangyujin/
Homepage: http://www.ee.tsinghua.edu.cn/zhangyujin/

CHAPTER 1

Introduction

IMAGES ARE ONE OF the main sources from which humans obtain information from the objective world. Images are obtained by observing the objective world with various different types of observation systems. They can act directly or indirectly on the human eye and then produce visual perception entities (Zhang 1996). There are many forms of images, including photos, drawings, animations, and even documents, etc. Videos are sequences of images that change regularly.

Humans use images to acquire and exploit meaning and information from the objective scene. People need not only to perceive the appearance and changes of the objective scene, but also to obtain characteristic information about objects of interest in it. They also need to use images to recognize the objective world, interpret the events in it, and determine their own actions to adapt, utilize, and transform the objective world. This involves the technology of image understanding.

Image understanding technology is a high-level image engineering technology, often based on low-level image-processing technology and middle-level image analysis technology.

The contents of each section of this chapter are arranged as follows:

Section 1.1 reviews and summarizes the development of image engineering over the past 25 years and more, and lists some statistical data from two special literature reviews to show the specific history and status quo of each branch of image engineering. Section 1.2 summarizes the research content of image understanding and its position within image engineering, discusses how it connects with and differs from computer vision and other related disciplines, and introduces some areas of application. Section 1.3 introduces the main points of Marr's visual computing theory that plays an important role in image understanding, as well as the improvements on Marr's theoretical framework. It also discusses the shortcomings of Marr's reconstruction theory, which not only aids understanding of the current situation of the whole field in general, but also promotes further in-depth research development. Section 1.4 explains the motivation, material selection, and structure, as well as how to use this book.

DOI: 10.1201/9781003362388-1

1.1 IMAGE ENGINEERING AND ITS DEVELOPMENT

An overview of the content of image engineering and its overall development is given first.

1.1.1 Basic Concepts and Overall Framework

Image is a physical form representing visual information. The human visual system is a typical system of observing and recognizing the world, and acquiring images from it. **Image technology** is the general name given to various technologies for treating images in a variety of ways to obtain the required information for human beings.

The comprehensive research and integrated application of image technology comes under the overall framework of image engineering (Zhang 1996). As we know, engineering is the general name of various disciplines that apply the principles of natural science to the industrial sector. Image engineering is an innovative subject that studies and applies the whole image field by using the principles of basic sciences such as mathematics and optics, combined with electronics technology, computer technology and technical experience accumulated in image application. In fact, the development and accumulation of image technology over the years have laid a solid foundation for the establishment of the discipline of image engineering, urgently required for a variety of applications (Zhang 1996, 2002, 2009a, 2015, 2018a, b).

Image engineering (IE) is very rich in content and widely used. It can be divided into three levels according to the degree of abstraction, research methods, operation objects and data volume, etc.: **image processing** (IP), **image analysis** (IA) and **image understanding** (IU) (see Figure 1.1). Image processing is a relatively low-level operation (Zhang 2017a) that occurs mainly at the level of image pixels, and the amount of data processed is very large. Image analysis is at the middle level (Zhang 2017b). Segmentation and feature extraction transform images originally described in pixels into simpler non-graphical descriptions. Image understanding mainly refers to high-level operations (symbolic operations): interpreting, judging, and making decisions based on more abstract descriptions (Zhang 2017c). Its operating process and methods have many similarities with human reasoning. Here, as the level of abstraction increases, the amount of data is gradually reduced. Specifically, the raw image data goes through a series of treatment processes and is gradually transformed into more organized and useful information. In this process, semantic information is continuously introduced, and the operation objects are gradually changed.

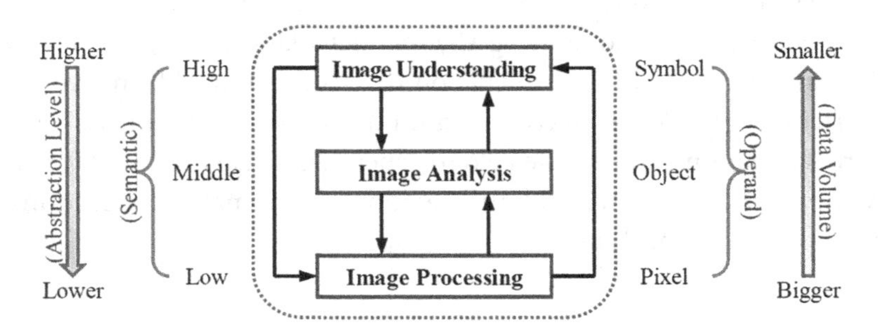

FIGURE 1.1 Schematic diagram of three levels of image engineering.

In addition, high-level operations can both guide and improve the efficiency of lower-level operations, and complete more complex tasks.

In a nutshell, image engineering is an organic combination of image processing, image analysis, and image understanding – which are all related but distinct. Image engineering also includes the engineering applications of image processing, image analysis, and image understanding. Conceptually, image engineering can not only better integrate many similar disciplines, but also emphasizes the application of image technology. This also makes image processing, image analysis and image understanding more closely related.

Image engineering is a new interdisciplinary subject that systematically studies various image theories, techniques, and applications. From the perspective of its research methods, it can learn from many disciplines such as mathematics, physics, biology, physiology (especially neurophysiology), psychology, electronics, and computer science; from the perspective of its research scope, it also intersects with multiple areas of study such as pattern recognition, computer vision, and computer graphics. In addition, the research progress of image engineering is closely related to artificial intelligence, neural networks, genetic algorithms, fuzzy logic, and other theories and technologies; its development and application are related to many fields, such as biomedicine, material science, remote sensing, communication, traffic management, military reconnaissance, document processing, and industrial automation.

Image engineering is a new discipline that comprehensively and systematically studies the theoretical methods of images, expounds the principles of image technology, promotes the application of image techniques, and summarizes practical experience in production. The main components of image engineering can be represented by the framework shown in Figure 1.2, where the dashed box is the basic module of image engineering. Various image techniques are used here to help people get information from the objective scene.

The first thing to do is to obtain images from the scene in various ways. Next, low-level processing of the image is mainly to improve the visual effect of the image or reduce the amount of data while maintaining the visual effect, and the processing result is mainly for

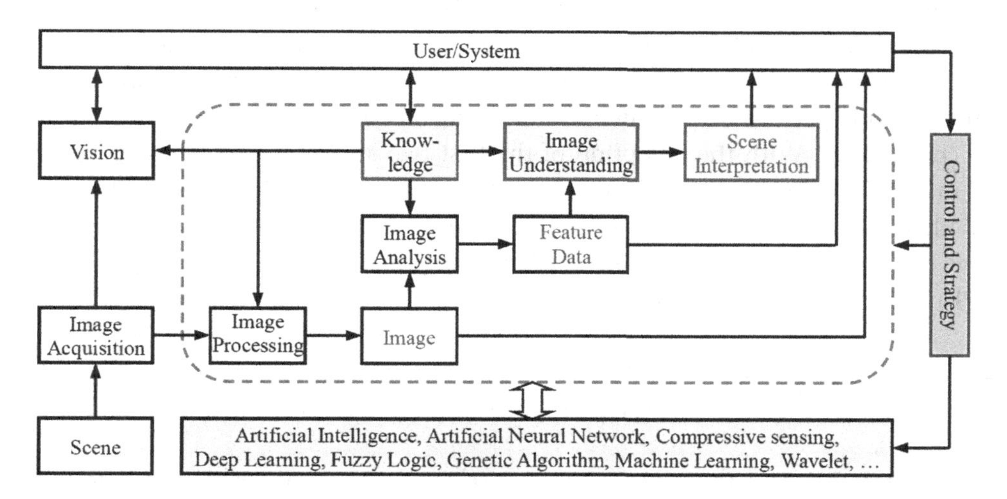

FIGURE 1.2 Overall framework for image engineering.

the user to display and watch. The purpose of the middle-level analysis is mainly to detect, extract, and measure the objects of interest in the image. The results of the analysis provide the user with data describing the characteristics and properties of the image object. Finally, the high-level understanding of the image is to grasp its meaning and explain the original objective scene through the study of the nature of the objects in the image and the relationship between them. The results of understanding provide the user with objective-world information that can guide and plan actions. These image technologies, from low level to high level, are strongly supported by new theories, new tools, and new technologies, including artificial intelligence, neural networks, genetic algorithms, fuzzy logic, image algebra, machine learning, and deep learning. Appropriate strategies are also required to control these tasks.

The content of this book will mainly involve high-level image understanding techniques, including (on the basis of processing and analysis) the acquisition and expression of 3-D objective scene information, scene reconstruction, scene interpretation, etc., as well as the applications, controls, and strategies used to accomplish these tasks.

1.1.2 Review of the Development of Image Technology

The development of research in a field can be analyzed through a review of relevant literature that discusses research results. The following will briefly introduce the development of some image technologies in the past half-century with the help of statistics from two series of surveys.

1.1.2.1 A Closed Survey Series of Image Technology

A series of surveys consisting of 30 papers on image technology dates from the last 30 or so years of the 20th century. The series was called "Image Processing" from its beginning until 1986, and "Image Analysis and Computer Vision" from 1987. The series was concluded by the author in 2000, as it was deemed no longer necessary due to advances in online access to information (Rosenfeld 2000). Details of the 30-year series are summarized in Table 1.1 (Zhang 2002). Note that there are both "big" and "small" years for publications, due to the large increase in the number of papers published in the biennial conference years.

The 34,293 articles cited in this survey series over 30 years were drawn from more than 40 journals (mostly US and some international journals) and more than ten major international conferences. With the exception of the first two survey papers, those papers published in the current year are for the relevant literature in the previous year (a literature list is provided). The first two survey papers in this series were published in *ACM Computing Surveys*, with the remaining 28 published in *Computer Graphics and Image Processing (CGIP)* and its renamed journals.

1.1.2.2 Image Engineering Survey Series in Progress

In 1996, a new survey series on image engineering started, which is now in its 27th year (Zhang 2022).

During these 27 years, the survey series has selected 17,535 papers belonging to the field of image engineering from 70,783 academic research and technical application papers

TABLE 1.1 Overview of a Closed Survey Series

#	Survey Title	Index Year	Paper Number	Journal for Publication	Publication Year
1	Picture Processing by Computer	~69	408	ACM Computing Surveys	1969
2	Progress in Picture Processing: 1969-71	69~71	580		1972
3	Picture Processing: 19xx (72-86)	72	350	Computer Graphics and Image Processing (CGIP)	1973
4		73	245		1974
5		74	341		1975
6		75	354		1976
7		76	461		1977
8		77	609		1978
9		78	819		1979
10		79	700		1980
11		80	897		1981
12		81	982		1982
13		82	1185	Computer Vision, Graphics and Image Processing (CVGIP)	1983
14		83	1138		1984
15		84	1252		1985
16		85	1063		1986
17		86	1436		1987
18	Image Analysis and Computer Vision: 19xx (87-99)	87	1412		1988
19		88	1635		1989
20		89	1187		1990
21		90	1611	CVGIP: Image Understanding	1991
22		91	1178		1992
23		92	1897		1993
24		93	1281		1994
25		94	1911	Computer Vision and Image Understanding (CVIU)	1995
26		95	1561		1996
27		96	2148		1997
28		97	1691		1998
29		98	2268		1999
30		99	1693		2000

published in 15 important Chinese journals (a total of 3,275 issues) about image engineering. Unlike the previous survey series, this survey series is characterized not only by classifying the selected articles, but also by comparisons and analyses of their statistics, so in addition to assisting in the literature search, it contributes to determining the research direction of image engineering and further formulating scientific research decisions.

In order to more easily and intuitively see the development and changes over the years, Figure 1.3 draws the curve of the total number of articles, the total number of selected articles, and the selection rate in the past 27 years. The horizontal axis indicates the year, and the left vertical axis indicates the number of articles (the total number of articles or

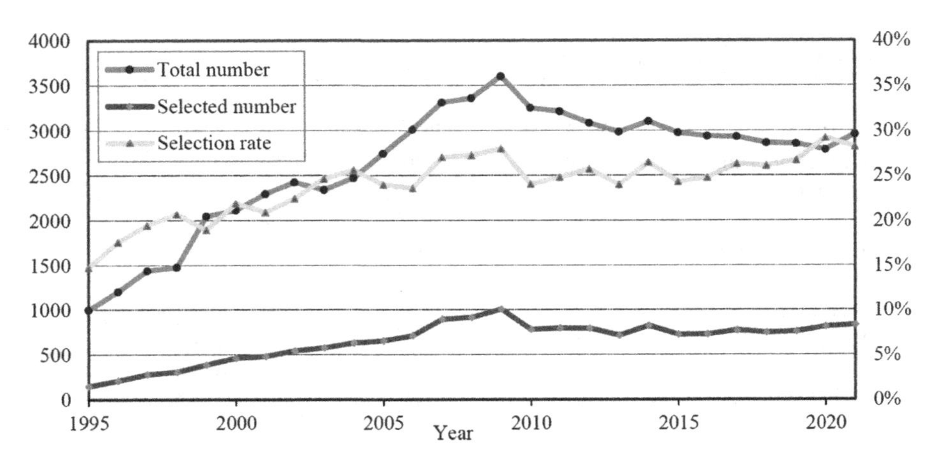

FIGURE 1.3 Selection curve of image engineering articles in the past 27 years.

total number of selected articles), and the vertical axis on the right indicates the selection rate of articles.

During the first 15 years, the survey series selected 8,217 papers belonging to the field of image engineering from 34,841 papers published in the 15 image engineering journals (a total of 1,537 issues). The total number of articles in each of the 15 years, the selected number of articles in each year, and the selection rate can be found in Table 1.2. The article selection rate reflects the relative importance of image engineering in the range of specialties covered by each journal. The selection rate for 1995 was only about 1/7, the selection rate for 1996 was about 1/6, and it remained around 1/5 until 2002. In 2003, the selection rate reached about 1/4, and in 2004 the selection rate exceeded 1/4 for the first time. In terms of the selected quantity, the number of image engineering articles selected for 2009 is nearly 7 times that for 1995 (the total number of articles in journals is only 3.6 times). This is the result of the increasing number of research results and submissions in image engineering over the years, and it is also a testament to the vigorous development of the image engineering discipline.

During the 12 years beginning 2010, the survey series selected 9,318 papers from the field of image engineering from 35,942 papers published in the 15 image engineering journals (a total of 1,738 issues). The total number of articles in each of these 12 years, the selected number of articles in each year, and the selection rate can be found in Table 1.3. The subject of image engineering has matured in the second decade of this century, and the total number of articles has fluctuated around 3,000 (with an improvement in the quality and an increase in the length of the articles, the total number has had a tendency to decrease slightly). With a total of 700~800 selected papers, the overall selection rate is relatively stable at a high level, with an average over 25%.

According to their main contents, the selected image engineering articles in the survey series are classified into five major categories – image processing, image analysis, image understanding, technical application, and review articles – and then further divided into 23 professional subcategories. Table 1.4 lists the first four major categories and their 22 subcategories.

TABLE 1.2 Selection Rates for the First 15 Years in the Image Engineering Survey Series

Articles Year	**1995**	**1996**	**1997**	**1998**	**1999**	**2000**	**2001**	**2002**	**2003**	**2004**	**2005**	**2006**	**2007**	**2008**	**2009**
Total Number	997	1205	1438	1477	2048	2117	2297	2426	2341	2473	2734	3013	3312	3359	3604
Selected Number	147	212	280	306	388	464	481	545	577	632	656	711	895	915	1008
Selection Rate (%)	14.74	17.59	19.47	20.72	18.95	21.92	20.94	22.46	24.65	25.60	23.99	23.60	27.02	27.24	27.97

TABLE 1.3 Selection Rates for the Last 12 Years in the Image Engineering Survey Series

Articles Year	**2010**	**2011**	**2012**	**2013**	**2014**	**2015**	**2016**	**2017**	**2018**	**2019**	**2020**	**2021**	**Average**
Total Number	3251	3214	3083	2986	3103	2975	2938	2932	2863	2854	2785	2958	2995
Selected Number	782	797	792	716	822	723	728	771	747	761	813	833	774
Selection Rate (%)	24.05	24.80	25.69	23.98	26.49	24.30	24.78	26.30	26.09	26.66	29.19	28.16	25.83

TABLE 1.4 Current Image Technology in the Three Levels of Image Processing, Analysis, and Understanding and in Technique Applications

Three Layers	Image Technology Categories and Names
Image Processing	A1: Image acquisition (including various imaging methods, image capture, representation and storage, camera calibration, etc.).
	A2: Image reconstruction (including image reconstruction from projection, indirect imaging, etc.).
	A3: Image enhancement/image restoration (including transformation, filtering, restoration, repair, replacement, correction, visual quality evaluation, etc.).
	A4: Image/video coding and compression (including algorithm research, implementation and improvement of related international standards, etc.).
	A5: Image information security (including digital watermarking, information hiding, image authentication and forensics, etc.).
	A6: Image multi-resolution processing (including super-resolution reconstruction, image decomposition and interpolation, resolution conversion, etc.).
Image Analysis	B1: Image segmentation and primitive detection (including edges, corners, control points, points of interest, etc.).
	B2: Object representation, object description, feature measurement (including binary image morphology analysis, etc.).
	B3: Object feature extraction and analysis (including color, texture, shape, space, structure, motion, saliency, attributes, etc.).
	B4: Object detection and object recognition (including object 2-D positioning, tracking, extraction, identification and classification, etc.).
	B5: Human body biological feature extraction and verification (including detection, positioning and recognition of human body, face and organs, etc.).
Image Understanding	C1: Image matching and fusion (including registration of sequence and stereo image, mosaic, etc.).
	C2: Scene restoration (including 3-D scene representation, modeling, reconstruction, etc.).
	C3: Image perception and interpretation (including semantic description, scene model, machine learning, cognitive reasoning, etc.).
	C4: Content-based image/video retrieval (including corresponding labeling, classification, etc.).
	C5: Spatial-temporal techniques (including high-dimensional motion analysis, object 3-D posture detection, spatial-temporal tracking, behavior judgment and behavior understanding, etc.).
Technique Applications	D1: System and hardware, fast/parallel algorithm implementation, etc.
	D2: Telecommunication, video transmission and broadcasting (including TV, network, radio, etc.).
	D3: Documents and text (including text, numbers, symbols, etc.).
	D4: Biology and medicine (physiology, hygiene, health, etc.).
	D5: Remote sensing, radar, sonar, surveying and mapping, etc.
	D6: Others (technical applications not directly/explicitly included in the above categories).

For the most recent year, 2021, the number of articles in all 23 subcategories are shown in Figure 1.4 (Zhang 2022). Compared with the number of articles in each subcategory of image processing and image analysis, the number of papers in each subcategory of image understanding is relatively small (only two subcategories, C1 and C5, can be compared). It can be seen that much remains to be done in the field of image understanding.

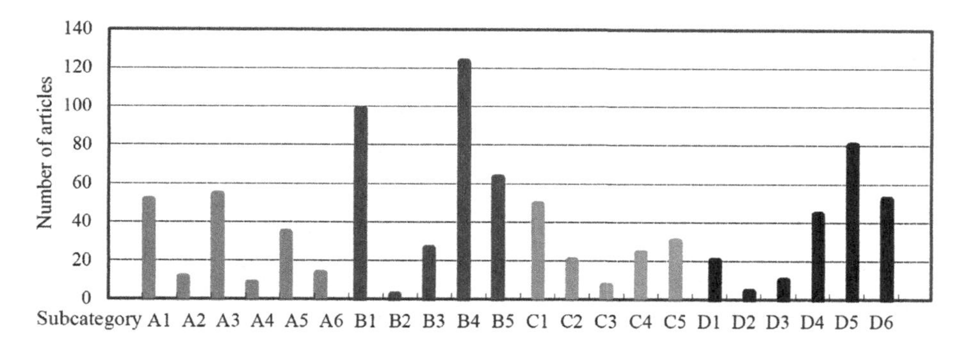

FIGURE 1.4 Statistical results of the number of articles in 23 subcategories in 2021.

1.2 IMAGE UNDERSTANDING AND RELATED DISCIPLINES

Image understanding is closely related to some other disciplines, which will be discussed briefly below.

1.2.1 Image Understanding

The high level of image engineering, which focuses on the combination of artificial intelligence and cognitive theory in the context of image analysis, studies the properties of various objects in the image and the relationship between them, and aims to understand the meaning of image content and interpret the corresponding objective scene to guide and plan actions. If image analysis mainly studies the objective world centered on the observer (i.e., mainly studies observable things), then image understanding is centered on the objective world to a certain extent, and uses knowledge, experience, etc. to grasp the entire objective world (including things that are not directly observed).

Image understanding is concerned with how to describe and judge the scene according to the image. It uses the computer to build a system to help explain the meaning of the image, so that image information can be used to explain the objective world. It needs to determine what information is to be obtained from the objective world through image acquisition to complete certain tasks, what information is to be extracted from the image through image processing and analysis, and what information is to be used to continue obtaining the required decision. It studies the mathematical model of comprehension ability, and realizes a computer simulation of comprehension ability by programming a mathematical model.

Many of these tasks cannot be fully automated, due to the current limitations of computer capabilities and image understanding technology. In most cases, it is the "system" that mainly performs lower-level work, and the human needs to do some of the higher-level work (Figure 1.5).

Without any systems, humans would have to perform tasks in all the layers. If the system has only low-level capabilities, people are needed to complete middle- and high-level system processing tasks. If the system has low-level and middle-level capabilities, then users only need to complete high-level system processing tasks. If the system has the ability

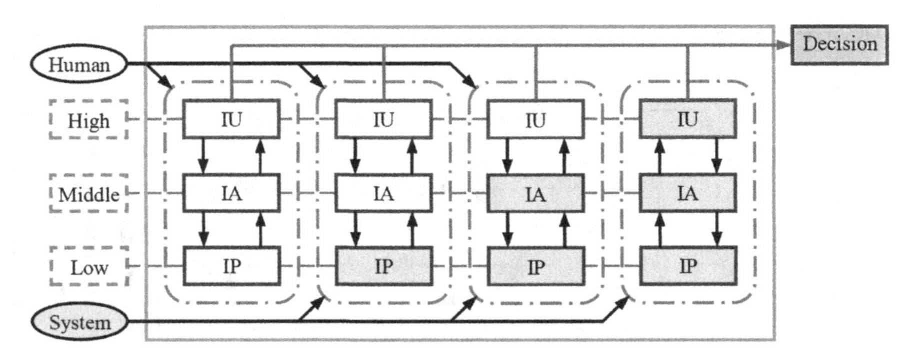

FIGURE 1.5 The system and user cover different levels.

to go from low level to high level, users can easily obtain decisions. At present, the bottleneck of research and development is mainly at the high level.

1.2.2 Computer Vision

The **human vision** process can be seen as a complex procedure from sensation (feeling an image obtained by 2-D projection of the 3-D world) to perception (perceiving the content and meaning of the 3-D world from the 2-D image) (Kong 2002). In a narrow sense, the ultimate purpose of vision is to make interpretations and descriptions of a scene meaningful to the observer, while in a broad sense, the ultimate purpose of vision is to formulate behavioral plans based on these interpretations and descriptions, according to the surrounding environment and the wishes of the observer. **Computer vision** refers to the use of computers to realize human vision functions, aiming to make meaningful judgments about actual objects and scenes based on perceived images (Shapiro and Stockman 2001). This is actually also the goal of image understanding.

1.2.2.1 Research Methods

Research on computer vision mainly adopts two kinds of methods.

(1) The bionics method

The essence of the bionics method is **imitation**, that is, referring to the structural principles of the human vision system, establishing corresponding processing modules or making equipment with visual capabilities to perform similar functions. Three related questions are involved here (Sonka et al., 2014): (i) empirical question: What is? It is needed to determine how to design the existing vision system; (ii) standardized question: What should be? It is needed to determine the properties to be expected from a natural or ideal vision system; (iii) theoretical question: What could be? It is needed to determine the mechanism in the intelligent vision system.

(2) The engineering method

The essence of the engineering method is **simulation**, that is, starting from the analysis of the function of the human vision process, without simulating the internal structure of the human vision system deliberately, but only considering the input and output of the system, and using any existing feasible technical means to achieve the required system functions. This is the main method discussed in this book, and it will be covered in detail in subsequent chapters.

1.2.2.2 Realization of Engineering Methods

Depending on the direction of the information flow and the amount of prior knowledge, there are two ways to implement human vision functions with engineering methods:

(1) *Bottom-up reconstruction.* The 3-D shape of the object needs to be reconstructed from a 2-D image or a set of 2-D images, where both luminance and depth images can be used. Marr's **theory of visual computation** (see Subsection 1.3.1) is a typical approach, which is strictly bottom up and requires little prior knowledge of the object.

(2) *Recognition from top to bottom.* It is also known as **model-based vision**, that is, prior knowledge about the object is represented by the model of the object, of which the 3-D model is more important. For example, in CAD model-based recognition, uncertain vision problems can in many cases be solved due to constraints embedded in the model.

In practice, these two methods are often used in combination, depending on the specific task, and this is reflected in subsequent chapters.

1.2.2.3 Research Objectives

The main research objectives of computer vision can be summarized as two complementary goals.

The first research goal, which is the main focus of this book, is to build computer vision systems to accomplish various vision tasks. In other words, the computer can obtain images of the scene with the help of various visual sensors (such as CCD and CMOS camera devices, etc.), so as to perceive and restore the geometric properties, posture structures, motion information, and mutual relations of objects in the 3-D environment. The objective scene can be described in such a way as to identify, describe, and explain it, and finally to make judgments and decisions. The main research consists of technical mechanisms. Current work in this domain is focused on building various specialized systems to complete specialized vision tasks proposed in various practical situations. In the long run, it is expected that more general systems will be built (Jain and Dorai 1997).

The second research goal is to use this research as a means to explore the working mechanism of human vision, and to further deepen understanding of human brain vision (such as computational neuroscience). The main research here concerns biological mechanisms. Much research has been done on the human brain visual system from the physiology, psychology, nerves, and cognition, etc. aspects, but it is far from revealing all the mysteries of the vision process. It should be pointed out that a full understanding of human brain vision will also promote in-depth research on computer vision (Finkel and Sajda 1994), and research on the powerful understanding of the human vision system can help people to develop new image understanding and computer vision algorithms.

This book will mainly consider and develop against the first research objective.

1.2.2.4 The Relationship between Image Understanding and Computer Vision

Image understanding and computer vision are closely related. Image is a physical form of representing visual information, and image understanding must be carried out with the help of computers, and based on image processing and analysis. As a discipline, computer vision has a very close connection and different degrees of intersection with many disciplines that take images as the main research object, especially image processing, image analysis, and image understanding. Computer vision mainly emphasizes the realization of human visual functions with computers, which actually requires the use of many technologies at the three levels of image engineering, although the current research content is mainly combined with image understanding.

The close connection between image understanding and computer vision can also be seen in the definition of computer vision (Sonka et al., 2014), the central problem of which is to understand the object and/or scene and its properties from a single or multiple monocular, moving or stationary observers. This definition basically matches the definition of image understanding. The complexity of the comprehension task is related to the specific application. If there is less prior knowledge, such as in the human vision of nature, then understanding is complex; but in many cases where the environment and goals are defined and constrained, since the possible explanations are limited, then the understanding can be less complicated.

In building an image/visual information system and using computers to assist humans in completing various visual tasks, both image understanding and computer vision require the use of theories of projective geometry, probability theory and stochastic processes, and artificial intelligence. For example, they all rely on two types of intelligent activities: (i) perception, such as perceiving the distance, orientation, shape, movement speed, interrelationship, etc. of the visible parts of the scene; and (ii) thinking, such as analyzing the behavior of objects according to the structure of the scene, inferring the development and changes of the scene, deciding and planning the main action, etc. The former is closely related to **visual feeling**, while the latter is closely related to **visual perception**.

In fact, **computer vision** was originally studied as an artificial intelligence problem, so it is also called **image understanding** (Shah 2002). In practice, the terms image understanding and computer vision are often used interchangeably. Essentially, they are interrelated, and in many cases their contents overlap, with no absolute boundaries in terms of concept or practicality. In many contexts and situations, they have different focuses but often complement each other, so it is more appropriate to think of them as distinct terms customarily used by people of different professions and/or backgrounds.

1.2.3 Other Related Disciplines

Image understanding is closely related to computer science (as are image processing and image analysis, which underlie image understanding). In addition to computer vision, other computer-related disciplines, such as (in alphabetical order) artificial intelligence, computer graphics, machine learning, machine vision/robot vision, pattern recognition, etc., have played and will continue to play an important role in the development of image understanding.

1.2.3.1 Artificial Intelligence

Human intelligence mainly refers to the ability of human beings to understand the world, judge things, learn about the environment, plan behavior, reason and think, and solve problems. **Artificial intelligence (AI)** refers to the ability and technology to simulate, perform or reproduce some functions related to human intelligence by computer (Nilsson 1980; Winston 1984; Dean et al. 1995).

Visual function is a manifestation of human intelligence, so image understanding and computer vision are closely related to artificial intelligence. Many artificial intelligence technologies are used in image understanding research. In turn, image understanding can also be regarded as an important application field of artificial intelligence, which needs to be realized with the help of theoretical research results and systems of artificial intelligence.

1.2.3.2 Machine Learning and Deep Learning

Machine learning (ML) refers to the process by which computer systems achieve self-improvement through self-learning, simulating or realizing human learning behaviors to acquire new knowledge or skills, or to reorganize existing knowledge structures to continuously improve their performance. Machine learning problems can often be reduced to search problems, and different learning methods are defined and distinguished by different search strategies and search space structures. A major research direction in machine learning is automatic learning to recognize complex patterns and make intelligent decisions based on data.

A number of methods for automatically analyzing structures in data have been proposed. The research work can be divided into two categories:

(1) Unsupervised learning, also known as descriptive modeling, where the goal is to discover patterns or structures of interest in the data.

(2) Supervised learning, also known as predictive modeling, where the goal is to predict the value of one or more variables (on the basis of some other values given).

These goals are similar to those of **data mining**, but here the focus is more on automated machine performance than on how people learn from data.

Deep learning (DL), which has received extensive attention in recent years, is a branch of machine learning of the unsupervised learning type. It attempts to mimic the workings of the human brain, building neural networks that learn to analyze, recognize, and interpret data such as **images**. By combining low-level features to form more abstract and high-level representative attribute categories, distributed feature representation of data can be found. This is similar to how humans first grasp simple concepts in learning and then use them to express more abstract semantics. With the help of deep learning, unsupervised or semi-supervised feature learning can be used to achieve efficient feature extraction.

1.2.3.3 Machine Vision/Robot Vision

Machine vision or **robotic vision** is inextricably linked to computer vision and is used synonymously in a number of cases. Specifically, it is generally believed that computer vision focuses more on theories and algorithms for scene analysis and image interpretation, while machine vision/robot vision pays more attention to image acquisition, system structure and algorithm implementation, which are closely related to the technical application of image engineering (Zhang 2009b).

1.2.3.4 Pattern Recognition

There can be a wide range of patterns, and images are one type of pattern. Recognition refers to the mathematics and technology that automatically establish symbolic descriptions or logical reasoning from objective facts, so **pattern recognition** (PR) is defined as the discipline of classifying and describing objects and processes in the objective world (Bishop 2006). At present, the recognition of image patterns mainly focuses on the representation, description, identification, and classification of the content (objects of interest) in the image, which has a considerable intersection with image analysis. Many concepts and methods of pattern recognition are also used in image understanding, but visual information has its particularity and complexity, and traditional pattern recognition (competitive learning model) cannot include all image understanding.

1.2.3.5 Computer Graphics

Computer graphics (CG), the study of how to generate "images" from a given description, is also closely related to computer vision. Computer graphics is generally referred to as the inverse of computer vision because computer vision extracts 3-D information from 2-D images, while computer graphics uses 3-D models to generate 2-D visual scenes. In fact, computer graphics is often more associated with image analysis. Some graphics can be considered as visualization of image analysis results, and the generation of computer-realistic scenes can be considered as the inverse process of image analysis (Zhang 1996). Graphics technology also plays a central role in the process of human–computer interaction and modeling of vision systems. **Image-based rendering**, a research domain that combines the two, is a good example. It should be noted that, compared with the many uncertainties in image understanding and computer vision, computer graphics deals with more deterministic problems that can be solved mathematically. In many practical applications, people are more concerned with the speed and accuracy of graphics generation, that is, with achieving some kind of compromise between real time and fidelity.

From a broader perspective, image understanding uses engineering methods to solve biological problems and complete the inherent functions of biology, so it also has a mutual learning and interdependence relationship with biology, physiology, psychology, neurology and other disciplines. In recent years, image understanding researchers have cooperated closely with visual psychophysiological researchers and have obtained a series of research results. Image understanding belongs to engineering applied science and is inseparable from electronics, integrated circuit design, communication engineering, etc. On the one hand, image understanding research makes full use of the achievements of these disciplines; on the other hand, the application of image understanding also greatly promotes the in-depth research and development of these disciplines.

1.3 THE THEORETICAL FRAMEWORK OF IMAGE UNDERSTANDING

Research on image understanding and computer vision lacked a comprehensive theoretical framework in the early days. In the 1970s, research on object recognition and scene understanding basically detected primitives (e.g., points, edges, etc.) first, and then combined them to form more complex structures. But in practice, primitive detection is difficult and unstable, so the understanding system can only input simple lines and corners to form the so-called "block world".

1.3.1 Marr's Theory of Visual Computation

Marr's 1982 book *Vision* (Marr 1982) summed up a series of results of his and his colleagues' research on human vision, proposed a theory of visual computing, and outlined a framework for understanding visual information. This framework is both comprehensive and refined, and is the key to making the study of visual information understanding rigorous and to moving visual research from the level of description to the level of mathematical sciences. Marr's theory points to understanding the purpose of vision before understanding the details. This is suitable for a variety of information processing tasks (Edelman 1999).

1.3.1.1 Vision is a Complex Information Processing Process

Marr believes that vision is a much more complex information processing task and process than people's imagination, and its difficulty is often ignored. One of the main reasons is that although it is difficult to understand images with computers, it is often easy for people.

To understand the complex process of vision, two problems must first be addressed. One is the representation of visual information; the other is the processing of visual information. Representation here refers to a formal system (such as the Arabic numeric system, the binary numeric system) that can clearly represent certain entities or certain types of information, and a number of rules that explain how the system works. Some messages in the representation are prominent and explicit, while others are hidden and vague. Representation has a great influence on the difficulty of subsequent information processing. As for visual information processing, it achieves its goal by continuously processing, analyzing, and understanding information, converting different forms of representation, and gradually abstracting it. There are several different levels and aspects to accomplishing vision tasks.

Recent biological studies have shown that when organisms perceive the external world, their visual system can be divided into two cortical visual subsystems, that is, there are two visual pathways: the "what" pathway and the "where" pathway. Information transmitted by the "what" channel is related to the object in the external world, and the "where" channel is used to transmit spatial information about the object. Combined with the attention mechanism, "what" information can be used to drive bottom-up attention to form perception and object recognition; "where" information can be used to drive top-down attention and process spatial information. This research result is consistent with Marr's point of view, because according to Marr's computational theory, the visual process is an information processing process, and its main purpose is to find the object existing in the external world and the spatial location of the object from the image.

1.3.1.2 Three Key Elements of Visual Information Processing

To fully understand and interpret visual information, three key elements need to be grasped at the same time: computational theory, algorithm implementation, and hardware implementation.

First, if a task is to be done by a computer, it should be computable. This is the computability question, which needs to be answered with computational theory. Generally, for a particular problem, if there is a program that can give output in finite steps for a given input, the problem is computable. There are three research objects in computable theory: decision problems, computable functions, and computational complexity. Decision problems are mainly to determine whether the equation has a solution. Computable functions mainly discuss whether a function is computable. For example, a mathematical model such as a Turing machine can be used to determine whether a function is a computable function. Computational complexity mainly discusses **NP-complete problems**, and generally considers whether there is an efficient algorithm whose time and space complexity is polynomial. (The category of all problems that can be solved by polynomial time algorithms can be called class P, P $\subseteq$ NP, and NP-complete problems are the most difficult problems in the NP class, but NP-complete does not mean that there is no way to solve them. For some problems, approximate solutions that meet specific applications can be obtained.)

The highest level of visual information understanding is abstract computational theory. There is no clear answer to the question of whether vision can be computed by modern computers. Vision is a process of feeling and perception. People still have very little understanding of the mechanism of human visual function in terms of microscopic anatomical knowledge and objective visual psychological knowledge, so discussion on visual computability is still relatively limited, mainly focusing on the ability to process numbers and symbols for performing certain specific visual tasks with existing computers. At present, visual computability often refers to whether a computer can obtain results similar to those that can be obtained by human vision for a given input. Here the calculation goal is clear, and the output requirements can also be determined after the input is given, so the focus is on the information understanding step in the transformation from input to output. For example, given an image of a scene (input), the computational goal is to obtain an interpretation of the scene (output). There are two main research contents in visual computation theory: (i) what is calculated and why; and (ii) certain constraints are proposed to uniquely determine the final calculation result.

Secondly, the objects operated by today's computers are discrete numbers or symbols, and the storage capacity of the computer is also limited to a certain extent. With computation theory, therefore, the realization of the algorithm must also be considered. For this reason, it is necessary to choose a suitable representation for the entity operated by the processing. Here, on the one hand, the input and output representations for the processing must be selected; on the other hand, the algorithm for completing the representation transformation must be determined. Representations and algorithms are mutually constrained, and the following three points should be noted:

(i) there can be many alternative representations in general;

(ii) the determination of the algorithm often depends on the selected representation;

(iii) given a representation, there can be a variety of algorithms to accomplish the task.

From this point of view, the chosen methods of representation and manipulation are closely related. The instructions and rules used for processing are generally referred to as algorithms.

Finally, with representations and algorithms, it is also necessary to consider how the algorithm is physically implemented. Especially with the continuous improvement of real-time requirements, the problem of dedicated hardware implementation is often raised. It should be noted that the determination of the algorithm often depends on the hardware characteristics of the physical realization of the algorithm, and the same algorithm can also be realized by different technical approaches.

Table 1.5 summarizes the above discussion.

There is a certain logical causal connection between the above three elements, but there is no absolute dependence. In fact, there are many different options for each element. In many cases, the issues involved in interpreting each element are largely independent of the other two elements (each element is relatively independent), or some visual phenomena can be explained by only one or two elements. It has been pointed out that the above three elements are at the three levels of visual information processing, and that different problems need to be explained at different levels. The relationship between the three elements is often

TABLE 1.5 The Meaning of the Three Key Elements of Visual Information Processing

Key Element	Meaning and the Problems Solved
Computational Theory	What is the goal of computation? Why should we make such a computation?
Representation and Algorithms	How to achieve the computational theory? What is the input and output representation? What algorithms are used to achieve the conversion between representations?
Hardware Implementation	How to implement representations and algorithms in physics? What are the details of the computational structure?

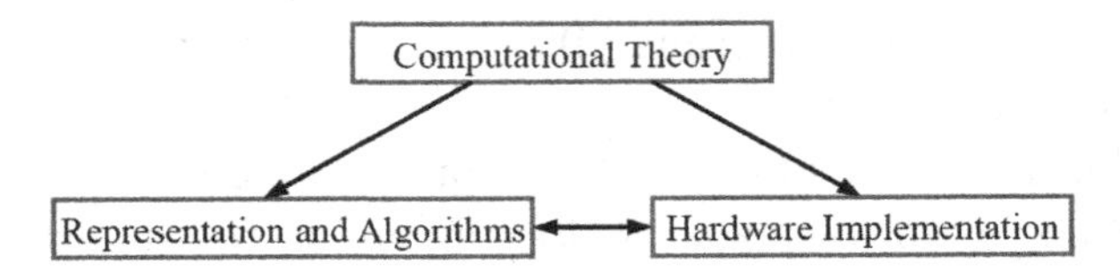

FIGURE 1.6 The links between the three key elements of visual information processing.

represented as in Figure 1.6 (actually it is more appropriate to regard it as two levels), in which the arrows indicate the meaning of guidance in the forward direction, and the meaning of the foundation in the reverse direction. Note that once there is the theory of computation, the representations and algorithms as well as hardware implementations interact.

1.3.1.3 Three-Level Internal Representation of Visual Information

According to the definition of visual computability, visual information processing can be decomposed into multiple transformation steps from one representation to another. Representation is the key to visual information processing. A basic theoretical framework for computer visual information understanding research is mainly composed of three-level representation structures of the visible world established, maintained, and explained by visual processing. For most philosophers, the nature of visual representations, how they relate to perception, and how they support action can all be interpreted differently. However, they tend to agree that the answers to these questions are related to the concept of "representation" (Edelman 1999).

(1) Primitive representation

Primitive representation refers to 2-D representation, which is a collection of image features that describe the contour parts of an object's surface properties that change. Primitive representation provides information about the outline of each object in the image, and is a sketch-like representation of 3-D objects. This kind of representation can be proved from the human visual process. When people observe the scene, they always pay attention to the dramatic part of the scene, so the primitive representation should be a stage of the human visual process.

It should be noted that using only primitive representation does not guarantee a unique interpretation of the scene. Taking the Necker cube illusion shown in Figure 1.7 as an example (Marr 1982), if the observer focuses on the intersection of the three lines at the top right of Figure 1.7(a), it will be interpreted as Figure 1.7(b),

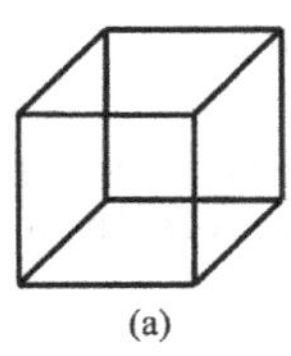
(a)

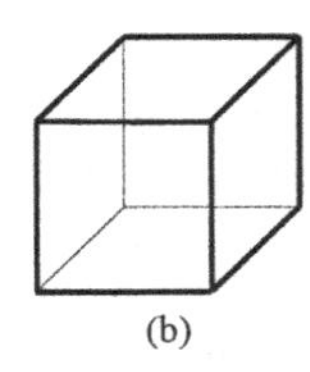
(b)

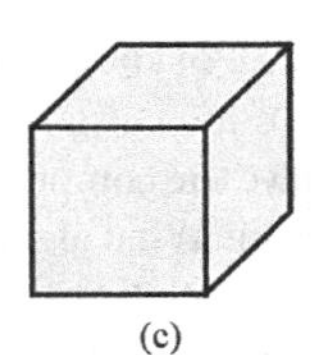
(c)

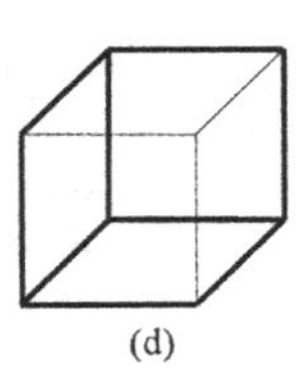
(d)

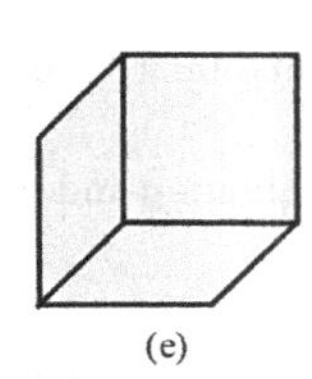
(e)

FIGURE 1.7 Necker cube illusion.

that is, the imaged cube is considered as shown in Figure 1.7(c). If the observer focuses on the intersection of the three lines at the lower left of Figure 1.7(a), it will be interpreted as Figure 1.7(d), that is, the imaged cube is considered as shown in Figure 1.7(e). This is because although Figure 1.7(a) gives people the clue of (part of) the 3-D object (cube), when they try to recover 3-D depth from it with the help of empirical knowledge, two different explanations can be obtained due to different comprehensive methods. One regards the intersection of the three lines at the top right of Figure 1.7(a) as the closest point to itself, while the other regards the intersection of the three lines at the bottom left of Figure 1.7(a) as the closest point to itself. Two different results are derived.

Incidentally, Necker's illusion can also be explained by **viewpoint reversal** (Davies 2005). When people observe a cube, they will intermittently regard the two middle vertices as the closest points to themselves, which is called **perception reversal** in psychology. Necker's illusion suggests that the brain makes different assumptions about the scene and even makes decisions based on incomplete evidence.

(2) 2.5-D representation

2.5-D representation is proposed entirely to adapt to the computing function of the computer. It decomposes the object according to the principle of **orthogonal projection** with a certain sampling density, so that the visible surface of the object is decomposed into many surface elements of a particular size and geometric shape, and each surface element has its own orientation. A 2.5-D representation graph (also called a needle graph) is formed by using a normal vector to represent the orientation of the surface element where it is located and to form a needle graph (the vector is represented by an arrow). The orientation of a normal vector is observer-centric. The specific steps to obtain the 2.5-D representation map are: (i) decompose the orthogonal projection of the visible surface of the object into a set of unit surfaces; (ii) use the normal line to represent the orientation of the unit surface; and (iii) draw each normal line and superimpose it on the visible surface inside the outline of the object. Figure 1.8 provides an example.

The 2.5-D map is actually a kind of "eigen-image" (see Subsection 1.3.1.2) because it represents the orientation of the surface elements of the object, thus giving information on the shape of the surface. It represents not only information on a part of the outline of the object (which is similar to the representation of the primitive representation), but also orientation information on the object surface that is visible and is centered on the observer.

Combining 2-D primitive representation and 2.5-D representation can provide 3-D information (including boundaries, depth, reflection properties, etc.). Such representations are also consistent with human understanding of 3-D objects.

(3) 3-D representation

3-D representations are object-centric (i.e., they also include invisible parts of objects). They describe the shape of a 3-D object and its spatial organization in an object-centric coordinate system.

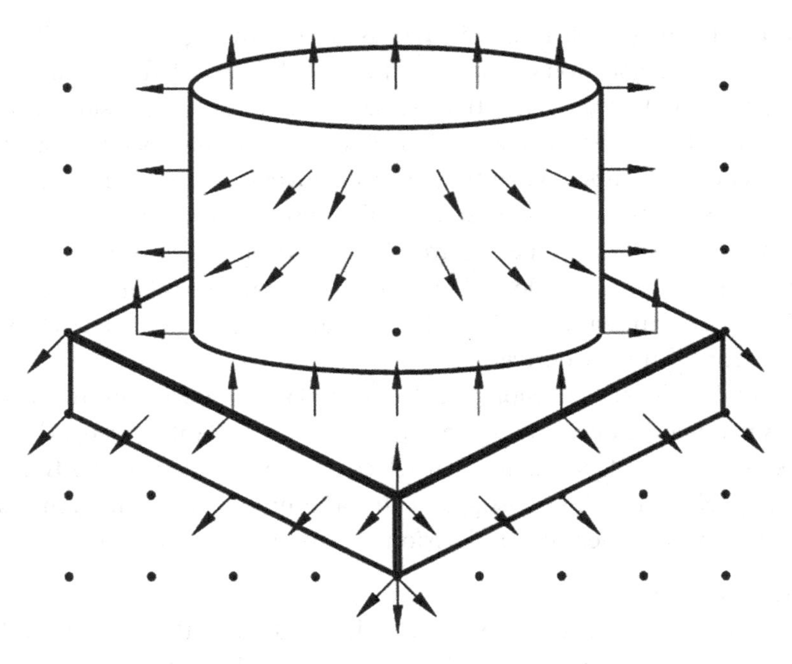

FIGURE 1.8 Example of 2.5-D representation.

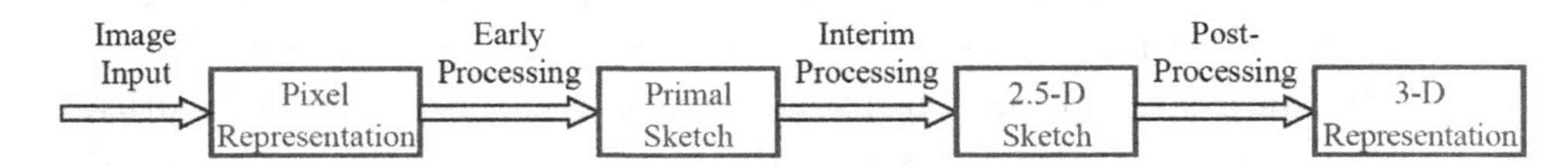

FIGURE 1.9 Three-level representation of Marr's framework.

Now we return to the problem of visual computability. From the point of view of computer or information processing, the problem of visual computability can be divided into several steps; between these steps a certain form of representation is required, and each step is a calculation/processing method that links the two forms of representation before and after (see Figure 1.9).

According to the above three-level representation point of view, the problem to be solved by visual computability is: how to start from the pixel representation of the original image, through the primitive representation map and the 2.5-D representation map, and finally obtain the 3-D representation map (Table 1.6).

1.3.1.4 Visual Information Understanding is Organized in the Form of Functional Modules

Not only is the idea of viewing a visual information system as a set of relatively independent functional modules supported by computational evolutionary and epistemological arguments, but some functional modules can also be isolated experimentally.

Psychological research also shows that people obtain various intrinsic visual information by using multiple cues or a combination. This suggests that the visual information system

TABLE 1.6 Representation Framework for Visual Computability Problems

Name	Purpose	Basic Element
Image	Represent the brightness of the scene or the illumination of the object.	Pixel (value).
Primal Sketch	Represent the location of the brightness change in the image, the geometrical distribution and the organization structure of the object contour.	Zero crossing point, endpoint, corner, inflection point, edge segment, boundary, etc.
2.5-D Sketch	Represent the orientation, depth, contour and other properties of visible object surface in the observer-centered coordinate system.	Local surface orientation ("needle" primitives), discontinue point of surface orientation, depth, depth discontinuity point.
3-D Map	Describe the shape and the spatial organization of shapes by voxels or sets of surface in an object-centered coordinate system.	3-D model, with the axis as the skeleton, attach the voxel or surface element to the axis.

should include many modules, each module obtains a specific visual cue and performs certain processing, so that different weights can be combined with different modules according to the environment to finally complete the visual information understanding task. According to this point of view, complex processing can be completed by some simple independent functional modules, which can simplify the research method and reduce the difficulty of specific implementation. This is also important from an engineering perspective.

1.3.1.5 The Formal Representation of Computational Theory Must Take Constraints into Account

During the process of image acquisition, various changes will occur to the information in the original scene, including:

(1) When a 3-D scene is projected as a 2-D image, information on object depth and invisible parts is lost.

(2) Images are always obtained from a specific perspective, and images of the same scene from different perspectives will be different. In addition, information will be lost due to objects or parts of them occluding each other.

(3) Imaging projection enables all factors, such as illumination, object geometry and surface reflection characteristics, camera characteristics, and the spatial relationship between light source and object and camera, to be integrated into a single image gray value, in which these factors are difficult to distinguish.

(4) Noise and distortion are inevitably introduced in the imaging process.

A problem/question is well-posed if its solution: (i) exists; (ii) is unique; and (iii) is continuously dependent on the initial data. If one or more of the above are not satisfied, it is an ill-posed problem. Due to the various changes of information in the above-mentioned original scenes, the method of solving the vision problem as the inverse problem of the optical imaging process becomes an ill-posed (ill-conditioned) problem that is difficult to solve. It is first necessary to find out the constraints of the problem according to the general characteristics of the external objective world, then turn them into precise hypotheses, so as to draw firm and testable conclusions. Constraints are generally obtained with the help of prior knowledge, and ill-conditioned problems can be changed by using constraints, because adding constraints to the computational problem can make its meaning clear and enable it to be solved.

1.3.2 Improvements to Marr's Theoretical Framework

Marr's theory of visual computing is the first theory to have a profound influence on visual research. This theory actively promotes research in this field and plays an important role in the development of image understanding and computer vision research.

Marr's theory also has its shortcomings. Four issues arise with the overall framework (see Figure 1.9):

(1) The input for the framework is passive, and the system will process whatever image is input.

(2) The purpose of processing in the framework remains unchanged, and the position and shape of objects in the scene are always restored.

(3) The framework lacks or does not pay enough attention to the guiding role of high-level knowledge.

(4) The information processing process in the entire framework is basically bottom up, that is, it is a unidirectional flow with no feedback.

In response to the above problems, a series of improvement ideas, corresponding to the framework of Figure 1.9, have been proposed in recent years, which can be integrated into new modules to obtain the framework in Figure 1.10. These include the following:

(1) Human vision is proactive, such as changing the line of sight or perspective as needed to aid observation and cognition. **Active vision** means that the vision system can determine the motion of the camera to obtain the corresponding image from the

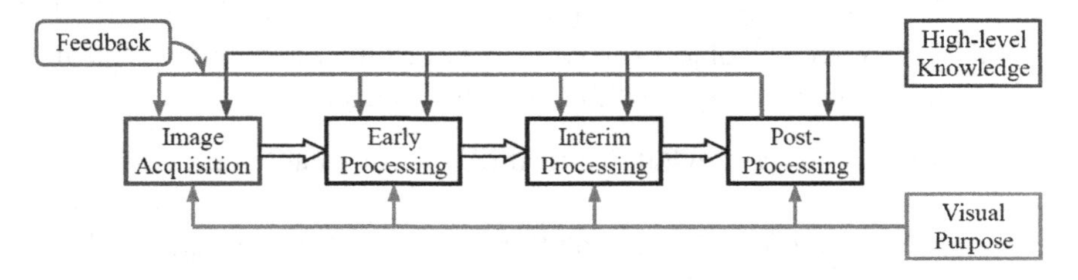

FIGURE 1.10 Improved Marr's framework.

appropriate position and perspective based on the existing analysis results and the current requirements of the vision task. Human vision is also selective, staring at one part (observing a region of interest at a higher resolution) and turning a blind eye to other parts of the scene. **Selective vision** means that the vision system can determine the camera's attention point to obtain the corresponding image based on the existing analysis results and the current requirements of the vision task. Taking these factors into consideration, an "image acquisition" module is added to the improved framework, which is also considered together with other modules in the framework. This module selects the image acquisition method according to the visual purpose.

Active and selective vision can also be viewed as two forms of active vision (Davies 2005): (i) moving the camera to focus on a specific object of interest in the current environment; and (ii) focusing on a specific region in the image and dynamically interacting with it for interpretation. Although these two forms of active vision look similar, in the first form, the activeness is mainly in the camera's observation; while in the second, the activeness is mainly in the processing level and strategy. Although there is interaction in both forms, i.e., vision is active, moving the camera to record and store the scene in its entirety is an expensive process, and the overall interpretation thus obtained is not necessarily all used. In contrast, collecting the most useful parts of the scene at the moment, narrowing it down, and enhancing its quality for useful explanations mimics the process of human interpretation of a scene.

(2) Human vision can be adjusted for different purposes. **Purposeful vision** means that the vision system makes decisions based on the purpose of vision, such as whether to completely and comprehensively recover information, such as the position and shape of objects in the scene, or just to detect whether there is an object in the scene. It has the potential to give simpler solutions to vision problems. The key issue here is to determine the purpose of the task, so a "visual purpose" box is added to the improved framework (Aloimonos 1992), which can be qualitatively or quantitatively analyzed according to the different purposes of understanding (in practice, there are quite a number of occasions where only qualitative results are sufficient, and quantitative results with high complexity are not required). However, the qualitative analysis still lacks complete mathematical tools at present. The motivation for purposeful vision is to clarify only part of the information that is needed. For example, collision avoidance of autonomous vehicles does not require precise shape descriptions; some qualitative results are sufficient. This line of thinking does not yet have a solid theoretical foundation, but the study of biological visual systems provides many examples.

Qualitative vision, closely related to purposeful vision, seeks a qualitative description of an object or scene. Its motivation is not to represent geometric information that is not needed for qualitative (non-geometric) tasks or decisions. The advantage of qualitative information is that it is less sensitive to various unwanted transformations (such as a slight change in perspective) or noise than quantitative information. Qualitative or invariant can allow easy interpretation of observed events at different levels of complexity.

(3) Humans have the ability to fully solve vision problems with only partial information from images due to the implicit use of various items of knowledge. For example, using CAD design data to obtain information about an object's shape (using the object model library) can help solve the difficulty of restoring the object shape from a single drawing. Using high-level knowledge can solve the problem of insufficient low-level information, so a "high-level knowledge" box is added to the improved framework (Huang and Stucki 1993).

(4) There is an interaction between the processing of "before" and "after" in human vision. Although the mechanism of this interaction is not fully understood, the important role of high-level knowledge and feedback from later results in early processing has been widely accepted. From this point of view, feedback control flow is added to the improvement framework.

1.3.3 Discussion on Marr's Reconstruction Theory

Marr's theory emphasizes the reconstruction of the scene and uses it as the basis for understanding the scene.

1.3.3.1 Problems Related to Reconstruction Theory

According to Marr's theory, the common core concept of different visual tasks/jobs is representation, and the common processing goal is to recover the scene from visual stimuli and incorporate it into the representation. If the vision system can recover the characteristics of the scene, such as the reflective properties of the surface of the object, the direction and speed of the object's movement, the surface structure of the object, etc., then there needs to be a representation that can help with various recovery tasks. In such a theory, different tasks should have the same conceptual core, understanding process, and data structure.

In his theory, Marr showed how people can extract from various cues the representations that construct the visual world. If the construction of such a unified representation is regarded as the ultimate goal of visual information processing and decision making, then vision can be viewed as a reconstruction process that starts with stimuli and is sequentially acquired and accumulated. This idea of reconstructing the scene first and then interpreting it can simplify the visual task, but it is not completely consistent with the human visual function. In fact, reconstruction and interpretation are not always serial and need to be adjusted for visual purposes.

The above assumptions have also been challenged. Some of Marr's contemporaries questioned the vision process as a hierarchical, single-pass data-processing process. One of the meaningful contributions, based on longstanding research in psychophysics and neuropsychology, is that the single-path hypothesis has been shown to be untenable. At the time Marr wrote *Vision*, there was little psychological research that took into account information about primates' higher-level vision, and little was known about the anatomy and functional organization of higher-level visual regions. As new data continue to be obtained and the understanding of the entire visual process deepens, it has been found that the visual process is less and less like a single-channel processing process (Edelman 1999).

Fundamentally, a correct representation of the objective scene should be available for any visual work. If this is not the case, then the visual world itself (which is an external appearance of internal representations) cannot support visual behavior. Nonetheless, further research has revealed that **reconstruction-based representations** are in many respects (see below) a poor explanation of vision, or involve a set of problems (Edelman 1999).

Let us first look at the implications of reconstruction for recognition or classification. If the visual world can be built inside, then the visual system is not necessary. In fact, acquiring an image, building a 3-D model, or even giving a list of locations of important stimulus features, does not guarantee recognition or classification. Of all the possible methods for interpreting the scene, the method involving reconstruction has the largest circle to run, since reconstruction does not directly contribute to the interpretation.

Secondly, it is also difficult to achieve reconstruction only by reconstruction from the original image. From a computer vision point of view, it is very difficult to recover scene representations from original images; there are now many findings in biological vision that support other representation theories.

Finally, reconstruction theory is also problematic conceptually. The source of the problem is the fact that theoretically, reconstruction can be applied to any representation work. Leaving aside the question of whether reconstruction is achievable in concrete terms, one might first ask whether it is worthwhile to seek a representation with universal unity. Since the best representation should be the one best suited to the task, a representation with universal uniformity may not be necessary. In fact, according to the theory of information processing, the importance of choosing the appropriate and correct representation for a given computational problem is self-evident. Marr himself has also pointed out this importance.

1.3.3.2 Representation Without Reconstruction

Several studies and experiments in recent years have shown that the interpretation of the scene does not necessarily have to be based on its 3-D reconstruction, or rather, it is not necessarily based on the complete 3-D reconstruction of the scene.

Since there are a series of problems with realizing representation according to reconstruction, other forms of representation methods have also been studied. For example, another representation first proposed by Locke in *An Essay Concerning Human Understanding* is now generally referred to as **mental representation semantics** (Edelman 1999). Locke suggests representing in a natural and predictable way. According to this view, a sufficiently reliable feature detector constitutes a primitive representation of the existence of a certain feature in the visual world. The representation of the entire goal and scene can then be constructed from these primitives (if there are enough of them).

In the theory of natural computing, the original concept of feature hierarchy was developed, influenced by the discovery of "insect detectors" in frog retinas. Recent computer vision and computational neuroscience research results suggest that modifications to the original feature-level representation hypothesis can serve as an alternative to the reconstruction theory. Today's feature detection differs from traditional feature detection in two ways. One is that a set of feature detectors can have much greater representative power than any one of them; the other is that many theoretical researchers realize that "symbols" are not the only elements that combine features.

Consider the representation for spatial resolution as an example. In a typical situation, the observer can see two straight line segments that are very close to each other (the offset distance between them may also be smaller than the distance between the photon receptors in the fovea). An early hypothesis was that at some stage of cortical processing, visual input is reconstructed with sub-pixel accuracy, making it possible to obtain distances in the scene that are smaller than pixels. Proponents of reconstruction theory do not believe that feature detectors can be used to build visual functions, Marr believes that "the world is so complex that it is impossible to analyze with feature detectors". Now this view is challenged. Taking the representation of spatial resolution as an example, a set of patterns covering the viewing field can contain all the information needed to determine the offset without the need for reconstruction.

As another example, consider the perception of **relative motion**. In monkeys' mid-cortical regions, receptor cells can be found that have movements aligned with a particular direction. The combined movement of these cells can be considered to represent the movement of the **field of view** (FOV). To illustrate this, note that a given mid-cortical region and determining movement in the FOV occur synchronously. Artificial simulations of cells produce similar behavioral responses to real moving stimuli, with the result that cells reflect motor events, but visual movements are difficult to reconstruct from movements in mid-cortical regions. This means that motion can be determined without reconstruction.

The above discussion shows that new thinking is needed for Marr's theory. A computationally hierarchical description of a task determines its input and output representations. For a low-level task, such as binocular vision, the input and output are well defined. A system with stereo vision must receive two different images of the same scene, and also need to produce a representation that unambiguously represents depth information. However, even in such a task, reconstruction is not entirely necessary. In stereoscopic viewing, qualitative information, such as the depth order of viewing surfaces, is useful and relatively easy to compute, and also approximates to what the human visual system actually does.

In high-level work, the choice of representation is less clear. A recognition system must be able to accept images of the object or scene to be recognized, but what should the representation of the desired recognition look like? It is not enough to store and compare raw images of objects or scenes. As many researchers have pointed out, the appearance of objects is related to the direction in which they are viewed, to the lighting on them, and to the presence and distribution of other objects. Of course, the appearance of an object is also related to its own shape. Can one recover the geometric properties of an object from its appearance and use it as its representation? Previous research has shown that this is also not feasible.

To sum up, on the one hand a complete reconstruction looks unsatisfactory for many reasons, and on the other hand it is unreliable to represent the object only with the original image. However, these relatively obvious methodological shortcomings do not imply that the entire theoretical framework based on representative concepts is wrong. It is only these shortcomings that suggest the need for further examination of the underlying assumptions behind this notion of representation.

1.3.4 Research on the New Theoretical Framework

For historical reasons, Marr did not study how to use mathematical methods to strictly describe visual information. Although he studied early vision more fully, he did not discuss the representation and utilization of visual knowledge, as well as the recognition based on visual knowledge. In recent years there have been many attempts to establish a new theoretical framework. Grossberg, for example, claimed to have established a new theory of vision: **apparent dynamic geometry** (Grossberg and Mingolia 1987). Challenging Marr's theory, it points out that the perceived surface shape is the aggregate result of multiple processing actions distributed over multiple spatial scales, so in practice the so-called 2.5-D graphs do not exist.

Another new vision theory is the **network-symbol model** (Kuvich 2004). Under this model framework, there is no need to precisely compute the 3-D model of the scene, but instead the image is transformed into an understandable relational format similar to the knowledge model. This is similar to the human visual system. In fact, it is very difficult to process natural images with geometric operations. The human brain constructs the relational network-symbol structure of the visual scene, and uses different cues to establish the relative order of the object surface relative to the observer and the interrelationships between various objects. In the network-symbol model, object recognition is performed not according to the field of view but according to the derived structure, which is not affected by local variations and object appearance.

Two other representative works are introduced below.

1.3.4.1 Knowledge-Based Theoretical Framework

Knowledge-based theoretical frameworks have been developed around the study of **perceptual feature clusters** (Lowe 1987, 1988; Goldberg 1987), their physiological basis being derived from research findings in psychology. This theoretical framework argues that the human visual process is only a recognition process and has nothing to do with reconstruction. For 3-D object recognition, human perception can be used to describe the object, which can be done directly through 2-D images under the guidance of knowledge, without the need for bottom-up 3-D reconstruction through visual input.

The process of understanding a 3-D scene from a 2-D image can be divided into the following three steps (see Figure 1.11):

(1) Using the process of perceptual organization, extract from image features those groupings and structures that remain unchanged over a large range with respect to the viewing direction.

(2) Building a model with the help of image features, and using the probabilistic queuing method to reduce the search space in this process.

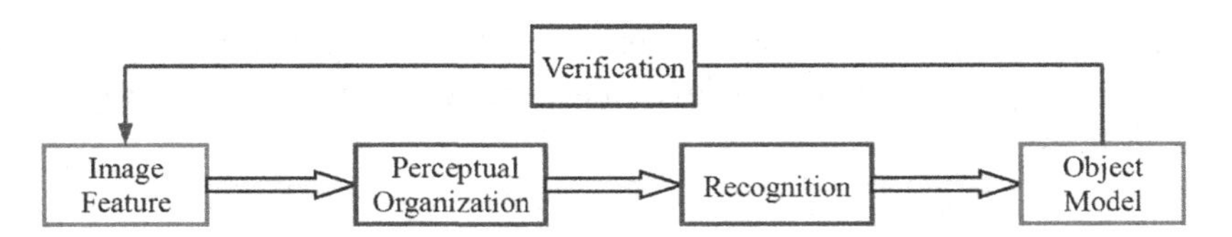

FIGURE 1.11 A knowledge-based theoretical framework.

(3) Finding the spatial correspondence by solving the unknown observation points and model parameters, so that the projection of the 3-D model directly matches the image features.

Throughout this process, there is no need to measure the 3-D object surface (no reconstruction), and information about the surface is deduced using the principle of perception. This theoretical framework shows high stability for handling occlusion and incomplete data. It introduces feedback, emphasizing the guiding role of high-level knowledge in vision. However, practice has shown that on some occasions recognition alone is not enough, such as when judging the size of the objects and estimating the distance among them, the 3-D reconstruction must be carried out. In fact, 3-D reconstruction still has a very wide range of applications. For example, in the virtual human project, a lot of human information can be obtained by 3-D reconstruction of a series of slices. As another example, the 3-D distribution of cells can be obtained by 3-D reconstruction of tissue slices, and the result has a good auxiliary effect on the localization of cells.

1.3.4.2 Active Vision Theory Framework

The active vision theoretical framework is mainly based on the initiative of human vision (or more generally biological vision). Human vision has two special mechanisms:

(1) *Selective attention mechanism.* Not all that the human eye sees is what people care about, and useful visual information is usually only distributed in a certain spatial range and time period, so human vision does not treat all parts of the scene equally, but selectively according to need. Special attention is paid to some parts, and for others just general observations are made or even a blind eye is turned. According to the **selective attention mechanism**, multi-azimuth and multi-resolution sampling can be performed when acquiring images, and information relevant to a specific task can be selected or retained.

(2) *Gaze control.* People can adjust their eyeballs so that they can "gaze" at different positions in the environment at different times according to their needs to obtain useful information, and this is called **gaze control**. Accordingly, the parameters of the camera can be adjusted so that it can always obtain visual information suitable for a specific task. Gaze control can be divided into **gaze stabilization** and **gaze change**. The former is a localization process, such as object detection; the latter is similar to the rotation of the eyeball, which controls the next fixation point according to the needs of a specific task.

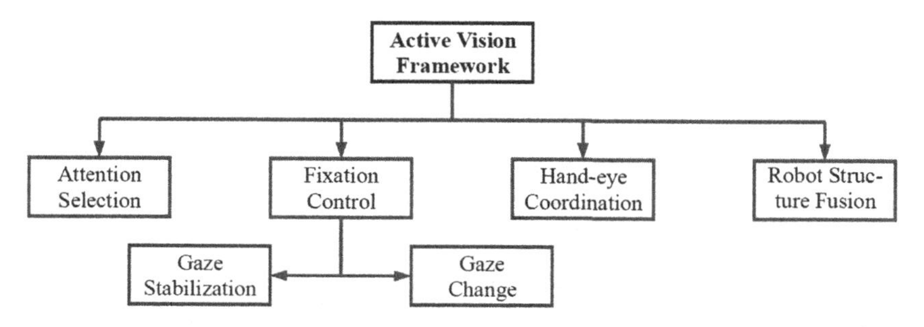

FIGURE 1.12 Active vision-based theoretical framework.

The theoretical framework of active vision proposed according to the human visual mechanism is shown in Figure 1.12.

The active vision theoretical framework emphasizes that the visual system should be **task oriented** and **purpose oriented**, and the vision system should have the ability to actively perceive. According to the existing analysis results and the current requirements of the vision task, the active vision system can control the motion of the camera through the mechanism of actively controlling the camera parameters, and coordinate the relationship between the processing task and the external signal. These parameters include camera position, orientation, focal length, aperture, etc. Active vision also incorporates the ability of "attention". By changing camera parameters or processing post-camera data, the "point of attention" can be controlled to achieve selective perception of space, time, resolution, etc.

Similar to knowledge-based theoretical frameworks, active vision theoretical frameworks also attach great importance to knowledge, arguing that knowledge belongs to high-level abilities that guide visual activities, and that these abilities should be utilized when completing visual tasks. However, feedback is lacking in current active vision theoretical frameworks. On the one hand, this feedback-free structure is not in line with the biological vision system; on the other hand, it often leads to problems including poor result accuracy, high computational complexity, significant noise effects, and lack of adaptability to some applications and environments.

1.4 CHARACTERISTICS OF THIS BOOK

There are many books about image technology, so what are the special characteristics of this book? In addition to structure and arrangement, we will consider the three aspects of writing motivation, material selection, and contents.

1.4.1 Writing Motivation

Image engineering covers a wide range of fields and contains many technologies. It is a huge project to fully understand and master image technology step by step. However, in many image applications and related scientific research and development work, it is often necessary to use specific and specialized image technology to complete the task as soon as possible. Many textbooks gradually introduce image technology starting with the basics

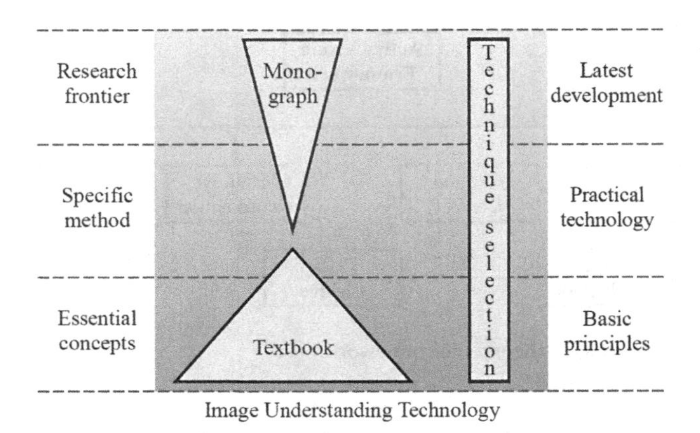

FIGURE 1.13 A complete introduction to the three layers of image understanding.

and progressing to more complex depths, but with this sequential approach it takes a long time for readers to reach a specific level of understanding. Although some monographs include an in-depth introduction to specific image technology, they require readers to have a better foundation at the beginning, so they are not suitable for readers having their first exposure to image technology and a particular task to perform.

As the schematic diagram in Figure 1.13 shows, a complete introduction to image engineering should consist of three parts (in fact, similar to any discipline and field), or three layers (corresponding to the three parts from bottom to top in the figure). The first layer is essential concepts and basic principles, the second layer covers specific methods and practical techniques, and the third layer is about research frontiers and the latest developments. Most textbooks focus on the first layer (as shown by the lower-middle triangle in Figure 1.13), and cover mostly essential concepts and basic principles. If one starts with the essential concepts and learns basic principles little by little, the foundations will be relatively solid, but it will take a long time to reach the second layer. It will be difficult for people working in different fields who only need certain skills and do not use many other concepts and principles. The monographs mainly focus on the third layer (as shown by the upper-middle triangle in Figure 1.13), in which the latest developments can be used as a reference for cutting-edge innovative scientific research, but the reader requires a higher level of technical understanding, and some technologies and methods may not be mature enough to solve current real-world problems.

This book attempts to combine the strengths of the textbook and the monograph, and to fill the gap between them, to meet the needs of readers who do not have a comprehensive foundation in image technology but need to use image techniques to solve specific tasks. To this end, according to the classification of image technology, we first select some of the more recently applied techniques to meet the needs of readers with specific applications; and then provide a step-by-step introduction to each type of technology, starting with the basic principles, so that readers with less fundamental knowledge can learn. We call it a selection of techniques, and for these selected techniques, all three layers are touched on, as shown in the rectangle on the right of Figure 1.13. An overview of the essential concepts and basic principles of each technique is given, providing sufficient explanation of current specific methods and practical techniques, as well as some discussion of research frontier trends and the latest results in conjunction with the development of technical methods.

1.4.2 Material Selection and Contents

This book focuses on IU (refer to Zhang 2017c) and selects seven technical fields that are currently receiving widespread attention and are commonly used in many applications: (i) camera model and calibration; (ii) stereo vision; (iii) generalized matching; (iv) scene analysis and semantic interpretation; (v) multi-sensor image information fusion; (vi) content-based visual information retrieval; and (vii) understanding spatial-temporal behavior. Definitions and explanations of relevant terms appearing in the book can be found in Zhang (2021).

Each chapter focuses on one type of technology. The following summarizes the contents of these seven chapters separately: Chapter 2 introduces the camera model and calibration techniques. Linear and nonlinear camera models are introduced, with basic camera calibration methods as well as internal and external parameters; and the classification of correction methods are also discussed. The traditional calibration method and the newer self-calibration method are described in detail, and some recent developments are introduced. Chapter 3 introduces stereo vision technology. Depth imaging methods, and the difference between depth image and grayscale image are presented, together with various methods of binocular stereo imaging. Region-based and feature-based stereo matching methods are introduced, both key techniques for obtaining depth information from stereo vision. Finally, some recent developments are introduced. Chapter 4 introduces generalized matching techniques and classification methods, and the connection and difference between matching and registration are discussed. Object matching principles and measurements are explained, together with a dynamic pattern matching technology, matching of various interrelationships between objects, basic definitions and concepts of graph theory, and how to use graph isomorphisms to match. Chapter 5 introduces scene analysis and semantic interpretation techniques. Among typical scene analysis techniques are fuzzy reasoning, predicate logic systems, and techniques for the labeling and classifying of scene objects. Chapter 6 introduces multi-sensor image information fusion technology, focusing on image fusion, especially the steps, levels, and effect evaluation of fusion. Pixel-level, feature-level and decision-level fusion technologies are analyzed and discussed. Rough set theory and its application in decision-level fusion are also presented in detail. Chapter 7 introduces content-based visual information retrieval techniques. Basic visual feature matching and retrieval methods are explained, and video retrieval based on motion information and two applications of video program retrieval are discussed in detail along with semantic-based retrieval. Chapter 8 discusses spatial-temporal behavior understanding technology, including space-time technology, the detection of spatial-temporal interest points, the analysis of spatial-temporal target dynamic trajectories, the classification and recognition of subject actions, the modeling of activities and behaviors, and the joint modeling of actors and actions.

This book assumes that the reader has a certain amount of background in science and engineering, and has some understanding of linear algebra, matrices, signal processing, statistics, and probability. Knowledge of some basic image concepts, such as pixels, image representation, image display, image transformation, image filtering, and so on is useful, along with elementary signal processing, because 2-D images can be seen as an augmentation of 1-D signals, and IU is an extension of signal processing. This main thrust of the book is the use of image technology to solve practical problems, so the experience and basic skills of practitioners in related industries are very useful.

This book does not give much consideration to the content from the comprehensive and systematic point of view; rather, it focuses on several specific technologies and provides information from basic to advanced levels. Although it is not written as a pure textbook, it can be used as a supplement to the textbook, especially for in-depth introductions to specific directions. This book is not a monograph in the traditional sense. It not only emphasizes advanced and real-time features, but also introduces some of the more mature technologies on the horizon (and also considers some of the latest scientific research results). This book attempts to cover the vertical range from introductory textbook to research monograph in selected technical directions to meet the specific needs of readers (Figure 1.13).

1.4.3 Structure and Arrangement

The styles of the following chapters of this book are relatively consistent. At the beginning of each chapter, in addition to the introduction of the basic concepts and overall content, some applications of the corresponding technologies are listed, which are reflected in the idea of application services; there is also an overview of each section to grasp the context of the whole chapter.

There are some similarities in the arrangement and structure of the body content of each chapter. Each chapter has multiple sections, which can be divided into the following three parts (corresponding to the three levels in Figure 1.13).

Principle and technology overview. The first section (or some second sections) at the beginning of each chapter has contents typical of textbooks. It introduces the principles, history, uses, methods and development of the image technology. Most of the information is drawn from professional textbooks (refer to Zhang 2017c).

Description of specific technical methods. The next few sections in the middle of each chapter combine the contents of textbooks and monographs. Typical relevant technologies are described in detail in terms of methods. The goal is to provide some ideas that can effectively and efficiently solve the problems faced by this type of image technology and introduce practical application solutions. These sections can have a progressive relationship or a relatively independent parallel relationship. Much of the content is drawn from literature in journals or conference papers that has been followed up and researched, but not made into professional textbooks or books.

Introduction to recent developments and directions. The last section of each chapter is more research oriented. It is based on the analysis and review of relevant new literature in recent important journals or conference proceedings. The goal is to provide some of the

TABLE 1.7 Classification Table of the Corresponding Sections of the Text of Each Chapter in this Book

#	Technology	Principle	Typical technique	Progress/trends
Chapter 2	Camera model and calibration	Sections 2.1–2.2	Sections 2.3–2.5	Section 2.6
Chapter 3	Stereo vision	Section 3.1–3.2	Sections 3.3–3.4	Section 3.5
Chapter 4	Generalized matching	Section 4.1	Sections 4.2–4.5	Section 4.6
Chapter 5	Scene analysis and semantic interpretation	Section 5.1	Sections 5.2–5.5	Section 5.6
Chapter 6	Multi-sensor image information fusion	Section 6.1–6.2	Sections 6.3–6.5	Section 6.6
Chapter 7	Content-based visual information retrieval	Section 7.1	Sections 7.2–7.5	Section 7.6
Chapter 8	Understanding spatial-temporal behavior	Section 8.1	Sections 8.2–8.6	Section 8.7

latest relevant information on focusing techniques and to help understand progress and trends in the corresponding technology.

The arrangement of the main text in sections of each chapter is shown in Table 1.7.

From the perspective of understanding the technical overview, one can only look at the sections of the principle introduction. If one wants to solve practical problems, one needs to learn some typical techniques. To master the technology more deeply, one can also refer to the recent progress/trends and look at more references.

REFERENCES

Aloimonos, Y. (ed.). 1992. Special issue on purposive, qualitative, active vision. *CVGIP-IU*, 56(1): 1–129.

Bishop, C.M. 2006. *Pattern Recognition and Machine Learning*. Berlin: Springer.

Bow, S.T. 2002. *Pattern Recognition and Image Preprocessing*. 2nd Ed. New York, NY: Marcel Dekker, Inc.

Davies, E.R. 2005. *Machine Vision: Theory, Algorithms, Practicalities*. 3rd Ed. Amsterdam: Elsevier.

Dean, T., J. Allen and Y. Aloimonos. 1995. *Artificial Intelligence: Theory and Practice*. Boston, MA: Addison Wesley.

Edelman, S. 1999. *Representation and Recognition in Vision*. Cambridge, MA: MIT Press.

Finkel, L.H. and P. Sajda. 1994. Constructing visual perception. *American Scientist*, 82(3): 224–237.

Gonzalez, R.C. and R.E. Woods. 2018. *Digital Image Processing*, 4th Ed. Cambridge, UK: Pearson.

Grossberg, S. and E. Mingolia. 1987. Neural dynamics of surface perception: Boundary webs, illuminants and shape-from-shading. *CVGIP*, 37(1): 116–165.

Kong, B. 2002. Comparison between human vision and computer vision. *Ziran Zazhi*, 24(1): 51–55.

Huang, T. and P. Stucki. (eds.). 1993. Special section on 3-D modeling in image analysis and synthesis. *IEEE-PAMI*, 15(6): 529–616.

Jain, A.K. and C. Dorai. 1997. Practicing vision: Integration, evaluation and applications. *PR*, 30(2): 183–196.

Kuvich, G. 2004. Active vision and image/video understanding systems for intelligent manufacturing. *SPIE*, 5605: 74–86.

Lowe, D.G. 1987. Three-dimensional object recognition from single two-dimensional images. *Artificial Intelligence*, 31(3): 355–395.

Lowe, D.G. 1988. Four steps towards general-purpose robot vision. *Proceedings of the 4th International Symposium on Robotics Research*, 221–228.

Marr, D. 1982. *Vision – A Computational Investigation into the Human Representation and Processing of Visual Information*. New York: W.H. Freeman.

Nilsson, N.J. 1980. *Principles of Artificial Intelligence*. Palo Alto: Tioga Publishing Co.

Rosenfeld, A. 2000. Classifying the literature related to computer vision and image analysis. *CVIU*, 79: 308–323.

Shah, M. 2002. Guest introduction: The changing shape of computer vision in the twenty-first century. *IJCV*, 50(2): 103–110.

Shapiro, L. and Stockman, G. 2001. *Computer Vision*. Upper Saddle River, NJ: Prentice Hall.

Sonka, M., V. Hlavac and R. Boyle. 2014. *Image Processing, Analysis, and Machine Vision*. 4th Ed. Singapore: Cengage Learning.

Winston, P.H. 1984. *Artificial Intelligence*. 2nd Ed. Reading, MA: Addison-Wesley.

Zhang, Y.-J. 1996. Image engineering and bibliography in China. *Technical Digest of International Symposium on Information Science and Technology*, 96: 158–160.

Zhang, Y.-J. 2002. Image engineering and related publications. *International Journal of Image and Graphics*, 2(3): 441–452.

Zhang, Y.-J. 2009a. A study of image engineering. *Encyclopedia of Information Science and Technology*, 2nd Ed., VII: Chapter 575 (3608–3615). Hershey, USA: Information Science Reference (an imprint of IGI Global).

Zhang, Y.-J. 2009b. *Image Engineering: Processing, Analysis, and Understanding*. Singapore: Cengage Learning.

Zhang, Y.-J. 2015. Statistics on image engineering literatures. *Encyclopedia of Information Science and Technology*, 3rd Ed., Chapter 595 (6030–6040). Hershey, USA: Information Science Reference (an imprint of IGI Global).

Zhang, Y.-J. 2017a. *Image Engineering, Vol. 1: Image Processing*. Germany: De Gruyter.

Zhang, Y.-J. 2017b. *Image Engineering, Vol. 2: Image Analysis*. Germany: De Gruyter.

Zhang, Y.-J. 2017c. *Image Engineering, Vol. 3: Image Understanding*. Germany: De Gruyter.

Zhang, Y.-J. 2018a. Development of image engineering in the last 20 years. *Encyclopedia of Information Science and Technology*, 4th Ed., Chapter 113 (1319–1330). Hershey, USA: Information Science Reference (an imprint of IGI Global)

Zhang, Y.-J. 2018b. An overview of image engineering in recent years. *Proceedings of the 21st IEEE International Conference on Computational Science and Engineering*, 119–122, 2018.

Zhang, Y.-J. 2021. *Handbook of Image Engineering*. Singapore: Springer Nature.

Zhang, Y.-J. 2022. Image engineering in China: 2021. *Journal of Image and Graphics*, 27(4): 1009–1022.

CHAPTER 2

Camera Model and Calibration

IMAGE UNDERSTANDING STARTS WITH the acquisition of images. The most commonly used image-capture device is a (video) camera.

The **camera model** expresses the relationship between the coordinates of the scene in the world coordinate system and the coordinates of the scene in the image coordinate system, that is, the projection relationship between the 3-D object point (spatial point) and the 2-D image point of the acquired image. Camera models can be divided into linear and nonlinear camera models.

Camera calibration, also called camera correction, uses the characteristic point coordinates (X, Y, Z) of a given 3-D object and its 2-D image coordinates (x, y) to calculate the internal and external parameters of the camera, in order to establish a quantitative connection between the objective scene and the collected images. Camera calibration is the basis of machine vision technology and photogrammetry.

This chapter is organized as follows. Section 2.1 introduces the linear camera model, explains the basic imaging model and analyzes the imaging transformation and perspective transformation. A variety of approximate projection modes are discussed and the basic imaging model is extended to the general camera model. Section 2.2 describes the nonlinear camera model and the various types of imaging distortion that cause nonlinearity. A more general imaging model that takes into account distortion factors is also discussed. Basic camera calibration procedures are explored in Section 2.3, focusing on the steps of nonlinear camera calibration. In Section 2.4, traditional camera calibration is explained, and a typical two-stage calibration method is introduced, along with methods for improving its accuracy. Section 2.5 discusses recent developments in self-calibration methods and introduces a method of realization that uses active vision technology. Section 2.6 provides a brief guide to recent technique developments and promising research directions.

DOI: 10.1201/9781003362388-2

2.1 LINEAR CAMERA MODEL

The **linear model** is also called the **pinhole model**. This type of model assumes that the image formed by any point in the 3-D space on the image coordinate system is formed according to the principle of small hole imaging.

2.1.1 Imaging Transformation

In image acquisition, a camera is used to **perspectively project** the scene of the 3-D objective world onto the 2-D image plane. This projection can be spatially described by imaging transformation (also called perspective transformation or geometric perspective transformation).

2.1.1.1 Various Coordinate Systems

Imaging transformation involves transformations between different spatial coordinate systems (Zhang 2017). Considering that the final result of image acquisition is to obtain a digital image that can be input into a computer, the coordinate systems involved in imaging a 3-D space scene mainly include the following:

(1) The **world coordinate system** is also called the real-world coordinate system, XYZ, which represents the absolute coordinates of the objective world (so it is also called the objective coordinate system). General 3-D scenes are represented by this coordinate system.

(2) The **camera coordinate system** is a coordinate system, xyz, formulated with the camera as the center. Generally, the optical axis of the camera is taken as the z-axis.

(3) The **image plane coordinate system** is the coordinate system, $x'y'$, on the imaging plane in the camera. Generally, the imaging plane is parallel to the xy plane of the camera coordinate system. In addition, the x-axis and the x'-axis, as well as the y-axis and y'-axis are coincident, respectively, so that the origin of the image plane is on the optical axis of the camera.

(4) The **computer image coordinate system** is the coordinate system, MN, used to express images inside the computer. The image is finally stored in the memory in the computer, so the projection coordinates of the image plane must be converted to the computer image coordinate system.

2.1.1.2 Imaging Model

According to the different interrelationships of the several coordinate systems described above, different (camera) imaging models can be obtained. Consider the most basic and simplest case first: the world coordinate system coincides with the camera coordinate system, and the camera coordinate system coincides with the image plane coordinate system (the computer image coordinate system is not considered first).

Figure 2.1 is a schematic diagram of the geometric perspective transformation model of the imaging process, where the image plane in the camera coordinate system xyz coincides

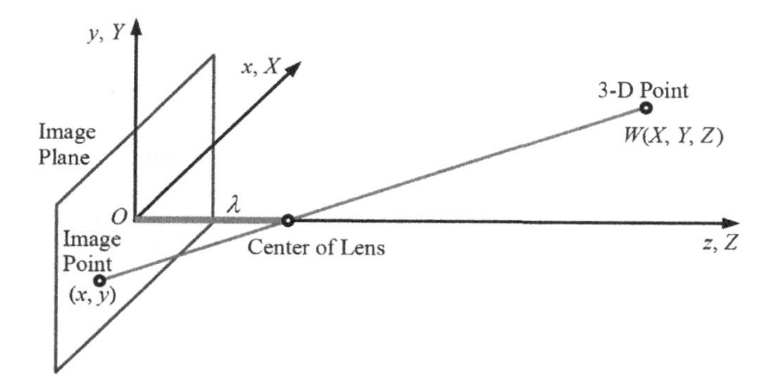

FIGURE 2.1 A basic image capturing model.

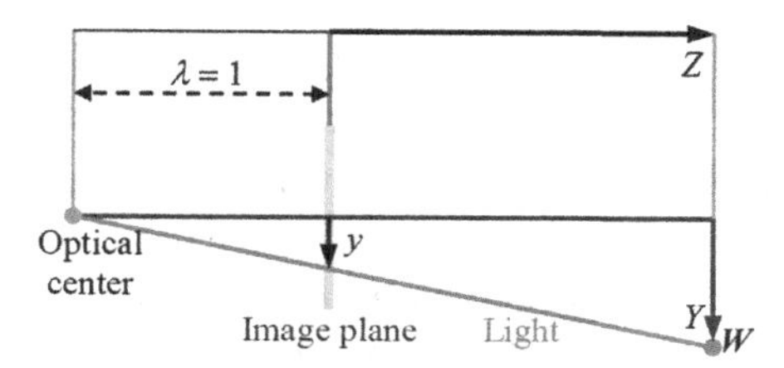

FIGURE 2.2 Normalized camera.

with the xy plane and the optical axis (through the lens center) is positively outward along the z-axis. In this way, the center of the image plane is at the origin, the coordinates of the lens center are $(0, 0, \lambda)$, and λ represents the focal length of the lens (Zhang 2017).

A special case of the above situation is the case where the focal length is 1. This particular camera is called a normalized camera, and also refers to a simplified camera model. Figure 2.2 shows a section of the model (X is a constant in YZ plane), in which the x-axis and X-axis go from the inside to the outside of the paper, the y-axis and Y-axis go from top to bottom, and the Z-axis goes from left to right. The y coordinate in the image corresponding to the point $\mathbf{W} = [X, Y, Z]^{\mathrm{T}}$ of the world coordinate system is Y/Z (the x coordinate is X/Z). It can be seen from the figure that for the farther (larger Z) target, its projection is closer to the center of the image (the smaller y is).

2.1.1.3 Perspective Transformation

Perspective transformation establishes the geometric relationship between space point coordinates (X, Y, Z) and image point coordinates (x, y). In the following discussion, assume that $Z > \lambda$, that is, all the points of interest in the objective scene are in front of the lens. According to Figure 2.2, the following two equations can be easily obtained with the help of the relationship of similar triangles (Zhang 2017):

$$\frac{x}{\lambda} = \frac{-X}{Z-\lambda} = \frac{X}{\lambda - Z} \tag{2.1}$$

$$\frac{y}{\lambda} = \frac{-Y}{Z-\lambda} = \frac{Y}{\lambda - Z} \tag{2.2}$$

In these equations, the negative signs before both X and Y mean that the image point is reversed. From these two equations, the image plane coordinates after 3-D point perspective projection can be obtained:

$$x = \frac{\lambda X}{\lambda - Z} \tag{2.3}$$

$$y = \frac{\lambda Y}{\lambda - Z} \tag{2.4}$$

The above perspective transformation projects the line segments in the 3-D space (except along the projection direction) to the line segments on the image plane. If the line segments are parallel to each other in the 3-D space and also parallel to the projection plane, these line segments are still parallel to each other after projection. The rectangle in the 3-D space may be any quadrilateral after being projected onto the image plane, which is determined by four vertices. Therefore, perspective transformation is often referred to as **4-point mapping**.

Perspective transformation is a special case of **projection transformation**, as are various affine transformations. The projection transformation matrix is non-singular, and can map three collinear points into three collinear points. Projection transformation is also called **collinearity** or **homography**.

The camera focal length used in practice is not always 1, and the position is represented by pixels rather than physical distance on the image plane. Considering these two factors, referring to Figure 2.2, the relationship between image plane coordinates and world coordinates is (S is the scale factor):

$$x = \frac{SX}{Z} \tag{2.5}$$

$$y = \frac{SY}{Z} \tag{2.6}$$

It should be noted that the change of focal length and the spacing of photon receiving units in the sensor will affect the relationship between both image plane coordinate points and world coordinate points. As shown in Figures 2.3(a) and 2.3(b), when the focal length is reduced to half, the imaging size (such as y) is also reduced to half. However, the field of view increases with the decrease in focal length. As shown in Figures 2.3(c) and 2.3(d), the imaging size determined in pixels decreases with the increase in sensor unit spacing, When the sensor density (corresponding number) is reduced to half, the number of imaging pixels is also reduced to half. Taken together, the focal length and sensor density change the mapping relationship from scene to pixel in the same way.

2.1.1.4 Telecentric Imaging and Supercentric Imaging

In a general standard optical imaging system, the light beam is convergent. This has obvious adverse effects on optical measurement (see Figure 2.4(a)). If the position of the

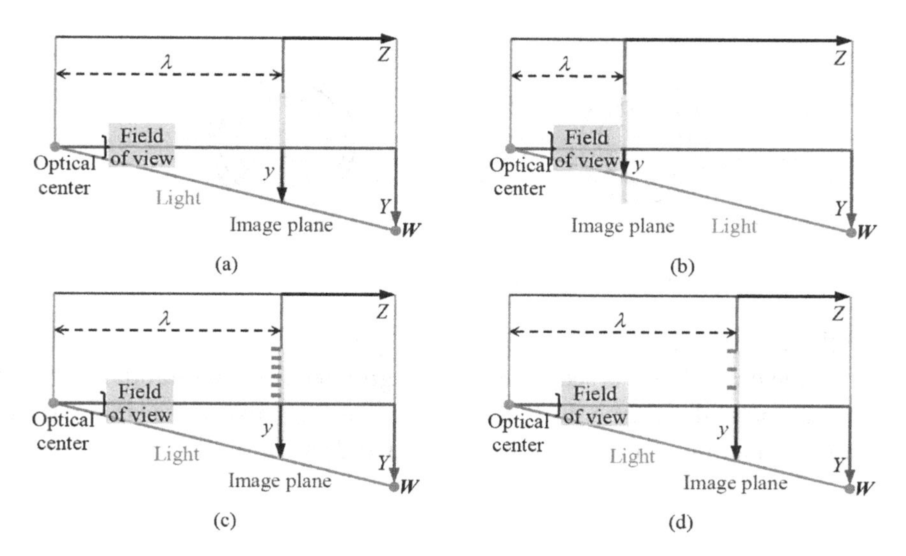

FIGURE 2.3 The effect of focal length and sensor unit pitch changes.

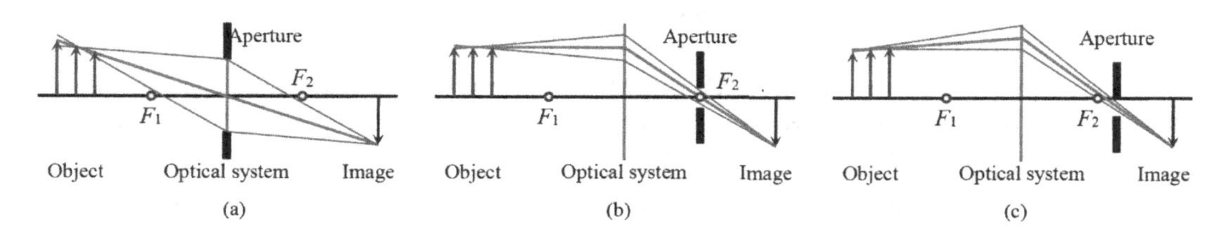

FIGURE 2.4 Moving the aperture position can change the property of the optical system.

target changes, its image will become larger when the target is close to the lens and smaller when it is far away from the lens. Because the depth of the target cannot be obtained directly from the image, measurement errors are inevitable unless the target is placed at a known distance. One way to solve this problem is to use a telecentric lens to shoot the object under test.

If the position of the aperture is moved to the convergence point (F_2) of the parallel light, a **telecentric imaging** system is obtained (see Figure 2.4(b)). At this time, the principal ray (the light passing through the center of the aperture) is parallel to the optical axis in the target space, and a small change in the target position does not change the size of the target image (Zhang 2017). Of course, the farther the target is from the focus position, the more blurred it will be. However, in this case, the center of the blurred disc does not change its position. The disadvantage of telecentric imaging is that the diameter of the telecentric lens must at least reach the size of the target to be imaged. Thus telecentric imaging of large-scale targets is very expensive.

If the aperture is placed closer to the image plane than the convergence point of the parallel light (Figure 2.4(c)), the principal ray becomes the convergence line of the target space. Contrary to standard imaging (Figure 2.4(a)), the far target looks bigger at this time! This imaging technique is called **supercentric imaging**. One of its features is that the

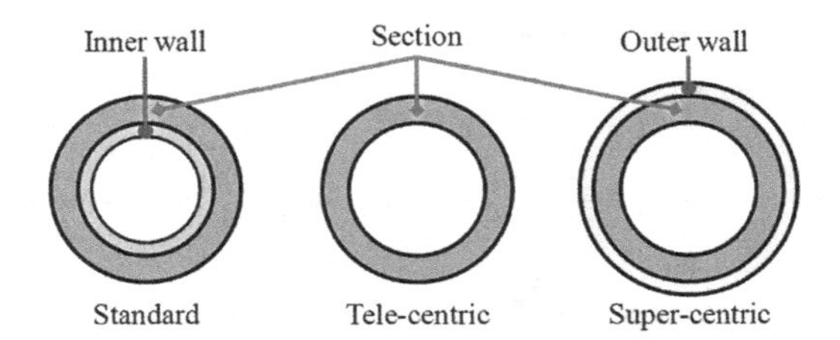

FIGURE 2.5 Comparison of three imaging methods.

surface parallel to the optical axis can be seen. Figure 2.5 illustrates how a thin-walled cylinder along the optical axis is seen in three imaging techniques. Standard imaging can see the cross-section and the inner wall, telecentric imaging can only see the cross-section, and supercentric imaging can see the cross-section and the outer wall (Jähne 2004).

2.1.1.5 Homogeneous Coordinates

Equations (2.3) and (2.4) are both nonlinear because they contain the variable Z in the denominator. In order to express them in the form of a linear matrix, they can be expressed homogeneously with the help of homogeneous coordinates (Zhang 2017).

The homogeneous coordinates corresponding to a point with coordinates $[X, Y, Z]$ in Cartesian space are defined as $[kX, kY, kZ, k]$, where k is an arbitrary non-zero constant. Obviously, a transformation from homogeneous coordinates back to Cartesian coordinates can be achieved by dividing the first three coordinate quantities with the fourth coordinate quantity. The 3-D space points in such a Cartesian world coordinate system can be expressed in vector form as

$$\boldsymbol{W} = \begin{bmatrix} X & Y & Z \end{bmatrix}^{\mathrm{T}} \tag{2.7}$$

Its corresponding homogeneous coordinates can be expressed as

$$\boldsymbol{W}_{\mathrm{h}} = \begin{bmatrix} kX & kY & kZ & k \end{bmatrix}^{\mathrm{T}} \tag{2.8}$$

If the perspective transformation matrix is defined as

$$\boldsymbol{P} = \begin{bmatrix} 1 & 0 & 0 & 0 \\ 0 & 1 & 0 & 0 \\ 0 & 0 & 1 & 0 \\ 0 & 0 & -1/\lambda & 1 \end{bmatrix} \tag{2.9}$$

The product $\boldsymbol{PW}_{\mathrm{h}}$ of $\boldsymbol{P}$ and $\boldsymbol{W}_{\mathrm{h}}$ gives a vector denoted by $\boldsymbol{c}_{\mathrm{h}}$:

$$\boldsymbol{c}_{\mathrm{h}} = \boldsymbol{PW}_{\mathrm{h}} = \begin{bmatrix} 1 & 0 & 0 & 0 \\ 0 & 1 & 0 & 0 \\ 0 & 0 & 1 & 0 \\ 0 & 0 & -1/\lambda & 1 \end{bmatrix} \begin{bmatrix} kX \\ kY \\ kZ \\ k \end{bmatrix} = \begin{bmatrix} kX \\ kY \\ kZ \\ -kZ/\lambda + k \end{bmatrix} \tag{2.10}$$

Here, the elements of $\boldsymbol{c}_\mathrm{h}$ are the camera coordinates in homogeneous form. These coordinates can be converted into Cartesian form by dividing the first three items of $\boldsymbol{c}_\mathrm{h}$ with the fourth item of $\boldsymbol{c}_\mathrm{h}$. Therefore, the Cartesian coordinates of any point in the camera coordinate system can be expressed in vector form:

$$\boldsymbol{c} = \begin{bmatrix} x & y & z \end{bmatrix}^\mathrm{T} = \begin{bmatrix} \dfrac{\lambda X}{\lambda - Z} & \dfrac{\lambda Y}{\lambda - Z} & \dfrac{\lambda Z}{\lambda - Z} \end{bmatrix}^\mathrm{T} \tag{2.11}$$

The first two items of $\boldsymbol{c}$ are the coordinates (x, y) that are the 3-D space point (X, Y, Z) projected to the image plane.

2.1.1.6 Inverse Perspective Transformation

Inverse perspective transformation refers to determining the coordinates of a 3-D objective scene according to its 2-D image coordinates, or inversely mapping an image point back to the 3-D space (Zhang 2017). Using matrix operation rules, it can obtain from Equation (2.10)

$$\boldsymbol{W}_\mathrm{h} = \boldsymbol{P}^{-1}\boldsymbol{c}_\mathrm{h} \tag{2.12}$$

where the inverse perspective transformation matrix $\boldsymbol{P}^{-1}$ is

$$\boldsymbol{P}^{-1} = \begin{bmatrix} 1 & 0 & 0 & 0 \\ 0 & 1 & 0 & 0 \\ 0 & 0 & 1 & 0 \\ 0 & 0 & 1/\lambda & 1 \end{bmatrix} \tag{2.13}$$

Can the coordinates of the corresponding 3-D objective scene points be determined from the 2-D image coordinate points by using the above-mentioned inverse perspective transformation matrix? Suppose the coordinates of an image point are $(x', y', 0)$, where 0 at the z position only means that the image plane is at $z = 0$. This point can be expressed in homogeneous vector form as

$$\boldsymbol{c}_\mathrm{h} = \begin{bmatrix} kx' & ky' & 0 & k \end{bmatrix}^\mathrm{T} \tag{2.14}$$

Substituting Equation (2.12) to get the homogeneous world coordinate vector:

$$\boldsymbol{W}_\mathrm{h} = \begin{bmatrix} kx' & ky' & 0 & k \end{bmatrix}^\mathrm{T} \tag{2.15}$$

The corresponding world coordinate vector in the Cartesian coordinate system is

$$\boldsymbol{W} = \begin{bmatrix} X & Y & Z \end{bmatrix}^\mathrm{T} = \begin{bmatrix} x' & y' & 0 \end{bmatrix}^\mathrm{T} \tag{2.16}$$

Equation (2.16) shows that the Z coordinate of a 3-D space point cannot be uniquely determined by the image point (x', y') (because it gives $Z = 0$ for any point). The problem here is caused by the many-to-one transformation that the 3-D objective scene is mapped to the image plane. The image point (x', y') now corresponds to the collection of all collinear 3-D space points on a straight line passing $(x', y', 0)$ and $(0, 0, \lambda)$ (see the connection between the image point and space point in Figure 2.1). In the world coordinate system, X and Y can be solved inversely by Equations (2.3) and (2.4):

$$X = \frac{x'}{\lambda}(\lambda - Z) \tag{2.17}$$

$$Y = \frac{y'}{\lambda}(\lambda - Z) \tag{2.18}$$

The above two equations indicate that unless there is some prior knowledge of the 3-D space point mapped to the image point (such as knowing its Z coordinate), it is impossible to completely recover the coordinates of a 3-D space point from its image point. In other words, to use the inverse perspective transformation to recover a 3-D space point from its image point, it is necessary to know at least one world coordinate of the point.

2.1.2 Approximate Projection Modes

Perspective projection is an accurate projection mode, but it is a nonlinear mapping, so calculation and analysis are more complicated. To simplify the calculation, some approximate projection modes can be used when the object distance is much larger than the scale of the scene itself.

2.1.2.1 Orthogonal Projection

Orthogonal projection, also called orthographic projection, is the simplest linear approximation. In orthogonal projection, the Z coordinate of the 3-D space point is not considered (the object distance information is lost), which is equivalent to directly/vertically mapping the 3-D space point to the image plane along the direction of the camera's optical axis. Figure 2.6 shows how a rod-shaped object is projected onto the Y-axis (corresponding to the YZ section in Figure 2.1) in two ways: orthogonal projection and perspective projection.

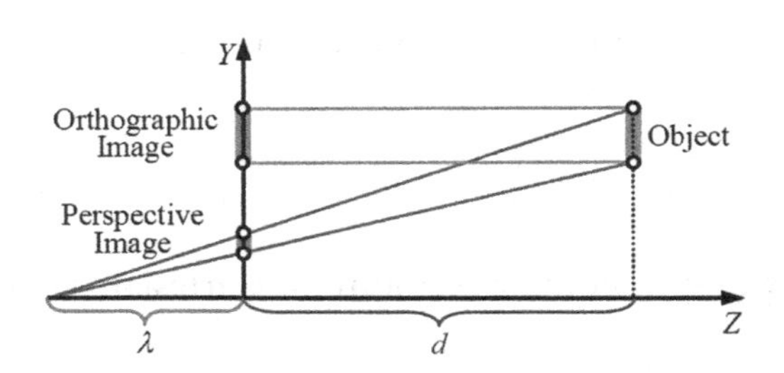

FIGURE 2.6 Comparison of orthogonal projection and perspective projection.

The result of orthogonal projection indicates the true scale of the cross-section of the scene, while the result of perspective projection is related to the object distance d. Orthogonal projection can be regarded as perspective projection when the focal length λ is infinite, so its projection transformation matrix can be written as

$$\boldsymbol{P} = \begin{bmatrix} 1 & 0 & 0 & 0 \\ 0 & 1 & 0 & 0 \\ 0 & 0 & 1 & 0 \\ 0 & 0 & 0 & 1 \end{bmatrix} \tag{2.19}$$

2.1.2.2 Weak Perspective Projection

Weak perspective projection (WPP) is an approximation of perspective projection. It is a special case when observing a specific scene, and the depth range ΔZ of the target itself is much smaller than the depth Z of the target in the scene. At this time, the depth of each point of the target itself can be approximated by the depth Z_0 of its center of mass. In this case, the image can be seen as the result of orthographic projection, where the depth information is removed, and a scale factor is also used to adjust the target to give the observed size. Therefore, in weak perspective projection, there are two steps:

(1) Orthogonally project the scene onto the plane parallel to the image plane;

(2) Project the result of the previous step to the image plane perspectively again.

Here, the perspective projection in step (2) can be achieved with equal scaling in the image plane (so weak perspective projection is also called **scaled orthographic projection**). Suppose the **perspective scaling factor** is S, then

$$S = \lambda / d \tag{2.20}$$

It is the ratio of focal length to object distance. If S is also considered, the weak perspective projection transformation matrix can be written as

$$\boldsymbol{P} = \begin{bmatrix} S & 0 & 0 & 0 \\ 0 & S & 0 & 0 \\ 0 & 0 & 1 & 0 \\ 0 & 0 & 0 & 1 \end{bmatrix} \tag{2.21}$$

Not only does orthogonal projection in weak perspective projection not consider the Z coordinate of the 3-D space point, but it also changes the relative position of each point of the scene in the projection plane (corresponding to the change in size). When the distance between the optical axis of the camera and the target in the scene is relatively large, it will have a greater impact (it can be proved that the error in image is the first-order infinitesimal of the error in scene). Generally, weak perspective projection can only be used when $Z_0 > 10|\Delta Z|$.

2.1.2.3 Parallel Perspective Projection

Parallel perspective projection (also known as **para-perspective projection**) is also a kind of projection method between orthogonal projection and perspective projection (Dean et al. 1995). A schematic diagram of the parallel perspective projection can be seen in Figure 2.7. In the figure, the world coordinate system coincides with the camera coordinate system, the camera focal length is λ, the image plane and the Z-axis perpendicularly intersect at point $(0, 0, \lambda)$, point C is the center of mass of the target set, and the distance from the origin in the Z direction is d.

Given a projection plane located at $Z = d$ and parallel to the image plane, the process of parallel perspective projection is divided into two steps:

(1) Given a specific target P (one of the target set), first project it parallel to the projection plane parallel to the image plane, and then project each projection line parallel to the straight line OC (not necessarily perpendicular to the projection plane);

(2) The projection result on the projection plane is then perspective projected onto the image plane. Since the projection plane and the image plane are parallel, the projection on the image plane is reduced to λ/d with respect to the projection on the projection plane (the same as weak perspective projection).

The first step above considers the effects of perspective and pose (retaining the relative position of each point of the target in the projection plane), while the second step considers the influence of distance and other positions. It can be proved that the image error at this time is the second-order infinitesimal of the target error, so the parallel perspective projection is closer to the perspective projection than the weak perspective projection.

By modifying the parallel perspective projection, it is also possible to obtain a linear approximation with less error, called orthogonal or orthographic perspective. **Orthogonal perspective** changes the first step of parallel perspective projection to parallel projection of the target onto a projection plane that is perpendicular to the line between the optical center and the center of mass, and passes through the center of mass. The second step remains the same. Orthogonal perspective is relatively complicated.

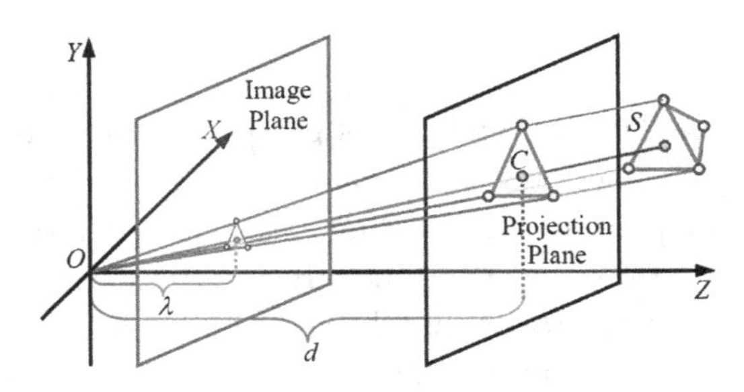

FIGURE 2.7 Para-perspective projection imaging diagram.

2.1.2.4 *Comparison of Various Approximate Modes and Perspective Projection*

The relationship between perspective projection and its two approximate modes (weak perspective projection and parallel perspective projection) can be illustrated with the help of Figure 2.8. The result of parallel perspective projection is closer to the result of perspective projection than the result of weak perspective projection. The gap between them is inversely proportional to the focal length and directly proportional to the object distance (Zhang 2017).

Orthogonal projection, weak perspective projection and parallel perspective projection all have the same form of projection matrix, which has eight degrees of freedom. A camera with such a projection matrix is called an **affine camera**. Two important properties of affine cameras are (not available in perspective projection):

(1) Preserving parallelism: parallel lines in 3-D space are still parallel after being projected into 2-D space;

(2) Centroid preservation: the centroid of the 3-D point set after projection is the centroid of the 2-D point set.

2.1.3 A General Camera Model

Now consider the situation when the camera coordinate system is separated from the world coordinate system, but the camera coordinate system coincides with the image plane coordinate system (the computer image coordinate system is still not considered). Figure 2.9 shows a schematic diagram of the geometric model of imaging at this time. The position deviation between the image plane center (origin) and the world coordinate system is recorded as a vector $\boldsymbol{D}$, and its components are D_x, D_y, and D_z. Here it is assumed that the camera **scans horizontally** at angle γ (the angle between the x and X axes) and **tilts vertically** at the angle α (the angle between the z and Z axes). If the XY plane is the equatorial plane of the earth, and the Z-axis points to the north pole of the earth, then the **sweep angle** corresponds to the longitude and the **tilt angle** corresponds to the latitude (Zhang 2017).

The above model can be converted from the basic camera model where the world coordinate system coincides with the camera coordinate system through the following series of

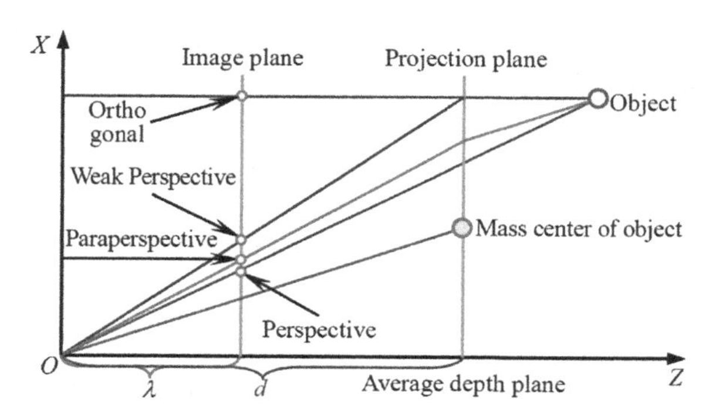

FIGURE 2.8 Different projection modes.

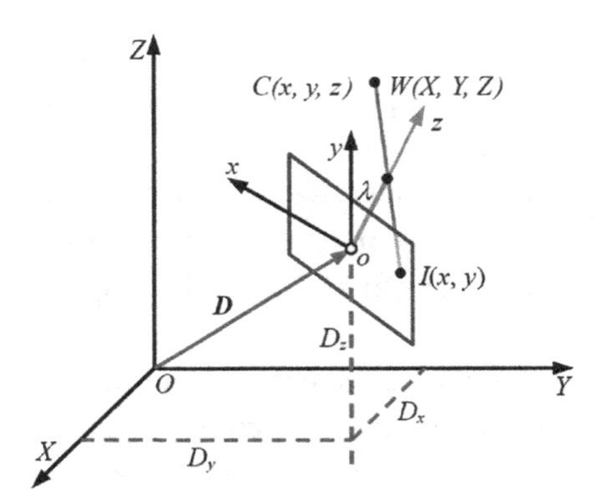

FIGURE 2.9 A general image-capture model.

steps: (i) move the origin of the image plane out of the origin of the world coordinate system by vector $\boldsymbol{D}$; (ii) use a certain sweep angle γ (around the z-axis) to scan the x-axis; and (iii) tilt the z-axis with a certain tilt angle α (rotate around the x-axis).

Moving the camera relative to the world coordinate system is also equivalent to moving the world coordinate system relative to the camera inversely. Specifically, the three steps taken in the above geometric relationship conversion can be performed on each point in the world coordinate system. Translating the origin of the world coordinate system to the origin of the image plane can be accomplished by the following transformation matrix:

$$\boldsymbol{T} = \begin{bmatrix} 1 & 0 & 0 & -D_x \\ 0 & 1 & 0 & -D_y \\ 0 & 0 & 1 & -D_z \\ 0 & 0 & 0 & 1 \end{bmatrix} \tag{2.22}$$

In other words, the homogeneous coordinate point $\boldsymbol{D}_\text{h}$ located at the coordinates (D_x, D_y, D_z) is located at the origin of the new coordinate system after the transformation $\boldsymbol{TD}_\text{h}$.

Now consider how to overlap the coordinate axes. The scanning angle γ is the angle between the x and X axes. In the normal (nominal) position, these two axes are parallel. In order to scan the x-axis at the required γ angle, you only need to rotate the camera counterclockwise (defined by looking at the origin from the positive direction of the rotation axis) around the z-axis by the γ angle:

$$\boldsymbol{R}_\gamma = \begin{bmatrix} \cos\gamma & \sin\gamma & 0 & 0 \\ -\sin\gamma & \cos\gamma & 0 & 0 \\ 0 & 0 & 1 & 0 \\ 0 & 0 & 0 & 1 \end{bmatrix} \tag{2.23}$$

The position without rotation ($\gamma = 0°$) corresponds to the parallel of the x-axis and the X-axis. Similarly, the tilt angle α is the angle between the z-axis and the Z-axis. The camera can be rotated counterclockwise around the x-axis to achieve the effect of tilting the camera axis α angle:

$$R_\alpha = \begin{bmatrix} 1 & 0 & 0 & 0 \\ 0 & \cos\alpha & \sin\alpha & 0 \\ 0 & -\sin\alpha & \cos\alpha & 0 \\ 0 & 0 & 0 & 1 \end{bmatrix} \tag{2.24}$$

The position without tilt ($\alpha = 0°$) corresponds to the parallel of the z-axis and the Z-axis.

The transformation matrices that complete the above two rotations can be cascaded to form one matrix:

$$R = R_\alpha R_\gamma = \begin{bmatrix} \cos\gamma & \sin\gamma & 0 & 0 \\ -\sin\gamma\cos\alpha & \cos\alpha\cos\gamma & \sin\alpha & 0 \\ \sin\alpha\sin\gamma & -\sin\alpha\cos\gamma & \cos\alpha & 0 \\ 0 & 0 & 0 & 1 \end{bmatrix} \tag{2.25}$$

Here R represents the effect of the camera's rotation in space.

Considering the translation and rotation transformations at the same time to coincide with the world coordinate system and the camera coordinate system, the perspective projection transformation corresponding to Equation (2.10) has the following homogeneous representation:

$$c_\mathrm{h} = PRTW_\mathrm{h} \tag{2.26}$$

Equation (2.26) is expanded and converted to Cartesian coordinates to get the coordinates in the image plane for the point (X, Y, Z) in the world coordinate system:

$$x = \lambda \frac{(X - D_x)\cos\gamma + (Y - D_y)\sin\gamma}{-(X - D_x)\sin\alpha\sin\gamma + (Y - D_y)\sin\alpha\cos\gamma - (Z - D_z)\cos\alpha + \lambda} \tag{2.27}$$

$$y = \lambda \frac{-(X - D_x)\sin\gamma\cos\alpha + (Y - D_y)\cos\alpha\cos\gamma + (Z - D_z)\sin\alpha}{-(X - D_x)\sin\alpha\sin\gamma + (Y - D_y)\sin\alpha\cos\gamma - (Z - D_z)\cos\alpha + \lambda} \tag{2.28}$$

2.2 NONLINEAR CAMERA MODEL

In real-world situations, a camera usually uses a lens (often containing multiple lenses) for imaging. Based on the current lens processing technology and camera manufacturing technology, the projection relationship of the camera cannot be simply described as a pinhole model. In other words, due to the influence of many factors such as lens processing and installation, the projection relationship of the camera is not a linear projection relationship; that is, the linear model cannot accurately describe the imaging geometric relationship of the camera.

The real optical system does not work exactly according to the idealized pinhole imaging principle, but there is **lens distortion**. Due to the influence of a variety of distortion factors, there is a deviation between the true position of the 3-D space point projected on the 2-D image plane and the ideal image point position without distortion. The optical distortion error is more obvious in the region close to the edge of the lens. Especially when using a wide-angle lens, there is often a lot of distortions in the image plane away from the center. This will cause deviations in the measured coordinates and reduce the accuracy of the obtained world coordinates. Therefore, a nonlinear camera model that takes into account the distortion must be used for camera calibration (Weng et al. 1992).

2.2.1 Type of Distortion

Due to the influence of various distortion factors, when projecting a 3-D space point onto a 2-D image plane, there is a deviation between the actual coordinates (x_a, y_a) and the undistorted ideal coordinates (x_i, y_i). The deviation can be expressed as

$$x_a = x_i + d_x \tag{2.29}$$

$$y_a = y_i + d_y \tag{2.30}$$

Among them, d_x and d_y are the total nonlinear distortion deviation values in the x and y directions, respectively. There are two common basic distortion types: **radial distortion** and **tangential distortion**, as shown in Figure 2.10, where d_r represents the deviation caused by radial distortion and d_t represents the deviation caused by tangential distortion. Other distortions are generally a combination of these two basic distortions. The most typical combined distortion is **eccentric distortion** (**centrifugal distortion**) and **thin prism distortion** (Zhang 2021).

2.2.1.1 Radial Distortion

Radial distortion is mainly caused by the irregularity of the lens shape (surface curvature error). The deviation caused by it is generally symmetrical about the main optical axis of the camera lens, and it is more obvious at the distance from the optical axis along the lens

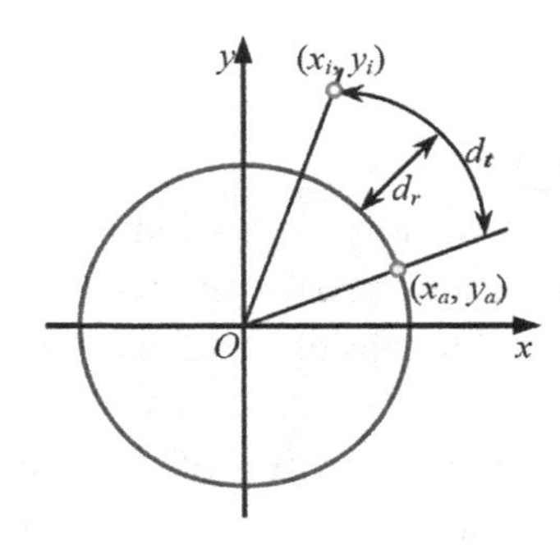

FIGURE 2.10 Schematic diagram of radial distortion and tangential distortion.

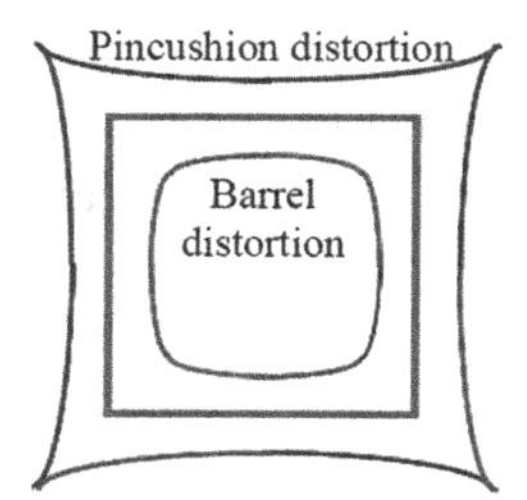

FIGURE 2.11 Schematic diagram of pincushion distortion and barrel distortion.

radius. Generally, the positive radial distortion is called **pincushion distortion**, and the negative radial distortion is called **barrel distortion**, as shown in Figure 2.11. The mathematical model is

$$d_{xr} = x_i \left(k_1 r^2 + k_2 r^4 + \cdots \right) \tag{2.31}$$

$$d_{yr} = y_i \left(k_1 r^2 + k_2 r^4 + \cdots \right) \tag{2.32}$$

where $r = (x_i^2 + y_i^2)^{1/2}$ is the distance from the image point to the image center, while k_1 and k_2 are radial distortion coefficients.

2.2.1.2 Tangential Distortion

Tangential distortion is mainly caused by the non-collinearity of the optical centers of the lens groups, resulting in the tangential movement of the actual image points on the image plane. Tangential distortion has a certain orientation in space, so there is the maximum distortion axis in a certain direction and the minimum distortion axis in the direction perpendicular to this direction, as shown in Figure 2.12, in which the solid line represents the situation without distortion and the dotted line represents the result caused by tangential distortion. Generally, the influence of tangential distortion is relatively small, and there are few cases of modeling alone.

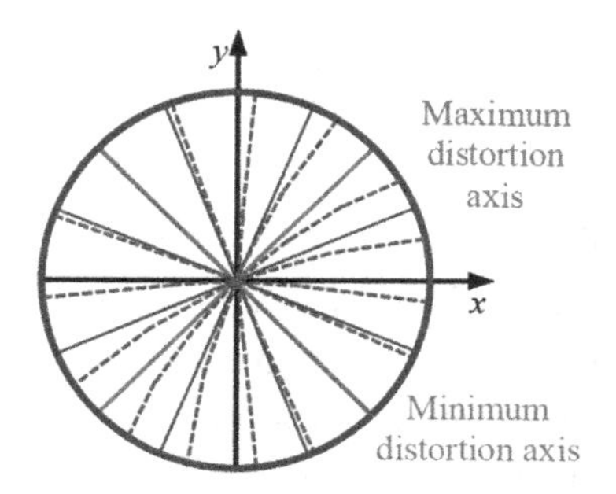

FIGURE 2.12 Tangential distortion.

2.2.1.3 Eccentric Distortion

Eccentric distortion is caused by the inconsistency between the optical center of the optical system and the geometric center, that is, the optical centers of the lenses are not strictly collinear. The mathematical model is

$$d_{xt} = l_1\left(2x_i^2 + r^2\right) + 2l_2x_iy_i + \cdots \tag{2.33}$$

$$d_{yt} = 2l_1x_iy_i + l_2\left(2y_i^2 + r^2\right) + \cdots \tag{2.34}$$

Among them, $r = (x_i^2 + y_i^2)^{1/2}$ is the distance from the image point to the image center, and l_1, l_2, etc. are the eccentric distortion coefficients.

2.2.1.4 Thin Prism Distortion

Thin prism distortion is caused by improper lens design and assembly. This kind of distortion is equivalent to adding a thin prism to the optical system, which not only causes radial deviation, but also tangential deviation. The mathematical model is

$$d_{xp} = m_1\left(x_i^2 + y_i^2\right) + \cdots \tag{2.35}$$

$$d_{yp} = m_2\left(x_i^2 + y_i^2\right) + \cdots \tag{2.36}$$

where m_1, m_2, etc. are the distortion coefficients of the thin prism.

The total distortion deviation d_x and d_y in considering radial distortion, eccentric distortion and thin prism distortion are

$$d_x = d_{xr} + d_{xt} + d_{xp} \tag{2.37}$$

$$d_y = d_{yr} + d_{yt} + d_{yp} \tag{2.38}$$

Ignoring terms higher than order 3, and let $n_1 = l_1 + m_1$, $n_2 = l_2 + m_2$, $n_3 = 2l_1$, $n_4 = 2l_2$, then

$$d_x = k_1xr^2 + \left(n_1 + n_3\right)x^2 + n_4xy + n_1y^2 \tag{2.39}$$

$$d_y = k_1yr^2 + n_2x^2 + n_3xy + \left(n_2 + n_4\right)y^2 \tag{2.40}$$

2.2.2 A Complete Imaging Model

The various types of distortion introduced above have their own characteristics. In actual imaging applications, it is often necessary to consider one or more of them according to specific conditions. In addition, the imaging model in practical applications should also consider two factors: (i) The camera coordinate system xyz is not only separate from the world coordinate system XYZ, but also separate from the image coordinate system $x'y'$; and

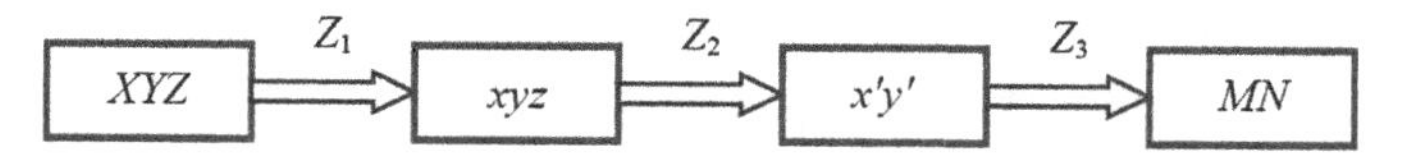

FIGURE 2.13 The coordinate system conversion schematic diagram of the complete imaging process.

(ii) the ultimate goal of imaging is to be used for computer processing, so it is necessary to establish a connection between the world coordinate system and the computer image coordinate system *MN*. Since the image coordinate unit used in the computer is the number of discrete pixels in the memory, it is necessary to round and transform the coordinates on the image plane. It has also been suggested that there are two types of image coordinate system: the image physical coordinate system and the image pixel coordinate system. The former corresponds to the coordinates on the image plane, while the latter corresponds to the coordinates in the computer.

From this point of view, a complete imaging process involves a total of three conversions between four non-coincident coordinate systems (distortion should be considered according to the actual situation), as shown in Figure 2.13 (Zhang 2017):

(1) Conversion Z_1 from the world coordinate system, *XYZ*, to the camera coordinate system, *xyz*. This conversion is expressed as

$$\begin{bmatrix} x \\ y \\ z \end{bmatrix} = \boldsymbol{R} \begin{bmatrix} X \\ Y \\ Z \end{bmatrix} + \boldsymbol{T} \tag{2.41}$$

Among them, $\boldsymbol{R}$ and $\boldsymbol{T}$ are respectively a 3×3 rotation matrix (actually a function of the angle between the three sets of corresponding coordinate axes of the two coordinate systems) and a 1×3 translation matrix:

$$\boldsymbol{R} \equiv \begin{bmatrix} r_1 & r_2 & r_3 \\ r_4 & r_5 & r_6 \\ r_7 & r_8 & r_9 \end{bmatrix} \tag{2.42}$$

$$\boldsymbol{T} \equiv \begin{bmatrix} T_x & T_y & T_z \end{bmatrix}^{\mathrm{T}} \tag{2.43}$$

(2) The conversion Z_2 from the camera coordinate system, *xyz*, to the image plane coordinate system, *x'y'*, can be expressed as

$$x' = \lambda \frac{x}{z} \tag{2.44}$$

$$y' = \lambda \frac{y}{z} \tag{2.45}$$

(3) The transformation Z_3 from the image plane coordinate system, $x'y'$, to the computer image coordinate system, MN, can be expressed as

$$M = \mu \frac{x' M_x}{S_x L_x} + O_m \tag{2.46}$$

$$N = \frac{y'}{S_y} + O_n \tag{2.47}$$

Among them, M and N are the total number of rows and total columns (computer coordinates) of the pixels in the computer memory respectively; O_m and O_n are the number of rows and columns where the central pixel of the computer memory is located; S_x is the distance between the centers of two adjacent sensors along the x direction (scan line direction), S_y is the distance between the centers of two adjacent sensors along the y direction; L_x is the number of sensor elements in the x direction; M_x is the number of samples (number of pixels) of the computer in a row; and μ is an uncertain image scale factor that depends on the camera (generally nonlinear, or can be regarded as a kind of distortion). According to the working principle of the sensor, the time difference between the image acquisition hardware and the camera scanning hardware during progressive scanning or the inaccuracy of the camera scanning itself in time will introduce certain uncertain factors. These uncertain factors can be described by introducing the uncertain image scale factor μ, and establishing the connection between the image plane coordinate system, $x'y'$, and the computer image coordinate system, MN, affected by the uncertain image scale factor.

2.3 CAMERA CALIBRATION

The purpose of camera calibration is to align the camera coordinate system with the world coordinate system.

2.3.1 Basic Calibration Procedure

According to the discussion of the general camera model in Subsection 2.1.3, if a series of transformations $\boldsymbol{PRTW}_\mathrm{h}$ are performed on the homogeneous coordinates $\boldsymbol{W}_\mathrm{h}$ of a space point, the world coordinate system can be overlapped with the camera coordinate system. Here, $\boldsymbol{P}$ is the imaging projection transformation matrix, $\boldsymbol{R}$ is the camera rotation matrix, and $\boldsymbol{T}$ is the camera translation matrix. Let $\boldsymbol{A} = \boldsymbol{PRT}$, the elements in $\boldsymbol{A}$ include camera translation, rotation and projection parameters, there is a homogeneous expression of image coordinates: $\boldsymbol{C}_\mathrm{h} = \boldsymbol{AW}_\mathrm{h}$. If we set $k = 1$ in the homogeneous expression, we obtain

$$\begin{bmatrix} C_{h1} \\ C_{h2} \\ C_{h3} \\ C_{h4} \end{bmatrix} = \begin{bmatrix} a_{11} & a_{12} & a_{13} & a_{14} \\ a_{21} & a_{22} & a_{23} & a_{24} \\ a_{31} & a_{32} & a_{33} & a_{34} \\ a_{41} & a_{42} & a_{43} & a_{44} \end{bmatrix} \begin{bmatrix} X \\ Y \\ Z \\ 1 \end{bmatrix} \tag{2.48}$$

According to the definition of homogeneous coordinates, the camera coordinates (image plane coordinates) in Cartesian form are

$$x = C_{h1} / C_{h4} \tag{2.49}$$

$$y = C_{h2} / C_{h4} \tag{2.50}$$

Substitute Equations (2.49) and (2.50) into Equation (2.48) and expand the matrix product to obtain

$$xC_{h4} = a_{11}X + a_{12}Y + a_{13}Z + a_{14} \tag{2.51}$$

$$yC_{h4} = a_{21}X + a_{22}Y + a_{23}Z + a_{24} \tag{2.52}$$

$$C_{h4} = a_{41}X + a_{42}Y + a_{43}Z + a_{44} \tag{2.53}$$

where the expansion of C_{h3} is omitted because it is related to z.

Substituting C_{h4} into Equations (2.51) and (2.52), two equations with a total of 12 unknowns can be obtained:

$$(a_{11} - a_{41}x)X + (a_{12} - a_{42}x)Y + (a_{13} - a_{43}x)Z + (a_{14} - a_{44}x) = 0 \tag{2.54}$$

$$(a_{21} - a_{41}y)X + (a_{22} - a_{42}y)Y + (a_{23} - a_{43}y)Z + (a_{24} - a_{44}y) = 0 \tag{2.55}$$

It can be seen that a calibration procedure should include the following steps: (i) obtain $M \geq 6$ space points with known world coordinates (X_i, Y_i, Z_i), $i = 1, 2, \ldots, M$ (in practical applications, often more than 25 points are taken, then least-squares fitting is used to reduce the error); (ii) use the camera to shoot these points at a given position to get their corresponding image plane coordinates (x_i, y_i), $i = 1, 2, \ldots, M$; and (iii) substitute these coordinates into Equations (2.54) and (2.55) to solve for the unknown coefficients.

In order to realize the calibration procedure described above, it is necessary to obtain corresponding spatial points and image points. In order to accurately determine these points, it is necessary to use a calibration object (also called a calibration target, that is, a standard reference object), which has a fixed pattern of marking points (reference points). The most commonly used 2-D calibration object has a series of regularly arranged square patterns (similar to a chess board), and the vertices of these squares (cross-hairs) can be used as reference points for calibration. If a coplanar reference point calibration algorithm is used, the calibration object corresponds to one plane; if a non-coplanar reference point calibration algorithm is used, the calibration object generally corresponds to two orthogonal planes.

2.3.2 Camera Internal and External Parameters

The calibration parameters (Zhang 2021) involved in camera calibration can be divided into external parameters (outside the camera) and internal parameters (inside the camera).

2.3.2.1 External Parameters

The first step of transformation in Figure 2.13 is to transform from the 3-D world coordinate system to the 3-D camera coordinate system with the center at the optical center of the camera. The transformation parameters are called **external parameters**, also called **camera pose parameters**. The rotation matrix $\boldsymbol{R}$ has a total of nine elements, but in fact there are only three degrees of freedom, which can be represented by the three Euler angles of the rigid body rotation. The Euler angles are shown in Figure 2.14 (the line of sight is inverse to the X-axis), where the intersection line AB of the XY plane and the xy plane is called the pitch line, and the angle θ between AB and the x-axis is the first Euler angle, which is called the yaw angle (also called the deflection angle), and which is the angle of rotation around the z-axis. The angle ψ between AB and the X-axis is the second Euler angle, which is called the tilt angle (also called the precession angle), and which is the angle of rotation around the Z-axis. The angle ϕ between the Z-axis and the z-axis is the third Euler angle, which is called the pitch angle (also called the roll angle), and which is the angle of rotation around the pitch line.

Using Euler angles, the rotation matrix can be represented as a function of θ, ψ, ϕ:

$$\boldsymbol{R} = \begin{bmatrix} \cos\psi\cos\theta & \sin\psi\cos\theta & -\sin\theta \\ -\sin\psi\cos\phi + \cos\psi\sin\theta\sin\phi & \cos\psi\cos\phi + \sin\psi\sin\theta\sin\phi & \cos\theta\sin\phi \\ \sin\psi\sin\phi + \cos\psi\sin\theta\cos\phi & -\cos\psi\sin\phi + \sin\psi\sin\theta\cos\phi & \cos\theta\cos\phi \end{bmatrix} \tag{2.56}$$

It can be seen that the rotation matrix has three degrees of freedom. The translation matrix also has three degrees of freedom (translation coefficients in three directions). In this way, the camera has six independent external parameters, namely the three Euler angles θ, ψ, ϕ in $\boldsymbol{R}$ and the three elements T_x, T_y, T_z in $\boldsymbol{T}$.

2.3.2.2 Internal Parameters

The last two steps of transformation in Figure 2.13 are from the 3-D camera coordinate system to the 2-D computer image coordinate system. The transformation parameters are

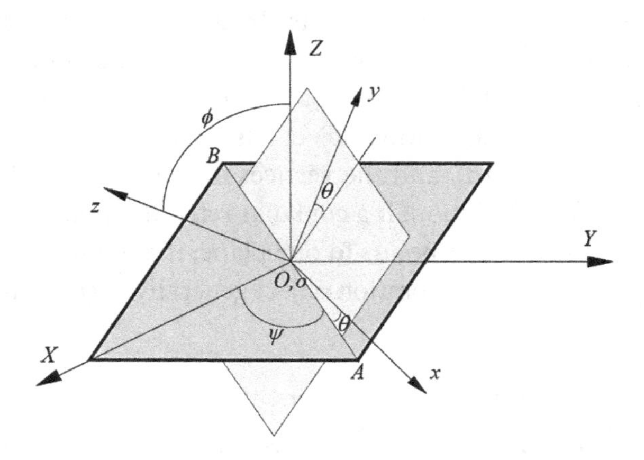

FIGURE 2.14 Euler angle schematic.

called **internal parameters**, also called **camera internal parameters**. There are four internal parameters: focal length λ, uncertainty image scale factor μ, and computer image coordinates, O_m and O_n, of the origin of the image plane.

The main significance of distinguishing external parameters and internal parameters is that when a camera is used to acquire multiple images at different positions and directions, the external camera parameters corresponding to each image may be different, but the internal parameters will not change, so after moving the camera, only the external parameters need to be re-calibrated, but not the internal parameters.

2.3.2.3 Another Description of Internal and External Parameters

From the viewpoint that the purpose of camera calibration is to align the camera coordinate system with the world coordinate system, the internal parameters and external parameters in the camera calibration can also be described as follows (Zhang 2021).

It is possible to decompose a complete **camera calibration transformation matrix** $\boldsymbol{C}$ into the product of the **internal parameter matrix** $\boldsymbol{C}_\mathrm{i}$ and the **external parameter matrix** $\boldsymbol{C}_\mathrm{e}$:

$$\mathrm{C} = \mathrm{C}_\mathrm{i}\mathrm{C}_\mathrm{e} \tag{2.57}$$

$\boldsymbol{C}_\mathrm{i}$ is a 4×4 matrix in general, but it can commonly be simplified to a 3×3 matrix:

$$\mathrm{C}_\mathrm{i} = \begin{bmatrix} S_x & P_x & T_x \\ P_y & S_y & T_y \\ 0 & 0 & 1/\lambda \end{bmatrix} \tag{2.58}$$

Among them, S_x and S_y are the scaling factors along the x-axis and y-axis, respectively; P_x and P_y are the skew factors along the x-axis and y-axis, respectively (derived from the non-strict orthogonality of the optical axis of the actual camera, that is, the row and column of pixels are not strictly 90° as reflected in the image); T_x and T_y are the translation coefficients along the x-axis and y-axis (to move the projection center of the camera to a suitable position); and λ is the focal length of the lens.

The general form of $\boldsymbol{C}_\mathrm{e}$ is also a 4×4 matrix, which can be written as

$$\mathrm{C}_\mathrm{e} = \begin{bmatrix} \boldsymbol{R}_1 & \boldsymbol{R}_1 \cdot \boldsymbol{T} \\ \boldsymbol{R}_2 & \boldsymbol{R}_2 \cdot \boldsymbol{T} \\ \boldsymbol{R}_3 & \boldsymbol{R}_3 \cdot \boldsymbol{T} \\ 0 & 1 \end{bmatrix} \tag{2.59}$$

where $\boldsymbol{R}_1$, $\boldsymbol{R}_2$, and $\boldsymbol{R}_3$ are three row vectors of a 3×3 rotation matrix (only three degrees of freedom), $\boldsymbol{T}$ is a 3-D translation column vector, and $\mathbf{0}$ is a 1×3 vector.

It can be seen from the above that the matrix $\boldsymbol{C}_\mathrm{i}$ has seven internal parameters and the matrix $\boldsymbol{C}_\mathrm{e}$ has six external parameters. However, both matrices have rotation parameters, so the rotation parameters of the inner parameter matrix can be included into the outer parameter matrix. Because rotation is a combination of scaling and skew, after removing the rotation factor from the internal parameter matrix, P_x and P_y become the same ($P_x = P_y = P$). When considering the linear camera model, $P = 0$. So there are only five parameters in the internal parameter matrix, namely λ, S_x, S_y, T_x, T_y. In this way, the two matrices have a total of 11 parameters that need to be calibrated, which can be done according to the basic calibration procedure. In special cases, if the camera is very accurate, then $S_x = S_y = S = 1$, and there are only three internal parameters at this time. Furthermore, if the cameras are aligned, $T_x = T_y = 0$. This leaves only one internal parameter λ.

2.3.3 Nonlinear Camera Calibration

In practical applications, the influence of the radial distortion of the camera lens is often relatively large, and the radial distortion is often proportional to the distance between a point in the image and this point on the optical axis of the lens (Zhang 2021). The transformation from the undistorted image plane coordinates, (x', y'), to the actual image plane coordinates, (x^*, y^*), offset by the influence of the lens radial distortion, is

$$x^* = x' - R_x \tag{2.60}$$

$$y^* = y' - R_y \tag{2.61}$$

where R_x and R_y represent the radial distortion of the lens. From Equations (2.31) and (2.32), we can get

$$R_x = x^*\left(k_1 r^2 + k_2 r^4 + \cdots\right) \approx x^* k r^2 \tag{2.62}$$

$$R_y = y^*\left(k_1 r^2 + k_2 r^4 + \cdots\right) \approx y^* k r^2 \tag{2.63}$$

Here only a single **lens radial distortion coefficient** k is introduced to simplify the approximation. On the one hand, it is because the high-order term of r can be ignored in practice; on the other hand, the factor that radial distortion is often symmetrical about the main optical axis of the camera lens is considered.

Considering the conversion from (x', y') to (x^*, y^*), the conversion from the 3-D world coordinate system to the computer image coordinate system based on the nonlinear camera model is shown in Figure 2.15. The original transformation Z_3 is now decomposed into two transformations (Z_{31} and Z_{32}), and Equations (2.46) and (2.47) can still be used to define Z_{32} (just replace x' and y' with x^* and y^*).

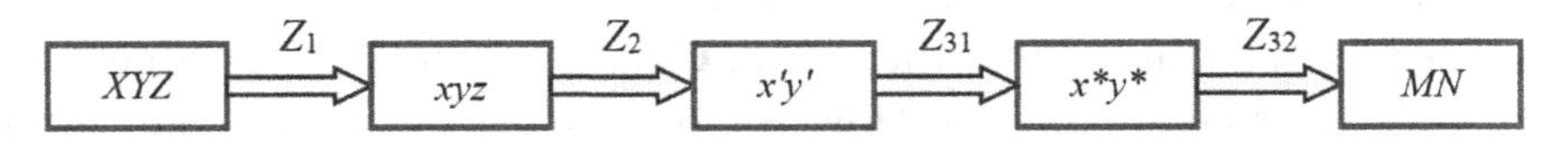

FIGURE 2.15 Schematic of the conversion from the 3-D world coordinate system to the computer image coordinate system under the nonlinear camera model.

Although Equations (2.60) and (2.61) only consider radial distortion, the forms of Equations (2.39) and (2.40) are actually applicable to various distortions. In this sense, the process in Figure 2.15 is suitable for any distortion, as long as the corresponding Z_{31} is selected according to the type of distortion. Comparing Figure 2.15 with Figure 2.13, the "nonlinearity" is reflected in the conversion from $x'y'$ to x^*y^*.

2.3.4 Classification of Calibration Methods

There are many camera calibration techniques, and there are different classification methods, according to different points of view. For example, according to the characteristics of the camera model, it can be divided into linear methods and nonlinear methods; according to whether calibration objects are needed, it can be divided into traditional camera calibration methods, camera self-calibration methods, and calibration methods based on active vision (the latter two methods are also combined by some). When using the calibration target, according to the dimension of the calibration target, it can be divided into the 2-D plane target method and the 3-D stereo target method. According to the result of solving the parameters, it can be divided into explicit and implicit calibration methods; according to whether the parameters of the camera can be changed, it can be divided into the methods with variable and with invariable internal parameters. According to the motion mode of the camera, it can be divided into limited motion mode and non-limited motion mode methods. According to the number of cameras used by the vision system, it can be divided into single-camera and multi-camera calibration methods. The classification of various calibration methods is shown in Table 2.1, which lists some classification criteria, categories and typical methods (Zhang 2021).

In Table 2.1, nonlinear methods are generally more complex, slower and require a good initial value; besides, nonlinear search cannot guarantee that the parameters converge to the global optimal solution. The implicit methods use conversion matrix elements as calibration parameters, and use a conversion matrix to represent the correspondence between 3-D space points and 2-D image plane points. Because the parameters themselves do not have a clear physical meaning, they are also called hidden parameter methods. Since the implicit parameter method only needs to solve linear equations, this method can obtain higher efficiency when the accuracy requirements are not very high. The **direct linear method** (DLT) takes the linear model as the object, and uses a 3 × 4 matrix to represent the correspondence between 3-D space points and 2-D plane image plane points, ignoring the intermediate imaging process (or comprehensively considering the process factor). The most common multi-camera calibration method is the dual-camera calibration method. Compared with single-camera calibration, dual-camera calibration not only needs to know the internal and external parameters of each camera itself, but also needs to measure the relative relationship (location and direction) between the two cameras through calibration.

2.4 TRADITIONAL CALIBRATION METHODS

Traditional camera calibration needs to use a known calibration target (2-D calibration board with known data, or 3-D calibration block), that is, the size and shape of the calibration target (position and distribution of calibration points) need to be known, and then

TABLE 2.1 Classifitcation of Calibration Methods

Classification criteria	Categories	Typical methods
Characteristics of the camera model	Linear	Two-stage calibration method
	Nonlinear	LM optimization method Newton Raphson (NR) optimization method Nonlinear optimization method for parameter calibration Method assuming only the condition of radial distortion
Whether calibration objects are required	Traditional calibration	Methods using optimization algorithms Methods using camera transformation matrix Two-step method considering distortion compensation Biplane method using camera imaging model Direct linear transformation (DLT) method Method using radial alignment constraint (RAC)
	Camera self-calibration	Method solving Kruppa's Equation directly Layered stepwise approach Method using absolute conic Method based on quadric surface
	Active vision-based calibration	Linear method based on two sets of three orthogonal motions Method based on four-group and five-group plane orthogonal motion Orthogonal movement method based on planar homography matrix Orthogonal motion method based on epi-pole
Dimension of calibration targets	2-D plane target	Black and white checkerboard calibration target (take grid intersection as calibration point) Arrange dots in a grid (take the center of the dot as the calibration point)
	3-D solid target	3-D objects of known size and shape
Results of solving the parameters	Implicit calibration	Consider calibration parameters with direct physical meaning (such as distortion parameters)
	Explicit calibration	Direct linear transformation (DLT) method to calibrate geometric parameters
Whether internal parameters of the camera can be changed	Variable internal parameters	During the calibration process, the optical parameters of the camera (such as focal length) can be changed
	Invariable internal parameters	During the calibration process, the optical parameters of the camera cannot be changed
Camera motion mode	Limited motion mode	Method in which camera only has a pure rotation Method for camera to perform orthogonal translation movement
	Non-limited motion	No limit to the movement of the camera during calibration
Number of cameras used by the vision system	Using a single camera	Calibrate only a single camera
	Using multi-cameras	Use 1-D calibration objects for multiple cameras (more than 3 collinear points with known distances), and use the maximum likelihood criterion to refine the linear algorithm

to establish the corresponding relationship between the point on the calibration target and the corresponding point on the captured image to determine the internal and external parameters of the camera. The advantage is that the theory is clear, the solution is simple, and calibration accuracy is high. The disadvantage is that the calibration process is relatively complicated, and the requirement for the accuracy of the calibration target is relatively high.

2.4.1 Basic Steps and Parameters

Calibration can be carried out along the conversion direction from 3-D world coordinates to computer image coordinates. As shown in Figure 2.16, the conversion from the world coordinate system to the computer image coordinate system has four steps, and each step has parameters to be calibrated.

Step 1: The parameters that need to be calibrated are the rotation matrix $\boldsymbol{R}$ and the translation matrix $\boldsymbol{T}$.

Step 2: The parameter that needs to be calibrated is the focal length of the lens λ.

Step 3: The parameters that need to be calibrated are the lens radial distortion coefficient k, eccentric distortion coefficient l, and thin prism distortion coefficient m.

Step 4: The parameter that needs to be calibrated is the uncertainty image scale factor μ.

2.4.2 Two-Stage Calibration Method

The **two-stage calibration method** is a typical traditional calibration method (Tsai 1987). It is so named because the calibration is divided into two steps: the first step is to calculate the external parameters of the camera (but the translation along the optical axis of the camera is not considered, yet), and the second step is to calculate other parameters of the camera. Since the radial alignment constraint (RAC) is used, it is also called the RAC method. Most of the equations in the calculation process are linear equations, so the process of solving parameters is relatively simple. This method has been widely used in industrial vision systems. The average accuracy of 3-D measurement can reach 1/4000, and the accuracy in the depth direction can reach 1/8000.

Calibration can be divided into two cases:

(1) If μ is known, only one image containing a set of coplanar reference points needs to be used for calibration. At this time, Step 1 calculates $\boldsymbol{R}$, T_x and T_y, and Step 2 calculates λ, k and T_z. Here, because k is the radial distortion coefficient of the lens, k may not be considered in the calculation of $\boldsymbol{R}$. Similarly, k may not be considered in the

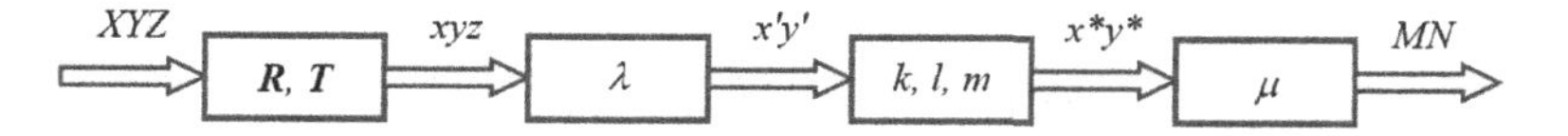

FIGURE 2.16 Camera calibration along the coordinate conversion direction.

calculation of T_x and T_y, but k needs to be considered in the calculation of T_z (the influence of T_z change on the image is similar to that of k), so it is placed in Step 2.

(2) If μ is unknown, an image containing a set of non-coplanar reference points is required for calibration. At this time, Step 1 calculates $\boldsymbol{R}$, T_x and T_y and μ, while Step 2 still calculates λ, k and T_z.

The specific calibration process is to calculate a set of parameters s_i (i = 1, 2, 3, 4, 5) or $\boldsymbol{s} = [s_1\ s_2\ s_3\ s_4\ s_5]^T$. With this set of parameters, the external parameters of the camera can be further calculated. Given M ($M \geq 5$) points with known world coordinates (X_i, Y_i, Z_i) and corresponding image plane coordinates (x_i, y_i), where i = 1, 2, …, M, matrix $\boldsymbol{A}$ can be constructed, and the row $\boldsymbol{a}_i$ can be expressed as follows:

$$a_i = \begin{bmatrix} y_i X_i & y_i Y_i & -x_i X_i & -x_i Y_i & y_i \end{bmatrix} \tag{2.64}$$

Let s_i have the following relationship with the rotation parameters r_1, r_2, r_4 and r_5 as well as the translation parameters T_x and T_y:

$$s_1 = \frac{r_1}{T_y} \quad s_2 = \frac{r_2}{T_y} \quad s_3 = \frac{r_4}{T_y} \quad s_4 = \frac{r_5}{T_y} \quad s_5 = \frac{T_x}{T_y} \tag{2.65}$$

Suppose the vector $\boldsymbol{u} = [x_1\ x_2\ \ldots\ x_M]^T$, $\boldsymbol{s}$ can be solved by the linear equations shown in Equation (2.66).

$$\boldsymbol{A}\boldsymbol{s} = \boldsymbol{u} \tag{2.66}$$

The rotation and translation parameters can be calculated according to the following steps:

(1) Set $S = s_1^2 + s_2^2 + s_3^2 + s_4^2$, calculate

$$T_y^2 = \begin{cases} \dfrac{S - \sqrt{\left[S^2 - 4\left(s_1 s_4 - s_2 s_3\right)^2\right]}}{4\left(s_1 s_4 - s_2 s_3\right)^2} & \left(s_1 s_4 - s_2 s_3\right) \neq 0 \\ \dfrac{1}{s_1^2 + s_2^2} & s_1^2 + s_2^2 \neq 0 \\ \dfrac{1}{s_3^2 + s_4^2} & s_3^2 + s_4^2 \neq 0 \end{cases} \tag{2.67}$$

(2) Set $T_y = (T_y^2)^{1/2}$, that is, take the positive square root and calculate

$$r_1 = s_1 T_y \quad r_2 = s_2 T_y \quad r_4 = s_3 T_y \quad r_5 = s_4 T_y \quad T_x = s_5 T_y \tag{2.68}$$

(3) Choose a point whose world coordinates are (X, Y, Z) and require its image plane coordinates (x, y) to be farther from the image center, then calculate

$$p_X = r_1 X + r_2 Y + T_x \tag{2.69}$$

$$p_Y = r_4 X + r_5 Y + T_y \tag{2.70}$$

This is equivalent to applying the calculated rotation parameters to the X and Y of the point (X, Y, Z). If the signs of p_X and x are the same, and the signs of p_Y and y are the same, it means that T_y has the correct sign, otherwise T_y needs to be negative.

(4) Calculate other rotation parameters.

$$r_3 = \sqrt{1 - r_1^2 - r_2^2} \quad r_6 = \sqrt{1 - r_4^2 - r_5^2} \quad r_7 = \frac{1 - r_1^2 - r_2 r_4}{r_3} \quad r_8 = \frac{1 - r_2 r_4 - r_5^2}{r_6} \quad r_9 = \sqrt{1 - r_3 r_7 - r_6 r_8} \tag{2.71}$$

Note that if the sign of $r_1 r_4 + r_2 r_5$ is positive, then r_6 must be negative, and the signs of r_7 and r_8 must be adjusted after the focal length λ is calculated.

(5) Establish another set of linear equations to calculate the focal length λ and the translation parameter T_z in the z direction. A matrix $\boldsymbol{B}$ can be constructed first, where the matrix $\boldsymbol{b}_i$ can be expressed as

$$b_i = \lfloor r_4 X_i + r_5 Y_i + T_y \quad y_i \rfloor \tag{2.72}$$

where $\lfloor \cdot \rfloor$ means rounding down.

Suppose the element v_i of vector $\boldsymbol{v}$ can be expressed as

$$v_i = (r_7 X_i + r_8 Y_i) y_i \tag{2.73}$$

Then, $\boldsymbol{t} = [\lambda \ T_z]^{\mathrm{T}}$ can be solved by the linear equations shown in Equation (2.74). Note that what we get here is only an estimate of $\boldsymbol{t}$.

$$\boldsymbol{Bt} = \boldsymbol{v} \tag{2.74}$$

(6) If $\lambda < 0$, to use the right-hand coordinate system, r_3, r_6, r_7, r_8, λ, T_z must be negative.

(7) Use the estimate of $\boldsymbol{t}$ to calculate the lens radial distortion coefficient k, and adjust the values of λ and T_z. Using the perspective projection equation including distortion here, the following nonlinear equation can be obtained:

$$\left\{ y_i (1 + kr^2) = \lambda \frac{r_4 X_i + r_5 Y_i + r_6 Z_i + T_y}{r_7 X_i + r_8 Y_i + r_9 Z_i + T_z} \right\} \qquad i = 1, 2, \cdots, M \tag{2.75}$$

Solve the above equation by nonlinear regression method to get the values of k, λ, T_z.

2.4.3 Precision Improvement

The above two-stage calibration method only considers the radial distortion of the camera lens. If the **tangential distortion** of the lens is further considered on this basis, it is possible to further improve the accuracy of camera calibration (Zhang 2021).

According to Equations (2.37) and (2.38), the total distortion deviations d_x and d_y considering radial distortion and tangential distortion are

$$d_x = d_{xr} + d_{xt} \tag{2.76}$$

$$d_y = d_{yr} + d_{yt} \tag{2.77}$$

Considering the fourth-order term for radial distortion and the second-order term for tangential distortion, there is

$$d_x = x_i\left(k_1 r^2 + k_2 r^4\right) + l_1\left(3x_i^2 + y_i^2\right) + 2l_2 x_i y_i \tag{2.78}$$

$$d_y = y_i\left(k_1 r^2 + k_2 r^4\right) + 2l_1 x_i y_i + l_2\left(x_i^2 + 3y_i^2\right) \tag{2.79}$$

The camera calibration can be divided into the following two steps:

(1) Assuming that the initial values of lens distortion coefficients k_1, k_2, l_1, and l_2 are all 0, calculate the values of $\boldsymbol{R}$, $\boldsymbol{T}$, and λ.

Referring to Equations (2.44) and (2.45), and refer to the derivation of Equation (2.75), we obtain

$$x = \lambda \frac{X}{Z} = \lambda \frac{r_1 X + r_2 Y + r_3 Z + T_x}{r_7 X + r_8 Y + r_9 Z + T_z} \tag{2.80}$$

$$y = \lambda \frac{Y}{Z} = \lambda \frac{r_4 X + r_5 Y + r_6 Z + T_y}{r_7 X + r_8 Y + r_9 Z + T_z} \tag{2.81}$$

From Equations (2.80) and (2.81):

$$\frac{x}{y} = \frac{r_1 X + r_2 Y + r_3 Z + T_x}{r_4 X + r_5 Y + r_6 Z + T_y} \tag{2.82}$$

Equation (2.82) is valid for all reference points, that is, an equation can be established by using the 3-D world coordinates and 2-D image coordinates of each reference point. There are eight unknowns in Equation (2.82), so if there are eight reference points, an equation system with eight equations can be constructed, and then the values of r_1, r_2, r_3, r_4, r_5, r_6, T_x, T_y can be calculated. Because $\boldsymbol{R}$ is an orthogonal matrix, the values of r_7, r_8, and r_9 can be calculated according to its orthogonality. Substituting the calculated values into Equations (2.80) and (2.81), and then arbitrarily taking the 3-D world coordinates and 2-D image coordinates of the two reference points, the values of T_z and λ can be calculated.

(2) Calculate the values of lens distortion coefficients k_1, k_2, l_1, and l_2.

According to Equations (2.29), (2.30), (2.76) ~ (2.79), we can get

$$\lambda\frac{X}{Z}=x=x_i+x_i\left(k_1r^2+k_2r^4\right)+l_1\left(3x_i^2+y_i^2\right)+2l_2x_iy_i \tag{2.83}$$

$$\lambda\frac{Y}{Z}=y=y_i+y_i\left(k_1r^2+k_2r^4\right)+2l_1x_iy_i+l_2\left(x_i^2+3y_i^2\right) \tag{2.84}$$

With the help of $\boldsymbol{R}$ and $\boldsymbol{T}$ which have been obtained, (X, Y, Z) can be calculated using Equation (2.82), and then substituting into Equations (2.83) and (2.84) to obtain

$$\lambda\frac{X_j}{Z_j}=x_{ij}+x_{ij}\left(k_1r^2+k_2r^4\right)+l_1\left(3x_{ij}^2+y_{ij}^2\right)+2l_2x_{ij}y_{ij} \tag{2.85}$$

$$\lambda\frac{Y_j}{Z_j}=y_{ij}+y_{ij}\left(k_1r^2+k_2r^4\right)+2l_1x_{ij}y_{ij}+l_2\left(x_{ij}^2+3y_{ij}^2\right) \tag{2.86}$$

where $j = 1, 2, \ldots, N$, N is the number of reference points. Using $2N$ linear equations and solving by the least-squares method, the values of four distortion coefficients k_1, k_2, l_1, and l_2 can be obtained.

2.5 SELF-CALIBRATION METHODS

The camera **self-calibration** method was proposed in the early 1990s. Camera self-calibration can calculate real-time, online camera model parameters from geometric constraints obtained from image sequences without resorting to high-precision calibration objects. This is especially suitable for cameras that often need to move. Since all the self-calibration methods are only related to the internal parameters of the camera, and have nothing to do with the external environment and the movement of the camera, the self-calibration method is more flexible than the traditional calibration method. However, the existing self-calibration methods are not very accurate and not very robust.

2.5.1 Basic Idea

The basic idea of the self-calibration method is to first establish a constraint equation about the parameter matrix in the camera through an absolute conic curve, which is called the Kruppa Equation. Then, the Kruppa Equation is solved to determine the matrix $\boldsymbol{C}$ ($\boldsymbol{C} = \boldsymbol{K}^{\mathrm{T}}\boldsymbol{K}^{-1}$, $\boldsymbol{K}$ is the internal parameter matrix). Finally, the matrix $\boldsymbol{K}$ is obtained by Cholesky decomposition (Faugeras 1993).

The self-calibration method can be realized with the help of active vision technology. However, some researchers put calibration methods based on active vision technology into a separate category. An active vision system means that the system can control the camera to obtain multiple images in motion, and then use the camera's motion trajectory and the corresponding relationship between the obtained images to calibrate the camera. The method based on active vision calibration is generally used when the motion parameters of the camera in the world coordinate system are known. It can usually be solved linearly and the results obtained have high robustness.

In practical applications, the method based on active vision calibration generally installs the camera accurately on a controllable platform, and actively controls the platform to perform special movements to obtain multiple images. The correspondence between these images and the camera motion parameters are used to determine the camera parameters. However, if the camera motion parameters are unknown or in situations where the camera motion cannot be controlled, this method cannot be used. In addition, the motion platform required by this method needs to be higher precision and it has higher cost.

2.5.2 A Practical Method

A typical self-calibration method (based on the active vision calibration method) is described in detail below (Zhang 2021). As shown in Figure 2.17, the camera's optical center is translated from O_1 to O_2, and the resulting two images are I_1 and I_2 (the origins of the coordinates are o_1 and o_2, respectively). A point P in space is imaged as point p_1 on I_1 and as point p_2 on I_2, and p_1 and p_2 form a pair of corresponding points. If a point p_2' is marked on I_1 according to the coordinate value of point p_2 on I_2, then the line between p_2' and p_1 is called the line of the corresponding point on I_1. It can be proved that when the camera performs pure translational motion, the lines of the corresponding points of all spatial points on I_1 intersect at the same point e, and it is the direction of the camera's motion (here e is on the line of O_1 and O_2, O_1O_2 is the translational motion trajectory).

According to the analysis of Figure 2.17, it can be known that by determining the intersection of the corresponding point lines, the translational movement direction of the camera under the camera coordinate system can be obtained. In this way, by controlling the camera to perform translational movement in three directions during calibration, and calculating the corresponding intersection $\boldsymbol{e}_i$ $(i = 1, 2, 3)$ before and after each movement, the directions of three translational movements can be obtained.

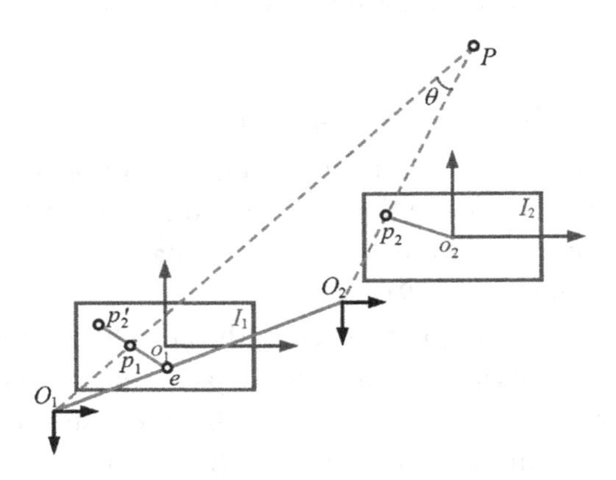

FIGURE 2.17 The geometric relationship between the images by camera translation.

With reference to Equations (2.46) and (2.47), consider the ideal situation of uncertain image scale factor μ being 1, and take each x-direction sensor to sample one pixel in each row, then Equations (2.46) and (2.47) can be written as

$$M = \frac{x'}{S_x} + O_m \tag{2.87}$$

$$N = \frac{y'}{S_y} + O_n \tag{2.88}$$

Equations (2.87) and (2.88) establish the conversion relationship between the image plane coordinate system $x'y'$ expressed in physical units (such as mm) and the computer image coordinate system MN expressed in pixels. According to Figure 2.17, the coordinates of the intersection point $\boldsymbol{e}_i$ ($i = 1, 2, 3$) on I_1 are (x_i, y_i), respectively. From Equations (2.87) and (2.88), we can see that the coordinates of $\boldsymbol{e}_i$ in the camera coordinate system are

$$\boldsymbol{e}_i = \left[(x_i - O_m)S_x \quad (y_i - O_n)S_y \quad \lambda \right]^{\mathrm{T}} \tag{2.89}$$

If the camera is translated three times, and the directions of movement of these three times are orthogonal, we can get $\boldsymbol{e}_i^{\mathrm{T}}\boldsymbol{e}_j = 0$ ($i \neq j$), and then get

$$(x_1 - O_m)(x_2 - O_m)S_x^2 + (y_1 - O_n)(y_2 - O_n)S_y^2 + \lambda^2 = 0 \tag{2.90}$$

$$(x_1 - O_m)(x_3 - O_m)S_x^2 + (y_1 - O_n)(y_3 - O_n)S_y^2 + \lambda^2 = 0 \tag{2.91}$$

$$(x_2 - O_m)(x_3 - O_m)S_x^2 + (y_2 - O_n)(y_3 - O_n)S_y^2 + \lambda^2 = 0 \tag{2.92}$$

Equations (2.90), (2.91) and (2.92) can be further rewritten as

$$(x_1 - O_m)(x_2 - O_m) + (y_1 - O_n)(y_2 - O_n)\left(\frac{S_y}{S_x}\right)^2 + \left(\frac{\lambda}{S_x}\right)^2 = 0 \tag{2.93}$$

$$(x_1 - O_m)(x_3 - O_m) + (y_1 - O_n)(y_3 - O_n)\left(\frac{S_y}{S_x}\right)^2 + \left(\frac{\lambda}{S_x}\right)^2 = 0 \tag{2.94}$$

$$(x_2 - O_m)(x_3 - O_m) + (y_2 - O_n)(y_3 - O_n)\left(\frac{S_y}{S_x}\right)^2 + \left(\frac{\lambda}{S_x}\right)^2 = 0 \tag{2.95}$$

Define two intermediate variables:

$$Q_1 = \left(\frac{S_y}{S_x}\right)^2 \tag{2.96}$$

$$Q_2 = \left(\frac{\lambda}{S_x}\right)^2 \tag{2.97}$$

Then Equations (2.93), (2.94) and (2.95) become three equations including four unknown quantities of O_m, O_n, Q_1, and Q_2. These equations are nonlinear. If Equation (2.93) is used to subtract Equations (2.94) and (2.95), two linear equations can be obtained:

$$x_1(x_2 - x_3) = (x_2 - x_3)O_m + (y_2 - y_3)O_nQ_1 - y_1(y_2 - y_3)Q_1 \tag{2.98}$$

$$x_2(x_1 - x_3) = (x_1 - x_3)O_m + (y_1 - y_3)O_nQ_1 - y_2(y_1 - y_3)Q_1 \tag{2.99}$$

Express O_nQ_1 in Equations (2.98) and (2.99) with intermediate variable Q_3:

$$Q_3 = O_nQ_1 \tag{2.100}$$

Then Equations (2.98) and (2.99) become two linear equations about three unknowns including O_m, Q_1, and Q_3. Since the two equations have three unknowns, the solutions of Equations (2.98) and (2.99) are generally not unique. In order to obtain a unique solution, the camera can be moved three times along the other three orthogonal directions to obtain three other intersection points $\boldsymbol{e}_i$ (i = 4, 5, 6). If these three translational motions have different directions from the previous three translational motions, two equations similar to Equations (2.98) and (2.99) can be obtained. In this way, a total of four equations are obtained, and any three equations can be selected or the least-squares method can be used to solve O_m, Q_1, and Q_3 from the four equations. Next, solve for O_n from Equation (2.100), and then substitute O_m, O_n, and Q_1 into Equation (2.95) to solve for Q_2. In this way, all the internal parameters of the camera can be obtained by controlling the camera to perform two sets of three orthogonal translational movements.

2.6 SOME RECENT DEVELOPMENTS AND FURTHER RESEARCH

In the following sections, technical developments and promising research directions from the last few years are briefly overviewed.

2.6.1 Calibration of Structured Light Active Vision System

The **structured light active vision system** can be regarded as mainly consisting of a camera and a projector, and the accuracy of the 3-D reconstruction of the system is mainly determined by their calibration. There are many methods for camera calibration, which are often realized by means of **calibration plates** and feature points. The **projector** is generally regarded as a camera with a reverse light path. The biggest difficulty in **projector calibration** is to obtain the world coordinates of the feature points. One common solution is to project the projection pattern onto the calibration plate used to calibrate the camera, and obtain the world coordinates of the projection point according to the known feature points on the calibration plate and the calibrated camera parameter matrix. This method requires the camera to be calibrated in advance, so the camera calibration error will be superimposed into the **projector calibration error**, resulting in an increase in the projector calibration error. Another commonly used method is to project the encoded structured light onto a calibration plate containing several feature points, and then use the phase technique to obtain the coordinate points of the feature points on the projection plane. This method does not need to calibrate the camera in advance, but needs to project the sinusoidal grating many times, and the total number of collected images will be relatively large.

In the following, a calibration method for active vision system based on color concentric circle array is introduced (Li and Yan 2021). The projector projects a pattern of colored concentric circles to a calibration plate drawn with an array of concentric circles, and separates the projected concentric circles and the calibration plate concentric circles from the captured image through **color channel filtering**. The pixel coordinates of the center of the circle on the image are calculated by the geometric constraints satisfied by the concentric circle projection, and the homographic relationship between the calibration plane, the projector projection plane and the camera imaging plane is established, and then the system calibration is realized. This method can achieve calibration by collecting at least three images.

2.6.1.1 Projector Model and Calibration

The projection process of the projector and the imaging process of the camera have the same principle but opposite directions, so the **reverse pinhole camera model** can be used as the mathematical model of the projector.

Like the imaging model of the camera, the projection model of the projector is designed as a conversion between three coordinate systems (respectively, the world coordinate system, the projector coordinate system, and the projection plane coordinate system – the coordinate system in the computer is not considered first). The world coordinate system is still represented by XYZ. The projector coordinate system is a coordinate system, xyz, centered on the projector, and the optical axis of the projector is generally taken as the z-axis. The projection plane coordinate system is the coordinate system, $x'y'$, on the projecting (imaging) plane of the projector.

For simplicity, the corresponding axes of the world coordinate system, XYZ, and the projector coordinate system, xyz, can be coincident (and the projector optical center is located at the origin), and then the xy plane of the projector coordinate system and the projecting plane of the projector can be coincident, so that the origin of projection plane is on the optical axis of the projector, and the z-axis of the projector coordinate system is perpendicular to the projection plane and points toward the projection plane, as shown in Figure 2.18. The spatial point (X, Y, Z) is projected to the projection point (x, y) of the projection plane through the optical center of the projector, and the connection between them is a spatial projection ray.

The coordinate system and transformation ideas in the calibration are as follows. First, use the projector to project the calibration pattern to the calibration plate (with the world coordinate system $\boldsymbol{W} = (X, Y, Z)$), and then use the camera (with the camera coordinate system $\boldsymbol{c} = (x, y, z)$) to collect the projected image, and separate the pattern of the

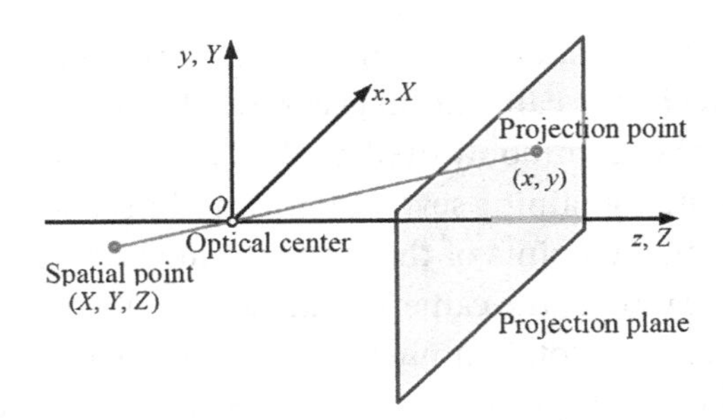

FIGURE 2.18 Basic projector projection model.

calibration plate and the projected pattern. By acquiring and matching the feature points on these patterns, the direct linear transformation (DLT) algorithm (Hartley and Zisserman 2004) can be used to calculate the **homography matrix** $\boldsymbol{H}_{wc}$ between the calibration plate and the camera imaging plane, as well as the homography matrix $\boldsymbol{H}_{cp}$ caused by the projection plate plane imaging plane, and between the camera imaging plane and projection planes (with the projector coordinate system $\boldsymbol{p} = (x', y')$). They are both 3 × 3 nonsingular matrices representing a 2-D projective transformation between two planes.

After obtaining $\boldsymbol{H}_{wc}$ and $\boldsymbol{H}_{cp}$, the virtual dots $\boldsymbol{I} = [1, \mathrm{i}, 0]^{\mathrm{T}}$ and $\boldsymbol{J} = [1, -\mathrm{i}, 0]^{\mathrm{T}}$ on the calibration plate plane, that is, the pixel coordinates $\boldsymbol{I}'_c$ and $\boldsymbol{J}'_c$ on the camera imaging plane, as well as the pixel coordinates $\boldsymbol{I}'_p$ and $\boldsymbol{J}'_p$ on the projector projection plane, can be obtained as follows:

$$\boldsymbol{I}'_c = \boldsymbol{H}_{wc}\boldsymbol{I} \qquad \boldsymbol{J}'_c = \boldsymbol{H}_{cp}\boldsymbol{J} \tag{2.101}$$

$$\boldsymbol{I}'_p = \boldsymbol{H}_{cp}\boldsymbol{I}'_c \qquad \boldsymbol{J}'_p = \boldsymbol{H}_{cp}\boldsymbol{J}'_c \tag{2.102}$$

By changing the position and direction of the calibration plate, the pixel coordinates of at least three groups of different plane imaginary dots on the camera and projector can be obtained, thus the image of the absolute conic curve in the camera imaging plane and projector projection plane, namely S_c and S_p, can be fitted. Then Cholesky decomposition is performed on S_c and S_p and the internal parameter matrices $\boldsymbol{K}_c$ and $\boldsymbol{K}_p$ of the camera and projector can be obtained, respectively. Finally, using $\boldsymbol{K}_c$ and $\boldsymbol{K}_p$, as well as $\boldsymbol{H}_{wc}$ and $\boldsymbol{H}_{cp}$, the extrinsic parameter matrices of the camera and projector can be obtained.

2.6.1.2 Pattern Separation

Using a projector to project a new pattern onto the calibration plate that has already drawn a pattern, and then using a camera to capture the projected calibration plate image, the two patterns in the captured image are overlapped and need to be separated. For this purpose, it is possible to consider using two patterns of different colors, with the aid of color filtering to separate them.

Specifically, a calibration plate with a magenta concentric circle array (7 × 9 concentric circles) on a white background can be used, and a blue-green concentric circle array

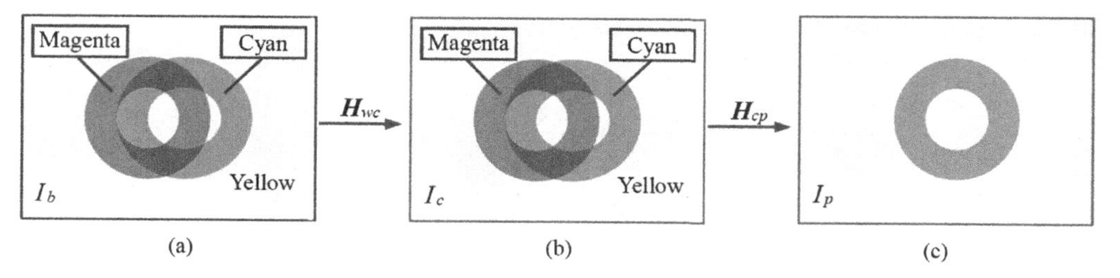

FIGURE 2.19 Extracting the projected pattern from the overlapping calibration plate pattern and projected pattern.

(also 7 × 9 concentric circles) with a yellow background is projected onto the calibration plate by a projector. When the patterns are projected onto the calibration plate I_b with the projector, the calibration plate pattern and the projected pattern are overlapped, as shown in Figure 2.19(a), in which only a pair of each of the two circular patterns is drawn as an example. The region where the two patterns overlap will change color, where the intersection of the magenta circle and the yellow background turns red, the intersection of the magenta circle and the cyan circle turns blue, and the intersection of the white background of the calibration board and the projected pattern turns into the color of the projected pattern. First convert it to the camera image I_c with the help of the homography matrix $\boldsymbol{H}_{wc}$ (as shown in Figure 2.19(b)), and then convert it to the projector image I_p with the help of the homography matrix $\boldsymbol{H}_{cp}$ (as shown in Figure 2.19(c)).

In the color filtering process, the image is first passed through the green, red and blue filtering channels, respectively. After passing through the green filtering channel, since the circle pattern on the calibration plate has no green component, it will appear black, and other regions will appear white, which can separate the calibration plate pattern. After passing through the red filtering channel, the projected circular pattern appears black because there is no red component in it, while the yellow background part and the calibration plate circle pattern appear close to white. After passing through the blue filtering channel, since the yellow background region projected onto the calibration plate and the red circle pattern on the calibration plate have no blue component, they will appear close to black, while the projected cyan circle pattern will appear close to white. Since the color difference of each pattern part is relatively large, the overlapping patterns can be separated more easily. Taking the centers of the separated concentric rings as feature points and obtaining their image coordinates, the homography matrix $\boldsymbol{H}_{wc}$ and the homography matrix $\boldsymbol{H}_{cp}$ can be calculated.

2.6.1.3 Calculation of Homography Matrix

In order to calculate the homography matrix between the calibration plate and the projection plane of the projector with the imaging plane of the camera, it is necessary to calculate the center of the concentric circles on the calibration plate and the image coordinates projected to the center of the concentric circles on the calibration plate. Here, consider that a plane in space has a pair of concentric circles $\boldsymbol{C}_1$ and $\boldsymbol{C}_2$ with the center O. The vector form of any point $\boldsymbol{p}$ on the plane relative to the epipolar line $\boldsymbol{l}$ of the circle $\boldsymbol{C}_1$ is $\boldsymbol{l} = \boldsymbol{C}_1\boldsymbol{p}$, and the

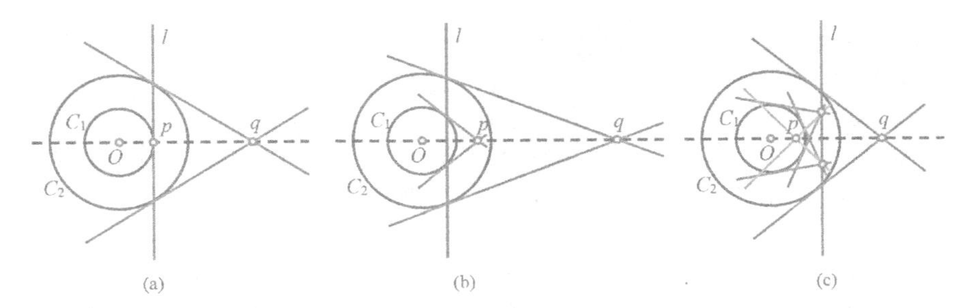

FIGURE 2.20 Constraints between epipolar lines and epipoles of concentric circles.

polar line $\boldsymbol{l}$ relative to the circle $\boldsymbol{C}_2$ is $\boldsymbol{q} = \boldsymbol{C}_2^{-1}\boldsymbol{l}$. The point $\boldsymbol{p}$ can be on the circumference of the circle $\boldsymbol{C}_1$ (as shown in Figure 2.20(a)), outside the circumference of the circle $\boldsymbol{C}_1$ (as shown in Figure 2.20(b)), or on the inside of the circumference of the circle $\boldsymbol{C}_1$ (as shown in Figure 2.20(c)). However, in these three cases, according to the constraint relationship between the epipoles and the epipolar lines of the conic, the line connecting the point $\boldsymbol{p}$ and the point $\boldsymbol{q}$ will pass through the center O.

The projection transformation maps the concentric circles $\boldsymbol{C}_1$ and $\boldsymbol{C}_2$ with the center O on the plane S to the camera imaging plane S_c, the corresponding point of the circle center O on S_c is O_c, and the corresponding conic curves of the concentric circles $\boldsymbol{C}_1$ and $\boldsymbol{C}_2$ on S_c are $\boldsymbol{G}_1$ and $\boldsymbol{G}_2$, respectively. If the epipolar line of any point $\boldsymbol{p}_i$ on the plane S_c relative to $\boldsymbol{G}_1$ is $\boldsymbol{l}_i'$, and the epipole of $\boldsymbol{l}_i'$ relative to $\boldsymbol{G}_2$ is $\boldsymbol{q}_i$, then according to the **projection invariance** of the collinear relationship and the epipolar line–epipole relationship, it can be known that the connection between $\boldsymbol{p}_i$ and $\boldsymbol{q}_i$ goes through O_c. If the connection between $\boldsymbol{p}_i$ and $\boldsymbol{q}_i$ is recorded as $\boldsymbol{m}_i$, then we have

$$\boldsymbol{m}_i = \begin{pmatrix} m_{i1} & m_{i2} & m_{i3} \end{pmatrix}^{\mathrm{T}} = \boldsymbol{q}_i \times \boldsymbol{G}_2^{-1}\boldsymbol{G}_1\boldsymbol{p}_i \tag{2.103}$$

If the normalized homogeneous coordinates of the center projection point are $\boldsymbol{u} = (u, v, 1)^{\mathrm{T}}$, the distance d_i from the center projection point to the straight line $\boldsymbol{m}_i$ can be written as:

$$d_i^2 = \frac{(\boldsymbol{m}_i \cdot \boldsymbol{u})^2}{m_{i1}^2 + m_{i2}^2 + m_{i3}^2} \tag{2.104}$$

One can take any n points on the conic curve $\boldsymbol{G}_1$, and use the Levenberg–Marquardt algorithm to search for the following cost function:

$$f(u,v) = \sum_{i=1}^{n} d_i^2 \tag{2.105}$$

The local minimum point of $f(u, v)$ provides the optimal projection position of the center of the circle.

In order to automatically extract and match concentric circle images, a Canny operator can be used for sub-pixel edge detection to extract circle boundaries and fit quadratic conic

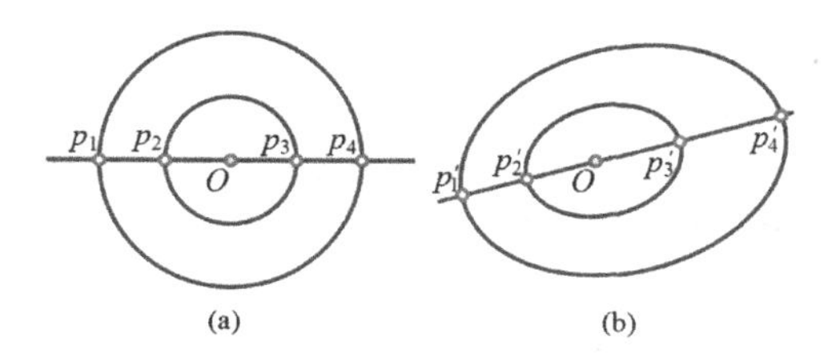

FIGURE 2.21 Use the intersection ratio constraint to match the circles and the curves.

curves. Among the large number of conic sections detected in each image, conic pairs from the same concentric circles are first found using the rank constraint of concentric circles (Kim et al. 2005). Consider two conic sections $\boldsymbol{G}_1$ and $\boldsymbol{G}_2$, whose generalized eigenvalues are λ_1, λ_2, and λ_3, respectively. If $\lambda_1 = \lambda_2 = \lambda_3$, then $\boldsymbol{G}_1$ and $\boldsymbol{G}_2$ are the same conic section; if $\lambda_1 = \lambda_2 \neq \lambda_3$, then $\boldsymbol{G}_1$ and $\boldsymbol{G}_2$ are the projections of a pair of concentric circles; if $\lambda_1 \neq \lambda_2 \neq \lambda_3$, then $\boldsymbol{G}_1$ and $\boldsymbol{G}_2$ come from different concentric circles.

After pairing the conic sections, one also needs to match the concentric circles on the calibration plate with the curve pairs in the image. Here, the cross-ratio invariance can be used for automatic matching of concentric circles. As shown in Figure 2.21(a), set the straight line where the diameter of the concentric circles is located and the concentric circles to intersect at four points $\boldsymbol{p}_1$, $\boldsymbol{p}_2$, $\boldsymbol{p}_3$, and $\boldsymbol{p}_4$, and map them to $\boldsymbol{p}_1'$, $\boldsymbol{p}_2'$, $\boldsymbol{p}_3'$, and $\boldsymbol{p}_4'$ after projective transformation (such as in Figure 2.21(b)). According to the invariance of the cross-ratio, the following relationship can be obtained (where $|\boldsymbol{p}_i \boldsymbol{p}_j|$ represents the distance from point $\boldsymbol{p}_i$ to point $\boldsymbol{p}_j$):

$$C(p_1, p_2, p_3, p_4) = \frac{|p_1p_2||p_3p_4|}{|p_1p_3||p_2p_4|} = \frac{|p'_1p'_2||p'_3p'_4|}{|p'_1p'_3||p'_2p'_4|} = C(p'_1, p'_2, p'_3, p'_4) \tag{2.106}$$

For concentric circles with different radius ratios, the intersection ratios formed by the straight line where the diameter is located with the four intersections of the concentric circles are different, so the radius ratio can be used to identify the concentric circles. When designing the calibration plate pattern and projection pattern, different radius ratios can be set according to the positions of different concentric circles to uniquely identify different concentric circles in the pattern. In practice, different radius ratios can be set only for some concentric circles, and after the corresponding homography matrix is obtained, the positions of other concentric circles and the projection point of the center of the circle can be obtained with the help of the homography matrix.

2.6.1.4 Calculation of Calibration Parameters

After the homography matrix $\boldsymbol{H}_{wc}$ and the homography matrix $\boldsymbol{H}_{cp}$ are determined, the internal and external parameters of the camera and the projector can be calculated. First, the homography matrix $\boldsymbol{H}_{wc}$ between the calibration plate plane and the camera imaging plane can be expressed as:

$$O'_i \sim H_{wc} O_i \tag{2.107}$$

Among them, $\boldsymbol{O}_i = (x_i, y_i, 1)^{\mathrm{T}}$ is the coordinates of the center of the concentric circles on the calibration plate in the calibration plate coordinate system, and $\boldsymbol{O}_i' = (u_i, v_i, 1)^{\mathrm{T}}$ is the image coordinates of the $\boldsymbol{O}_i$ projection point. $\boldsymbol{H}_{wc}$ can be calculated by calculating the image coordinates of the center of the concentric circles of four or more calibration plates and using the DLT algorithm.

Similar to the above process, the homography matrix $\boldsymbol{H}_{wc}$ between the projection plane of the projector and the imaging plane of the camera can also be calculated. Then, with the help of Equations (2.101) and (2.102), the internal parameter matrices $\boldsymbol{K}_c$ and $\boldsymbol{K}_p$ of the camera and the projector can be calculated.

Further, the extrinsic parameter matrices of the camera and projector are computed. Set the calibration board plane to coincide with the X_wY_w plane of the world coordinate system, the homogeneous coordinate of the previous point $\boldsymbol{X}$ in the world coordinate system is $\boldsymbol{X}_w = (x_w, y_w, 0, 1)^{\mathrm{T}}$, its image point on the camera $\boldsymbol{x}_c = (u_c, v_c, 1)^{\mathrm{T}}$ satisfies (where $\boldsymbol{R}_p$ and $\boldsymbol{T}_p$ are the rotation matrix and translation vector of the calibration plate plane relative to the world coordinate system, respectively):

$$\boldsymbol{x}_c \sim \boldsymbol{K}_c\left[\boldsymbol{R}_c \mid \boldsymbol{T}_c\right]\boldsymbol{X}_w \tag{2.108}$$

Denote the 2-D coordinate plane corresponding to the point $\boldsymbol{X}_w = (x_w, y_w, 0, 1)^{\mathrm{T}}$ as $\boldsymbol{x}_w = (x_w, y_w, 1)^{\mathrm{T}}$, and use $\boldsymbol{r}_{c1}$ and $\boldsymbol{r}_{c2}$ to represent the first two columns of $\boldsymbol{R}_c$, respectively, then there is $\boldsymbol{K}_c[\boldsymbol{R}_c|\boldsymbol{T}_c]\boldsymbol{X}_w = \boldsymbol{K}_c[\boldsymbol{r}_{c1}, \boldsymbol{r}_{c2}, \boldsymbol{T}_c]\boldsymbol{X}_w$, substitute it into Equation (2.108) to get:

$$\boldsymbol{x}_c \sim \boldsymbol{K}_c\left[\boldsymbol{r}_{c1},\ \boldsymbol{r}_{c2},\ \boldsymbol{T}_c\right]\boldsymbol{X}_w \tag{2.109}$$

If $\boldsymbol{r}_{c1}$, $\boldsymbol{r}_{c2}$, and $\boldsymbol{T}_c$ are not coplanar, that is, the plane of the calibration plate does not pass through the optical center of the camera, there is a homography matrix $\boldsymbol{H}_{wc}$ between the plane of the calibration plate and the image plane of the camera, which can be known from Equation (2.109)

$$\boldsymbol{H}_w \sim \boldsymbol{K}_c\left[\boldsymbol{r}_{c1},\ \boldsymbol{r}_{c2},\ \boldsymbol{T}_c\right] \tag{2.110}$$

From the above equation, $\boldsymbol{r}_{c1}$, $\boldsymbol{r}_{c2}$, and $\boldsymbol{T}_c$ can be obtained. Because $\boldsymbol{R}_c$ is a unit orthogonal matrix, so

$$\boldsymbol{r}_{c3} = \boldsymbol{r}_{c1} \times \boldsymbol{r}_{c2} \tag{2.111}$$

Similar to the above process, since the corresponding relationship is also satisfied between the calibration plate plane and the projector projection plane, the rotation matrix $\boldsymbol{R}_p$ and translation vector $\boldsymbol{T}_p$ of the projector coordinate system relative to the world coordinate system can be obtained. The rotation matrix $\boldsymbol{R}$ and translation vector $\boldsymbol{T}$ between the camera coordinate system and the projector coordinate system can be expressed as $\boldsymbol{R} = \boldsymbol{R}_c^{-1}\boldsymbol{R}_p$ and $\boldsymbol{T} = \boldsymbol{R}_c^{-1}(\boldsymbol{T}_p - \boldsymbol{T}_c)$, respectively.

2.6.2 Online Camera External Parameter Calibration

In the field of advanced driving-assistance systems (ADAS) or autonomous driving, in-vehicle cameras are required to detect and recognize road signs as well as to detect and track objects around the vehicle. The cameras' internal and external parameters have a large impact on the accuracy of these tasks. Among them, in addition to the traditional method based on the calibration plate (such as in the previous sections), the calibration of the parameters in the camera can also be based on the principle of stillness of the features in the environment. The constraint relationship between the feature points is first established, and then according to this relationship, the camera internal parameter calibration is performed in real time independent of specific calibration plates (Civera et al. 2009).

The external parameter calibration of the camera is to determine the coordinate system relationship between the camera and the vehicle. The general method is to establish a high-precision calibration field for auxiliary calibration. The high-precision calibration field is equipped with pose tracking equipment and specific calibration plates, and the robot hand–eye calibration method (Daniilidis, 1999) is used to determine the external parameters of the camera. The hand–eye calibration method needs the spatial pose relationship between the calibration plate and the camera in the process of solving the external parameters. According to the different dimensions of the calibration plate, it can be divided into 3-D, 2-D, and 1-D calibration methods (Zhang, 2004). However, these methods usually rely on ground signs that meet specific constraints such as points, lines or surfaces, and are mainly suitable for off-line calibration. In addition, due to maintenance and structural deformation, the external parameters of the camera may change significantly in the life cycle of the vehicle. How to calibrate and adjust the external parameters online is also very important.

Aiming at these problems, a method for online real-time camera external parameter calibration is proposed by matching the camera with the high-precision map without using the precise and expensive high-precision calibration field (Liao et al. 2021).

The basic idea of this method is: first, use deep learning technology to detect the lane lines in the image, by assuming an initial external parameter matrix $\boldsymbol{T}$, and project the lane line points $\boldsymbol{P}_w$ in the world coordinate system $\boldsymbol{W}(XYZ)$ to the camera coordinate system $\boldsymbol{C}(xyz)$ to obtain a 3-D image point $\boldsymbol{P}_c$ for matching with the map. Then, the projection error $L(\boldsymbol{T}_{cv})$ between $\boldsymbol{P}_c$ and the lane point $\boldsymbol{D}_c$ detected by the camera is evaluated by reasonably designing the error function L, and the idea of minimizing the reprojection error of the lane line curve to the image plane by using bundle adjustment (BA) is adopted (Triggs et al. 1999) to solve for the extrinsic parameter matrix $\boldsymbol{T}_{cv}$. Here, $\boldsymbol{T}_{cv}$ determines the coordinate system transformation between the camera coordinate system $\boldsymbol{C}(xyz)$ and the vehicle coordinate system $\boldsymbol{V}(x'y'z')$. $\boldsymbol{T}_{cv}$ consists of a rotation matrix $\boldsymbol{R}$ and a translation vector $\boldsymbol{T}$; the three degrees of freedom of $\boldsymbol{R}$ can be represented by three Euler angles (rotation angles) (Zhou et al., 2018). Considering that the vehicle-mounted camera needs to detect obstacles such as pedestrians and vehicles within a range of 200m, its detection accuracy is about 1m. Assuming that the horizontal field of view of the camera is about 57°, the accuracy requirement for the camera's external parameters is about 0.2° for the rotation angle, and about 0.2m for the translation.

2.6.2.1 Lane Line Detection and Data Screening

If the coordinates of lane line points on the image plane collected by the camera are (x', y'), it can be obtained according to the pinhole imaging model:

$$z_c \begin{bmatrix} x' \\ y' \\ 1 \end{bmatrix} = \boldsymbol{M}\boldsymbol{P}_c = \boldsymbol{M}\boldsymbol{T}_{cv}\boldsymbol{T}_{vw}\boldsymbol{P}_w \tag{2.112}$$

where z_c is the distance between the lane line point $\boldsymbol{P}_c$ and the camera, $\boldsymbol{M}$ is the internal parameter matrix of the camera, and $\boldsymbol{T}_{vw}$ is the coordinate transformation matrix between the world coordinate system $\boldsymbol{W}(XYZ)$ and the vehicle coordinate system $\boldsymbol{V}(x'y'z')$, which expresses the pose (position and attitude) of the vehicle.

The detection of **lane lines** can be performed with the help of a deep learning method based on the network structure U-Net++ (Zhou et al. 2018). After obtaining the lane line features in the image plane, the 3-D world coordinate system position cannot be directly recovered from the 2-D features in the image plane, so it is necessary to project the true value of the lane lines to the image plane, and to set the loss function for optimization in the image plane.

To prevent over-optimization and improve computational efficiency, the detected features need to be screened. Lane lines are usually composed of curves and straight lines, and the actual curvature is relatively small. When the vehicle is driving normally, in most cases, the lane line does not provide useful information for translating T_x, and it is necessary to select the scene of the vehicle steering for calibration. Therefore, the video captured by the vehicle camera can be divided into **useless frames**, **data frames** and **key frames** according to the following rules:

(1) When the number of lane line pixels detected in the frame image is less than a certain threshold, it is regarded as a useless frame, so as to avoid vehicles passing through intersections and traffic jams without obvious lane lines in the image.

(2) The frame images, when the vehicle driving distance from the previous key frame and the vehicle yaw angle are both less than a certain threshold, are classified as useless frames to avoid repeated collection of lane line information.

(3) When Rules (1) and (2) are not satisfied and the angle between the vehicle and the true value of the lane line (map data) is greater than a certain threshold, the frame image is classified as a key frame.

(4) The frame images collected in other cases are classified as data frames.

Since the useless frame does not contain lane line information, or only contains the lane line information that has been counted, it can be ignored in the optimization of the loss function to reduce the amount of data.

In actual driving, because the vehicle is parallel to the lane line most of the time, the number of key frame images collected is less than the number of data frames. As pointed out above, the lane lines in the data frame do not provide useful information for translating T_x, so not distinguishing between key frames and data frames may over-optimize other external parameters. To this end, a threshold can be set. If the number of collected key frames is small, only parameters other than T_x are optimized; if the number of collected key frames is sufficient, all external parameters are optimized.

2.6.2.2 Optimizing Reprojection Error

Defining the **reprojection error** of the observation points on the lane line and the reference points on the map as loss, the loss function can be expressed as:

$$L(T_{cv}) = \int \left\| \left[\frac{MT_{cv}T_{vw}P_w}{z_c} - (x', y', 1)^{\mathrm{T}} \right] \right\| \mathrm{d}P_w \tag{2.113}$$

$\boldsymbol{P}_w$ is the position of the lane line in the high-precision map in the world coordinate system, and $\boldsymbol{T}_{vw}$ can be obtained through a global positioning system (GPS) or the like. In this way, the loss function can be determined by determining $\boldsymbol{T}_{cv}$. When the losses under different poses of the vehicle traversing the lane line during driving are combined, the camera external parameter calibration problem can be reduced to an optimization problem that minimizes the loss:

$$\hat{T}_{cv} = \mathrm{argmin}\left[L(T_{cv}) \right] \tag{2.114}$$

In practice, the lane line has no obvious texture features in the direction of the vehicle, so it is impossible to establish a one-to-one mapping between $\boldsymbol{P}_w$ and $(x', y', 1)^{\mathrm{T}}$ to solve Equation (2.114). To do this, convert the point-to-point error in Equation (2.113) to a point-set-to-point-set error:

$$L(T_{cv}) = \int \left\| \left[\frac{MT_{cv}T_{vw}P_w}{z_c} - (x', y', 1)^{\mathrm{T}} \right] \right\| \mathrm{d}t \tag{2.115}$$

In this way, Equation (2.114) can be estimated and solved by numerical solution.

Suppose the position of the detected lane line point in the image plane is (x_i', y_i'), and the normal direction is ϕ, then Equation (2.115) can be converted into

$$L = \sum_{i}^{n} \left[k_1 \left\| \left(x'_i - x_n^w, y'_i - y_n^w \right) \right\| + \left\| \phi_i - \phi_n^w \right\| \right] \tag{2.116}$$

Among them, (x_n^w, y_n^w) is the projection of the lane line in the map on the image plane. The calculation of the normal direction can be found in Ouyang and Feng (2005).

The reprojection error calculation process includes the following steps:

(1) Project the lane line point set in the map (within a range of 200 m from the vehicle) into the camera coordinate system based on the camera external parameter matrix $\boldsymbol{T}_{cv}$ and the vehicle pose matrix $\boldsymbol{T}_{vw.}$

(2) Project the point set that has undergone coordinate system transformation into the image plane according to the camera's internal parameter matrix.

(3) Calculate the lane line point set on the projected map and the normal direction of the detected lane line point set.

(4) Determine the association between the lane line points on the map and the detected lane line points by matching.

(5) Determine the reprojection error according to Equation (2.116) (for example, a simple steepest descent method can be used).

REFERENCES

Civera, J., D.R. Bueno, A.J. Davison, et al. 2009. Camera self-calibration for sequential Bayesian structure from motion. *Proceedings of International Conference on Robotics and Automation*, 403–408.

Daniilidis, K. 1999. Hand-eye calibration using dual quaternions. *The International Journal of Robotics Research*, 18(3): 286–298.

Dean, T., J. Allen and Y. Aloimonos. 1995. *Artificial Intelligence: Theory and Practice*. Boston, MA: Addison Wesley.

Faugeras, O. 1993. *Three-dimensional Computer Vision: A Geometric Viewpoint*. Cambridge, MA: MIT Press.

Hartley, R. and A. Zisserman. 2004. *Multiple View Geometry in Computer Vision*, 2nd Ed. Cambridge, UK: Cambridge University Press.

Jähne, B. 2004. *Practical Handbook on Image Processing for Scientific and Technical Applications*, 2nd Ed. England: CRC Press.

Kim, J.-S., P. Gurdjos and I.-S. Kweon. 2005. Geometric and algebraic constraints of projected concentric circles and their applications to camera calibration. *IEEE Transaction on Pattern Analysis and Machine Intelligence*, 25(4): 78–81.

Li, Y. and Y.C. Yan. 2021. A novel calibration method for active vision system based on array of concentric circles. *Acta Electronica Sinica*, 49(3): 536–541.

Liao, W.L., H.Q. Zhao and J.C. Yan. 2021. Online extrinsic camera calibration based on high-definition map matching on public roadway. *Journal of Image and Graphics*, 26(1): 208–217.

Ouyang, D.S. and H.Y. Feng. 2005. On the normal vector estimation for point cloud data from smooth surfaces. *Computer-Aided Design*, 37(10): 1071–1079.

Triggs, B., P.F. McLauchlan, R.I. Hartley, et al. 1999. Bundle adjustment—a modern synthesis. *Proceedings of the International Workshop on Vision Algorithms: Theory and Practice*, 298–372.

Tsai, R.Y. 1987. A versatile camera calibration technique for high-accuracy 3D machine vision metrology using off-the shelf TV camera and lenses. *Journal of Robotics and Automation*, 3(4): 323–344.

Weng, J.Y., P. Cohen and M. Hernion. 1992. Camera calibration with distortion models and accuracy evaluation. *IEEE Transaction on Pattern Analysis and Machine Intelligence*, 14(10): 965–980.

Zhang, Y.-J. 2017. *Image Engineering, Vol. 1: Image Processing*. Germany: De Gruyter.

Zhang, Y.-J. 2021. *3D Computer Vision: Principles, Algorithms and Applications*. Singapore: Springer Nature.

Zhang, Z.Y. 2004. Camera calibration with one-dimensional objects. *IEEE Transactions on Pattern Analysis and Machine Intelligence*, 26(7): 892–899.

Zhou, Z.W., M.M.R. Siddiquee, N. Tajbakhsh, et al. 2018. UNet++: A nested U-net architecture for medical image segmentation. *Proceedings of the 4th International Workshop on Deep Learning in Medical Image Analysis and Multimodal Learning for Clinical Decision Support*, 3–11.

CHAPTER 3

Stereo Vision

THE GENERAL CAMERA IMAGING method obtains a 2-D image from a 3-D physical space, in which information on the plane perpendicular to the optical axis of the camera is retained in the image, but the depth information along the optical axis of the camera is lost. However, image understanding often needs to obtain 3-D information about the objective world or comprehensive information about higher dimensions.

Stereo vision mainly studies how to use (multi-image) imaging technology to obtain distance (depth) information about objects in a scene from (multiple) images, and pioneering work began as early as the mid-1960s (Roberts 1965). **Stereo vision** observes the same scene from two or more viewpoints, collects a set of images from different perspectives, and then obtains the **disparity** between corresponding pixels in different images through the principle of triangulation (that is, when the same 3-D point is projected on two 2-D images, the position difference between its two corresponding points on these images), from which depth information is obtained, and then the shape of the objects in the scene and the spatial position between them are calculated. The working process of stereo vision has many similarities with the perception process of the human visual system. In fact, the human visual system is a natural stereo vision system.

The chapter is organized as follows. Section 3.1 introduces various ways of depth imaging, compares depth images with common grayscale images and conducts an in-depth analysis from the perspective of intrinsic and extrinsic images. Section 3.2 introduces different imaging modes in binocular stereo vision due to the difference of the relative poses of the two cameras, including binocular lateral mode, binocular lateral convergence mode and binocular axial mode. Section 3.3 discusses the principle of binocular stereo matching based on regional gray-level correlation and introduces the various constraints used in matching, the essential matrix and the fundamental matrix, and the calculation of optical properties. Section 3.4 discusses two prominent (point) features that have been widely used in recent years in feature-based binocular stereo matching: scale-invariant feature transformation (SIFT) and speed-up robustness features (SURF). Section 3.5 explores recent technique developments and promising research directions.

DOI: 10.1201/9781003362388-3

3.1 DEPTH IMAGING AND DEPTH IMAGE

In order to better understand the image, not only the common 2-D projection image must be obtained, but also the third-dimensional information between the imaging device and the objective scene, that is, the distance information or depth information.

3.1.1 Depth Image and Grayscale Image

In the representation of image pixels using $f(x, y)$, x and y represent the position of a coordinate point in the 2-D space XY, and f represents the value of a certain property F of the image at the point (x, y). In the **grayscale image**, f represents the gray value, which often corresponds to the observed brightness of the objective scene. Text images are often **binary images**. Binary images are a special case of grayscale images. There are only two values for f, corresponding to text and blank space. The image at the point (x, y) can also have multiple properties at the same time. In this case, it can be represented by a vector $\boldsymbol{f}$. For example, a **color image** has three values of red, green and blue at each image point, which can be recorded as $\boldsymbol{f}(x, y) = [f_r(x, y), f_g(x, y), f_b(x, y)]$.

The depth image reflects the depth of the object, and its attribute is distance, that is, f in $f(x, y)$ represents the depth value. This is different from common grayscale (color) images.

Consider a profile on the object in Figure 3.1. Compared with the grayscale image, the depth collected from this profile has the following two characteristics:

(1) The pixel value of the same external surface on the corresponding object in the depth image changes at a certain rate (the surface is inclined relative to the image plane). This value changes with the shape and orientation of the object, but has nothing to do with the external lighting conditions. On the other hand, the corresponding pixel value in grayscale image depends not only on the illuminance of the surface, but also on the reflection coefficient of the surface. That is, it is related not only to the shape and orientation of the object, but also to the external lighting conditions.

(2) There are two types of boundary lines in depth images: one is the (distance) step edge between the object and the background; the other is the ridge-like edge at the intersection of two regions inside the object (corresponding to extreme value, depth is continuous). In the grayscale image, both places are step edges.

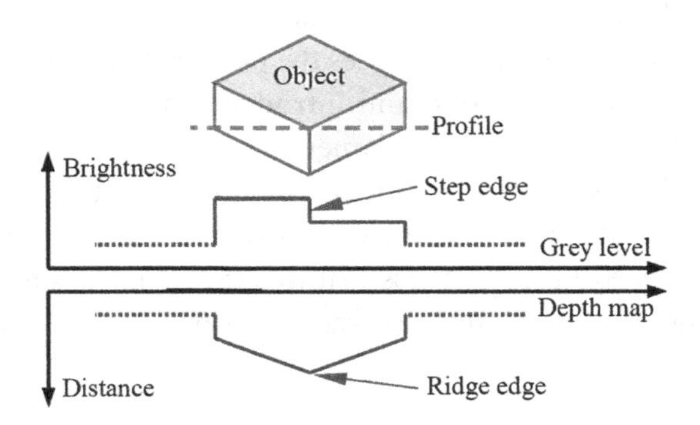

FIGURE 3.1 The difference between depth image and gray image.

3.1.2 Intrinsic Image and Non-Intrinsic Image

Image is a description form of objective scene, which can be divided into two categories: **intrinsic image** and **non-intrinsic image** according to the nature of the scene described (Ballard and Brown 1982). The image is the image of the scene captured by the observer or collector. The scene and the objects in the scene have some objectively existing characteristics (properties) that have nothing to do with the nature of the observer and the collector, such as surface reflectivity, transparency, surface orientation, movement speed of the scene, and the relative distance and orientation in space between the sceneries in the scene, etc. These characteristics are called intrinsic characteristics (of the scene), and the images representing the physical quantities of these intrinsic characteristics are called intrinsic images.

There are many types of intrinsic images, and each intrinsic image can only represent an intrinsic characteristic of the scene without the influence of other characteristics. If the intrinsic image can be obtained, it is very useful for correctly interpreting the scene represented by the image. For example, a depth image is one of the most commonly used intrinsic images, where each pixel value represents the distance between the scene point represented by the pixel and the camera (depth, also known as the elevation of the scene), and these pixel values actually reflect directly the shape of the visible surface of the scene (intrinsic characteristics).

The physical quantity represented by the non-intrinsic image is related not only to the scene, but also to the nature of the observer/collector or the conditions of image collection or the surrounding environment. A typical representative of non-intrinsic images is a common intensity image (luminance image), which is generally represented as a grayscale image. The intensity image is an image that reflects the intensity of the radiation received by the observation site. The intensity value is often the result of multiple factors such as the intensity of the radiation source, the orientation of the radiation mode, the reflection properties of the surface of the scene, as well as the location and performance of the collector. Many image understanding problems require the use of non-intrinsic images to restore intrinsic characteristics; that is, to obtain intrinsic images, which can further explain the scene.

3.1.3 Depth Imaging Modes

There are many modes of depth imaging, which are mainly determined by the mutual position and movement of the light source, collector, and scene (scenery). The most basic imaging method is monocular imaging, which uses a collector to take an image of the scene at a fixed position. Although the depth information of the scene is not directly reflected in the image at this time, this information is also implicit in the imaged geometric distortion, brightness (shadow), texture, surface contour, and other factors. If two collectors are used to take images of the same scene at one location each (one collector can also be used to take images of the same scene in two positions one after the other, or one collector and an optical imaging system can be used to obtain two images), it is binocular imaging. At this time, the parallax (disparity) generated between the two images (similar to the human eye) can be used to help calculate the distance between the collector and the scene, so this method is generally called stereo vision. If more than two collectors are used to take images of the same scene at different positions (and one collector can also be used to take images of the same scene at multiple positions), it is multi-eye imaging (essentially stereo vision). Monocular, binocular, or multi-eye methods can obtain not only still images, but also sequential images through continuous shooting. Monocular imaging is simpler than binocular imaging, but it is more complicated to obtain depth information from it. Conversely, binocular imaging increases the complexity of acquisition, but can reduce the complexity of acquiring depth information.

The above discussion assumes a fixed light source. If the collector is fixed relative to the scene and the light source moves around the scene, this imaging method is called light shift imaging (also called stereo photometric imaging). Since the surface of the same scene has different brightness under different lighting conditions, the surface orientation of the object can be obtained from the light shift image (but absolute depth information cannot be obtained in this case). If you keep the light source fixed and let the collector move to track the scene or let the collector and the scene move at the same time, it constitutes active visual imaging (referring to the initiative of human vision, that is, people will move their body or head according to the needs of observation to change the perspective and selectively pay special attention to part of the scene), the latter of which is also called active vision self-motion imaging. In addition, if a controllable light source is used to illuminate the scene, the structured light imaging method is used to explain the surface shape of the scene through the collected projection mode. In this way, the light source and the collector can be fixed while the scene rotates, or the scene can be fixed but the light source and the collector can rotate around the scene together.

Some of the characteristics of the light source, collector and scene in the above modes are summarized in Table 3.1.

From the perspective of image acquisition, these methods of obtaining intrinsic images (here, depth images) can be divided into two categories: (i) to directly collect intrinsic images (as in the structured light imaging mode in Table 3.1), and (ii) to first collect non-intrinsic images containing intrinsic information, and then restore intrinsic characteristics through image technology (as in other modes in Table 3.1). To use the former method, specific image acquisition equipment/imaging devices are required. Below we will discuss the stereo vision method. Other methods can be found, for example, in Zhang (2021).

TABLE 3.1 Characteristics of Common Imaging Modes

Imaging Modes	**Light Source**	**Collector**	**Scenery**
Monocular imaging	Fixed	Fixed	Fixed
Binocular imaging	Fixed	Two positions	Fixed
Multi-eye imaging	Fixed	Multiple-positions	Fixed
Video/sequence imaging	Fixed/Moving	Fixed/Moving	Moving /Fixed
Light shift (stereo photometric) imaging	Moving	Fixed	Fixed
Active vision imaging	Fixed	Moving	Fixed
Active vision (self-motion) imaging	Fixed	Moving	Moving
Structured light imaging	Fixed/Rotating	Fixed/Rotating	Rotating/Fixed

3.2 BINOCULAR IMAGING MODES

In this section, we mainly consider several typical imaging modes in the stereo vision method of basic binocular (stereo) imaging.

Binocular imaging can obtain two images of the same scene with different viewpoints (similar to human eyes). The **binocular imaging model** can be regarded as a combination of two monocular imaging models. In actual imaging, either two monocular systems can be used to collect at the same time, or one monocular system can be used in two poses to collect one after the other.

Depending on the relative positions of the two cameras, binocular imaging can have a variety of modes. Here are a few common situations.

3.2.1 Binocular Horizontal Mode

Figure 3.2 shows a schematic diagram of **binocular horizontal mode** imaging. The focal lengths of the two lenses are both λ, and the line between the centers of two lenses is called the baseline B of the system. The corresponding axes of the two camera coordinate systems are completely parallel (X-axis coincides), and both image planes are parallel to the XY plane of the world coordinate system. The Z coordinate of a 3-D space point W is the same for both camera coordinate systems.

3.2.1.1 Parallax and Depth

It can be seen from Figure 3.2 that the same 3-D space point corresponds to two image plane coordinate points, and the position difference between them is called parallax. The relationship between parallax and depth (object distance) in the binocular horizontal mode is discussed below with the help of Figure 3.3. It is a schematic diagram of the plane (XZ plane) where the two lenses are connected. Among them, the world coordinate system coincides with the first camera coordinate system and only has a translation amount B in the X-axis direction with the second camera coordinate system.

Consider first the geometric relationship between the coordinate X of the point W in the 3-D space and the coordinate x_1 of the projected point on the first image plane, which gives

$$\frac{|X|}{Z-\lambda} = \frac{x_1}{\lambda} \tag{3.1}$$

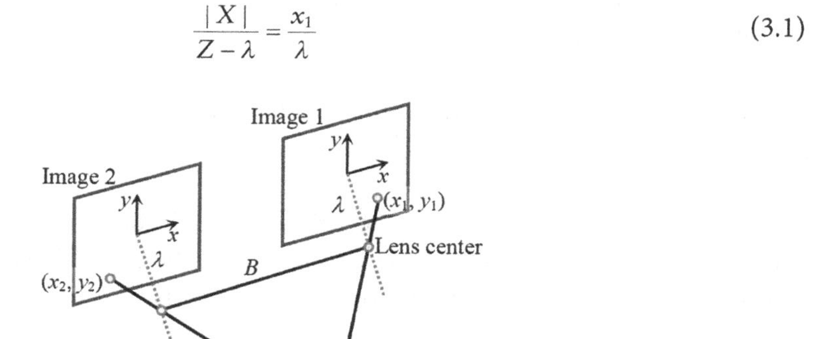

FIGURE 3.2 Schematic diagram of binocular horizontal mode imaging.

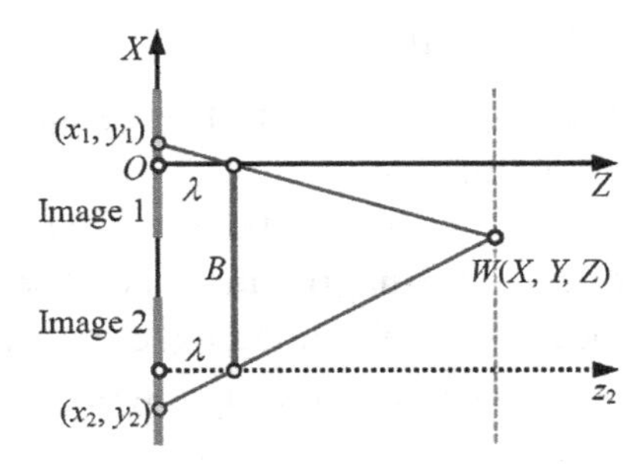

FIGURE 3.3 Parallax in parallel binocular imaging.

Then, consider the geometric relationship between the coordinate X of the point W in the 3-D space and the coordinate x_2 of the projected point on the second image plane, which gives

$$\frac{B-|X|}{Z-\lambda}=\frac{|x_2|-B}{\lambda} \tag{3.2}$$

Combine these two equations, eliminate X, and get the parallax:

$$d=x_1+|x_2|-B=\frac{\lambda B}{Z-\lambda} \tag{3.3}$$

Solve Z from Equation (3.3):

$$Z=\lambda\left(1+\frac{B}{d}\right) \tag{3.4}$$

Equation (3.4) directly relates the distance Z between the object and the image plane (that is, the depth in the 3-D information) and the parallax d. Conversely, it also shows that the size of the parallax is related to the depth, that is, the spatial information of the 3-D object is included in the parallax. According to Equation (3.4), when the baseline and focal length are known, it is very simple to calculate the Z coordinate of point W after determining the parallax d. In addition, after the Z coordinate is determined, the world coordinates X and Y of point W can be calculated with (x_1, y_1) or (x_2, y_2) referring to Equations (3.1) and (3.2).

Now let's look at the ranging accuracy. From Equation (3.4), we can see that the depth information is related to the parallax, and the parallax is related to the imaging coordinates. Suppose x_1 produces a deviation e, that is, $x_{1e} = x_1 + e$, then $d_{1e} = x_1 + e + |x_2| - B = d + e$, so the distance deviation is

$$\Delta Z=Z-Z_{1e}=\lambda\left(1+\frac{B}{d}\right)-\lambda\left(1+\frac{B}{d_{1e}}\right)=\frac{\lambda Be}{d(d+e)} \tag{3.5}$$

Substituting Equation (3.3) into Equation (3.5):

$$\Delta Z = \frac{e(Z-\lambda)^2}{\lambda B + e(Z-\lambda)} \approx \frac{eZ^2}{\lambda B + eZ} \tag{3.6}$$

The last step is to consider the simplification of $Z >> \lambda$ in the general case. It can be seen from Equation (3.6) that the accuracy of distance measurement is related to the focal length of the camera, the baseline length between the cameras, and the object distance. The longer the focal length and the longer the baseline, the higher the accuracy; but the larger the object distance, the lower the accuracy.

3.2.1.2 Angular Scanning Imaging

In the binocular horizontal mode imaging described above, in order to determine the information of a 3-D space point, the point needs to be in the common field of view of the two cameras. If the two cameras are rotated (around the *X*-axis), the common field of view is increased and panoramic images can be collected. This can be called **stereoscopic imaging** with an **angular scanning camera**. This is called a **binocular angular scanning mode**, in which the coordinates of the imaging point are determined by the **azimuth angle** and **elevation angle** of the camera. In Figure 3.4, θ_1 and θ_2, respectively, give the azimuth angle (corresponding to the saccade movement around the *Y* axis); and the elevation angle ϕ is the angle between the *XZ* plane and the plane determined by the two optical centers and the point *W*.

Generally, the azimuth angle of the lens can be used to indicate the spatial distance between objects. Using the coordinate system shown in Figure 3.4, there are

$$\tan\theta_1 = \frac{|X|}{Z} \tag{3.7}$$

$$\tan\theta_2 = \frac{B-|X|}{Z} \tag{3.8}$$

Eliminate *X* using two equations simultaneously, the *Z* coordinate of point *W* is

$$Z = \frac{B}{\tan\theta_1 + \tan\theta_2} \tag{3.9}$$

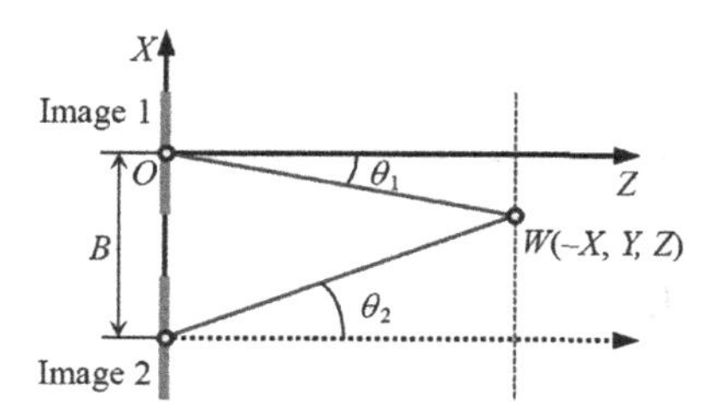

FIGURE 3.4 Angular scanning camera for stereoscopic imaging.

Equation (3.9) actually connects the distance Z between the object and the image plane (that is, the depth in the 3-D information) directly with the tangent of the two azimuth angles. Comparing Equation (3.9) and Equation (3.4), we can see that the effects of parallax and focal length are both implicit in the azimuth angle. According to the Z coordinate of the space point W, the X and Y coordinates can be obtained, respectively:

$$X = Z \tan \theta_1 \tag{3.10}$$

$$Y = Z \tan \varphi \tag{3.11}$$

3.2.2 Binocular Convergence Horizontal Mode

In order to obtain an even larger field of view overlap, you can place the two cameras side by side but let the two optical axes converge. This **binocular convergence horizontal mode** can be regarded as the extension of the **binocular horizontal mode** (at this time the **vergence** between the binoculars is not zero).

3.2.2.1 Parallax and Depth

Consider only the situation shown in Figure 3.5, which is obtained by rotating the two monocular systems in Figure 3.3 toward each other around their respective centers. Figure 3.5 shows the plane (XZ plane) where the two lenses connect. The distance between the centers of the two lenses (i.e., the baseline) is B. The two optical axes intersect at point $(0, 0, Z)$ in the XZ plane, and the angle of intersection is 2θ. Now let's look at how to find the coordinates (X, Y, Z) of point W in 3-D space if two image plane coordinate points (x_1, y_1) and (x_2, y_2) are known.

First of all, it can be seen from the triangle enclosed by the two world coordinate axes and the camera's optical axis:

$$Z = \frac{B}{2} \frac{\cos\theta}{\sin\theta} + \lambda \cos\theta \tag{3.12}$$

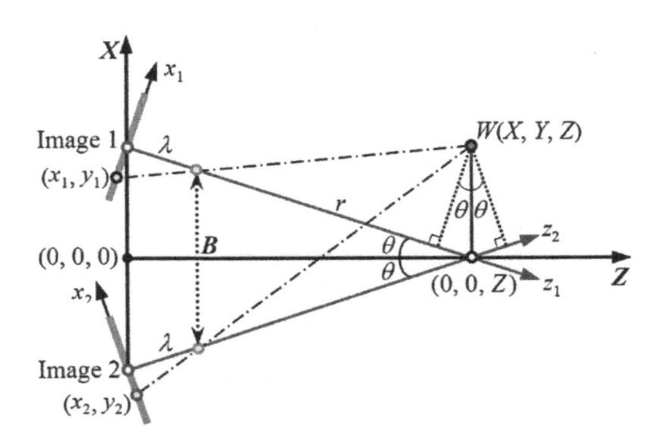

FIGURE 3.5 Parallax in convergent binocular imaging.

Now draw perpendicular lines from point W to the optical axis of the two cameras, because the angle between the two perpendicular lines and the X-axis is θ, so according to the relationship of similar triangles, we can get

$$\frac{|x_1|}{\lambda} = \frac{X\cos\theta}{r - X\sin\theta} \tag{3.13}$$

$$\frac{|x_2|}{\lambda} = \frac{X\cos\theta}{r + X\sin\theta} \tag{3.14}$$

where r is the distance from the (any) center of the lens to the convergence point of the two optical axes.

Combine Equation (3.13) and Equation (3.14), and eliminate r and X to get (refer to Figure 3.5)

$$\lambda\cos\theta = \frac{2|x_1|\cdot|x_2|\sin\theta}{|x_1| - |x_2|} = \frac{2|x_1|\cdot|x_2|\sin\theta}{d} \tag{3.15}$$

Substituting Equation (3.15) into Equation (3.12), we can get

$$Z = \frac{B\cos\theta}{2\sin\theta} + \frac{2|x_1|\cdot|x_2|\sin\theta}{d} \tag{3.16}$$

Equation (3.16), like Equation (3.4), also directly relates the distance Z between the object and the image plane with the parallax d. In addition, it can be obtained from Figure 3.5

$$r = \frac{B}{2\sin\theta} \tag{3.17}$$

Substituting Equation (3.13) or Equation (3.14) to get the X coordinate of point W

$$|X| = \frac{B}{2\sin\theta}\frac{|x_1|}{\lambda\cos\theta + |x_1|\sin\theta} = \frac{B}{2\sin\theta}\frac{|x_2|}{\lambda\cos\theta - |x_2|\sin\theta} \tag{3.18}$$

3.2.2.2 Image Rectification

The case of binocular convergence can also be converted to the case of binocular parallelism. **Image rectification** is the process of geometrically transforming the image obtained by the camera with the optical axis converging to obtain the image obtained by the camera with the optical axis parallel (Goshtasby 2005). Consider the images before and after rectification in Figure 3.6. The light from the object point W intersects the left image at (x, y) and (X, Y) before and after correction. Each point on the image before correction can be connected to the center of the lens and extended to intersect the image after rectification. Therefore, for each point on the image before rectification, the corresponding point on the image after rectification can be determined. The coordinates of the points before and after

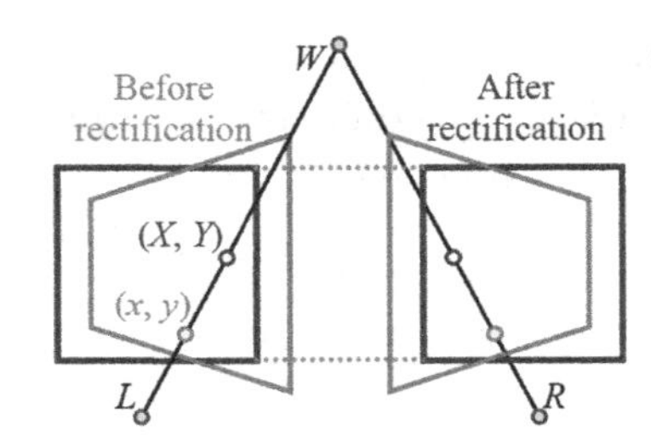

FIGURE 3.6 Using projection transformation to rectify the image obtained by two cameras converging with the optical axis.

rectification are related by the projection transformation (a_1 to a_8 are the coefficients of the projection transformation matrix):

$$x = \frac{a_1 X + a_2 Y + a_3}{a_4 X + a_5 Y + 1} \tag{3.19}$$

$$y = \frac{a_6 X + a_7 Y + a_8}{a_4 X + a_5 Y + 1} \tag{3.20}$$

The eight coefficients in the above two equations can be determined with the help of four sets of corresponding points on the images before and after rectification. Here we can consider using the horizontal epipolar line (the intersection of the plane formed by the baseline and a point in the scene and the imaging plane). For this reason, two epipolar lines in the image must be selected before rectification and mapped to the two horizontal lines in the image after rectification, as shown in Figure 3.7. The corresponding relationship is

$$X_1 = x_1 \quad X_2 = x_2 \quad X_3 = x_3 \quad X_4 = x_4 \tag{3.21}$$

$$Y_1 = Y_2 = \frac{y_1 + y_2}{2} \qquad Y_3 = Y_4 = \frac{y_3 + y_4}{2} \tag{3.22}$$

The above correspondence can maintain the width before and after image rectification, but there will be scale changes in the vertical direction (in order to map non-horizontal polar lines to horizontal polar lines). In order to obtain the rectified image, for each point (X, Y) on the rectified image, Equation (3.19) and Equation (3.20) are used to find the corresponding point (x, y) on the image before rectification. Moreover, the gray level at point (x, y) should be assigned to point (X, Y).

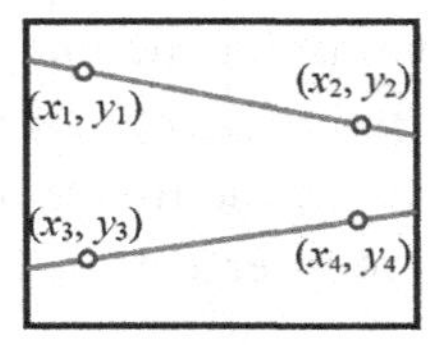

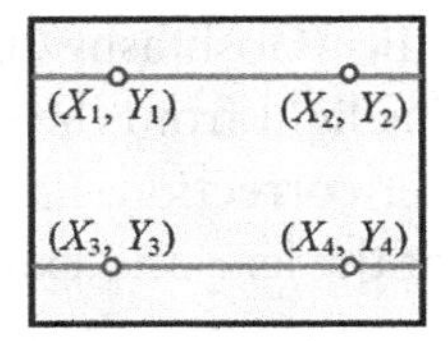

FIGURE 3.7 Schematic diagram of images before and after rectification.

The above process is also repeated for the right image. In order to ensure that the corresponding epipolar lines on the left and right images after rectification represent the same scan line, it is necessary to map the corresponding epipolar lines on the image before rectification to the same scan line on the image after rectification, so when rectifying both the left image and the right image, the Y coordinate in Equation (3.22) should be used.

3.2.3 Binocular Axial Mode

Binocular horizontal mode or binocular convergence horizontal mode need to be calculated according to the triangle method, so the baseline should not be too short; otherwise it will affect the accuracy of the depth calculation. However, when the baseline is longer, the problems caused by the mismatch (no overlap) of the field of view will be more serious. At this time, the **binocular axial mode**, also called **binocular longitudinal mode**, can be considered. In this mode, two cameras are arranged in sequence along the optical axis. This situation can also be seen as moving the camera along the optical axis and collecting the second image closer to the subject than the first image, as shown in Figure 3.8. In Figure 3.8, only the XZ plane is drawn, and the Y axis goes out from the inside of the paper. The origins of the two camera coordinate systems for the first image and the second image are only different in the Z direction by B, and B is also the distance between the optical centers of the two cameras (baseline).

According to the geometric relationship in Figure 3.8, there are

$$\frac{X}{Z-\lambda}=\frac{|x_1|}{\lambda} \tag{3.23}$$

$$\frac{X}{Z-\lambda-B}=\frac{|x_2|}{\lambda} \tag{3.24}$$

If Equation (3.23) and Equation (3.24) are solved simultaneously, it can be obtained (only x is considered, which is similar to y)

$$X=\frac{B}{\lambda}\frac{|x_1|\cdot|x_2|}{|x_2|-|x_1|}=\frac{B}{\lambda}\frac{|x_1|\cdot|x_2|}{d} \tag{3.25}$$

$$Z=\lambda+\frac{B|x_2|}{|x_2|-|x_1|}=\lambda+\frac{B|x_2|}{d} \tag{3.26}$$

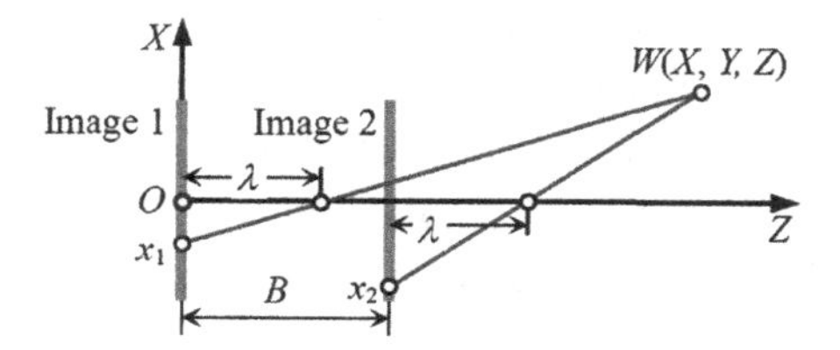

FIGURE 3.8 Binocular axial mode imaging.

Compared with the binocular horizontal mode, the common field of view of the two cameras in binocular axis mode is the field of view of the previous camera (the camera that acquired the second image in Figure 3.8), so the boundary of the common field of view can easily be determined, and the problem that the 3-D space point caused by occlusion is only seen by one camera can be basically eliminated. However, since the binoculars basically use the same angle to observe the scene at this time, the benefit of lengthening the baseline to the depth calculation accuracy cannot be fully reflected. In addition, the accuracies of the parallax and depth calculations are both related to the distance between the 3-D space point and the camera's optical axis (for example, in Equation (3.26), the depth Z and $|x_2|$ that is the distance between projection of the 3-D space point and the optical axis), which is different from the binocular horizontal mode.

3.3 BINOCULAR STEREO MATCHING BASED ON REGION

With binocular imaging, the relationship between parallax and distance can be established. In order to determine the parallax, it is necessary to determine the corresponding points of the same spatial points in the two images of the binocular imaging. Therefore, determining the relationship between the corresponding points in the binocular image is a key step to obtain the depth image. The following discussion only takes the **binocular horizontal mode** as an example. If we consider the unique geometric relationships among various modes, the results obtained from the binocular horizontal mode can also be extended to other modes.

To determine the relationship between corresponding points, a point-to-point correspondence matching method can be used. However, the direct use of single-point grayscale search will be affected by factors such as many points in the image having the same grayscale, the image noise, etc. Current practical technologies are mainly divided into two categories, namely grayscale correlation and feature matching. The former category is based on the region, that is, the neighborhood properties of each point that needs to be matched are considered. The latter category is based on feature points, that is, the points with unique or special properties in the image are selected as matching points. The features used by the latter method are mainly the inflection point and corner point coordinates in the image, the edge line segment, the contour of the object, and so on. The above two methods are similar to the region-based and edge-based methods in image segmentation.

3.3.1 Template Matching

The region-based method needs to consider the nature of the neighborhood of the point, and the neighborhood is often determined with the help of a template (also called a subimage or mask or window). When a point in the left image is given and the corresponding point in the right image needs to be searched for, the neighborhood centered on the point in the left image can be extracted as a template, and it can be translated onto the right image and calculated with respect to each position, and based on the correlation value to determine whether to match. If it matches, it is considered that the center point of the matching position in the right image and that point in the left image constitute a corresponding point pair. Here, the point with the maximum correlation value can be selected as the matching position. Or, a threshold value can be given first, and the points satisfying the correlation value greater than the threshold value can be extracted. The selection of matching position can also be made according to other factors.

3.3.1.1 Basic Method

The basic method of **template matching** is to use a smaller image (template) to match a part (sub-image) of a larger image. The result of the matching is to determine whether there is a small image in the large image, and if so, the position of the small image in the large image is further determined. In template matching, the template is often square, but it can also be rectangular or other shapes. Now consider finding the matching position of a template image $w(x, y)$ of size $J \times K$ and a large image $f(x, y)$ of $M \times N$, set $J \leq M$ and $K \leq N$. In the simplest case, the correlation function between $f(x, y)$ and $w(x, y)$ can be written as

$$c(s,t) = \sum_x \sum_y f(x,y) w(x-s, y-t) \tag{3.27}$$

where $s = 0, 1, 2, \ldots, M - 1$; $t = 0, 1, 2, \ldots, N - 1$.

The summation in Equation (3.27) is performed on the image region where $f(x, y)$ and $w(x, y)$ overlap. Figure 3.9 shows a schematic diagram of related calculations, assuming that the origin of $f(x, y)$ is at the upper left corner and the origin of $w(x, y)$ is at its center. For any given position (s, t) in $f(x, y)$, a specific value of $c(s, t)$ can be calculated according to Equation (3.27). When s and t change, $w(x, y)$ moves in the image region and gives all the values of the function $c(s, t)$. The maximum value of $c(s, t)$ indicates the position that best matches $w(x, y)$.

In addition to determining the matching position according to the maximum correlation criterion, the minimum mean square error function can also be used:

$$M_{\text{me}}(s,t) = \frac{1}{MN} \sum_x \sum_y \left[f(x,y) w(x-s, y-t) \right]^2 \tag{3.28}$$

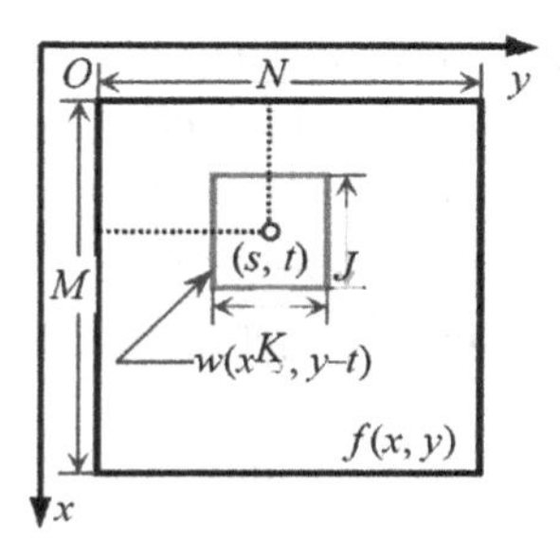

FIGURE 3.9 Schematic diagram of template matching.

In VLSI hardware, the square operation is more difficult to implement, so the absolute value can be used instead of the square value to obtain the minimum average difference function:

$$M_{ad}(s,t)=\frac{1}{MN}\sum_{x}\sum_{y}|f(x,y)w(x-s,y-t)| \tag{3.29}$$

The correlation function defined by Equation (3.27) has a disadvantage, that is, it is more sensitive to changes in the amplitude of $f(x, y)$ and $w(x, y)$. For example, when the value of $f(x, y)$ is doubled, the value of $c(s, t)$ will also be doubled. In order to overcome this problem, the following correlation coefficient can be defined:

$$C(s,t)=\frac{\sum_{x}\sum_{y}\left[f(x,y)-\bar{f}(x,y)\right]\left[w(x-s,y-t)-\bar{w}\right]}{\left\{\sum_{x}\sum_{y}\left[f(x,y)-\bar{f}(x,y)\right]^{2}\sum_{x}\sum_{y}\left[w(x-s,y-t)-\bar{w}\right]^{2}\right\}^{1/2}} \tag{3.30}$$

where $s = 0, 1, 2, \ldots, M - 1$; $t = 0, 1, 2, \ldots, N - 1$; $\bar{w}(x,y)$ is the mean value of $w(x, y)$ (you only need to calculate it once); and $\bar{f}(x,y)$ represents the mean value of the region corresponding to the current position of w in $f(x, y)$.

The sum in Equation (3.30) is performed on the common coordinates of $f(x, y)$ and $w(x, y)$. Because the correlation coefficient has been scaled to the interval [–1, 1], the change in its value has nothing to do with the amplitude change of $f(x, y)$ and $w(x, y)$.

Another method is to calculate the gray level difference between the template and the sub-image, and establish the correspondence between the two sets of pixels that meet the **mean squared difference** (MSD). The advantage of this type of method is that the matching result is not easily affected by the gray-level detection accuracy and density of the template, so it can get a high positioning accuracy and a dense parallax surface (Kanade et al. 1996). The disadvantage of this type of method is that it relies on the statistical characteristics of the image gray level, so it is more sensitive to the surface structure of the scene and the reflection of light. Therefore, there are certain difficulties when the scene surface in the space lacks sufficient texture details and the imaging distortion is relatively large (such as the baseline length being too large). In actual matching, some grayscale-derived quantities can also be used, but experiments have shown that in matching comparisons using grayscale, grayscale differential size and direction, grayscale Laplacian value, and grayscale curvature as matching parameters, the effect of using grayscale parameters is still the best (Lew et al. 1994).

As a basic matching technique, template matching has been applied in many aspects, especially when the image is only shifted. Using the calculation of the correlation coefficient above, the correlation function can be normalized to overcome the problems caused by the amplitude change. However, it is more difficult to normalize the image size and rotation.

The normalization of the size requires a spatial scale transformation, and this process requires a lot of calculations. Normalizing the rotation is more difficult. If the rotation angle of $f(x, y)$ is known, just rotate $w(x, y)$ by the same angle to align it with $f(x, y)$. But without knowing the rotation angle of $f(x, y)$, to find the best match, we need to rotate $w(x, y)$ at all possible angles. In practice, this method is not feasible, so in the case of arbitrary rotation or no restriction on the rotation, the region-related method is rarely used directly.

The method of using a template representing matching primitives for image matching must solve the problem that the amount of calculation will increase exponentially with the number of primitives. If the number of primitives in the image is n and the number of primitives in the template is m, then there are $O(n^m)$ possible correspondences between the primitives of the template and the image, where the number of combinations is $C(n, m)$, or $C_m{}^n$.

In order to reduce the amount of calculation for template matching, on the one hand, some prior knowledge (such as the epipolar constraint during stereo matching, see the next section) can be used to reduce the positions that need to be matched; on the other hand, it is possible to use the feature of considerable overlap in the template coverage at adjacent matching positions to reduce the number of recalculations of correlation values (Zhang and Lu, 2002). By the way, the correlation can also be calculated in the frequency domain by using fast Fourier transform (FFT), so it can be matched based on the frequency domain transform. If the sizes of f and w are the same, the calculation in the frequency domain will be more efficient than the direct spatial calculation. In fact, w is generally much smaller than f. Someone once estimated that if the non-zero items in w are less than 132 (approximately equivalent to a 13×13 sub-image), then directly using Equation (3.27) to calculate in the spatial domain may be more efficient than using FFT to calculate in the frequency domain. Of course, this number is related to the computer and programming algorithm used. In addition, the calculation of the correlation coefficient in Equation (3.30) is difficult to achieve in the frequency domain, so it is generally carried out directly in the space domain.

By the way, the generalized Hough transform (for example, see Zhang (2017)) can also be regarded as an improved template matching method (spatial matching filter), and its calculation amount is also less than template matching.

3.3.1.2 Using Geometric Hashing

In order to achieve efficient template matching, **geometric hashing** can also be used. Its basis is that three points can define a 2-D plane. That is, if you choose three non-collinear points P_1, P_2, P_3, you can use the linear combination of these three points to represent any point:

$$Q = P_1 + s(P_2 - P_1) + t(P_3 - P_1) \tag{3.31}$$

The above equation will not change under the affine transformation, that is, the value of (s, t) is only related to the three non-collinear points, and has nothing to do with the affine transformation itself. In this way, the value of (s, t) can be regarded as the affine coordinates of point Q. This feature also applies to line segments: three non-parallel line segments can be used to define an affine datum.

Geometric hashing requires the construction of a hash table, which can help the matching algorithm to quickly determine the potential position of a template in the image. The hash table can be constructed as follows. For any three non-collinear points (reference point group) in the template, calculate the affine coordinates (s, t) of other points. The affine coordinates (s, t) of these points will be used as the index of the hash table. For each point, the hash table retains the index (serial number) of the current reference point group. Searching for multiple templates in an image requires more template indexes to be kept.

To search for a template, randomly select a set of reference points in the image, and calculate the affine coordinates (s, t) of other points. Using this affine coordinate (s, t) as the index of the hash table, the index of the reference point group can be obtained. In this way, a vote for the occurrence of this reference point group in the image is obtained. If the randomly selected points do not correspond to the reference point group on the template, there is no need to accept voting. However, if the randomly selected point corresponds to the set of reference points on the template, the vote is accepted. If many votes are accepted, it means that this template is likely to be in the image, and the index of the reference point group can be obtained. Because there will be a certain probability that the selected reference point group is inappropriate, the algorithm needs to iterate to increase the probability of finding the correct match. In fact, it is only necessary to find a correct set of reference points to determine the matching template. Therefore, if k points of the N template points are found in the image, the probability that the reference point group is correctly selected at least once in m attempts is:

$$p = 1 - \left[1 - (k/N)^3\right]^m \tag{3.32}$$

If the ratio of the number of points in the template to the number of points in the image appears in the image, k/N is 0.2, and the probability of the template matching is 99% (that is, $p = 0.99$), then the number of attempts m is 574.

3.3.2 Stereo Matching

Using the principle of template matching, the similarity of regional gray levels can be used to search for the corresponding points of two images. Specifically, in the stereo image pair, first select a window centered on a specific pixel in the left image, build a template based on the grayscale distribution in the window, and then use the template to search in the right image to find the position of the best matching window, then the pixel in the center of the matching window corresponds to the pixel to be matched in the left image.

In the above search process, if there is no prior knowledge or any restriction on the position of the template in the right image, the range to be searched may cover the entire right image. It is very time-consuming to search in this way for each pixel in the left image.

Constraints that can be considered to reduce the scope of the search include the following (Forsyth and Ponce 2012):

(1) *Compatibility constraints.* The **compatibility constraint** means that black dots can only match black dots. More generally speaking, only features that originate from the same type of physical properties in the two images can be matched. It is also called **luminosity compatibility constraint**.

(2) *Uniqueness constraint.* The **uniqueness constraint** means that a single black point in one image can only be matched with a single black point in another image.

(3) *Continuity constraints.* The **continuity constraint** means that the disparity change near the matching point is smooth (gradual) in most points except the occluded region or the discontinuous region in the entire image, which is also called the **disparity smoothness constraint.**

When discussing stereo matching, in addition to the above three constraints, the epipolar constraints introduced below can also be considered.

3.3.2.1 Epipolar Line Constraint

The **epipolar line constraint** can help reduce the search range during the search process and speed up the search process.

First, with the help of the binocular horizontal convergence mode diagram in Figure 3.10, two important concepts of epipoles and epipolar lines are introduced. In Figure 3.10, the origin of the coordinates is the left eye center, the X-axis connects the left eye and right eye centers, the Z axis points to the observation direction, the spacing between left eye and right eye is B (also often called the system baseline), and the optical axes of the left and right image planes are both in the XZ plane, and the angle of intersection is θ. Consider the connection between the left and right image planes. O_1 and O_2 are the optical centers of the left and right image planes, and the connecting line between them is called the optical center line. The intersection points $\mathbf{e}_1$ and $\mathbf{e}_2$ of the optical center line with the left and right image planes are respectively called the epipoles of the left and right image planes (the epipole coordinates are respectively $\mathbf{e}_1$ and $\mathbf{e}_2$). The optical center line and the spatial point

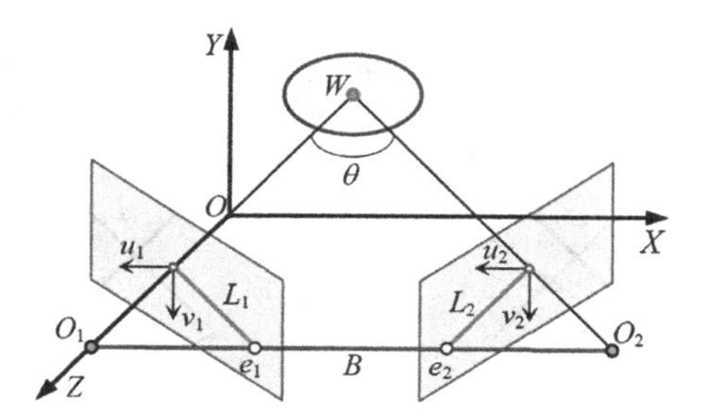

FIGURE 3.10 Schematic diagram of epipoles and epipolar lines.

W are in the same plane. This plane is called the **epipolar plane**. The intersection lines L_1 and L_2 of the epipolar plane with the left and right image planes are, respectively, called the epipolar lines of the projection points of the spatial point W on the left and right image planes. The epipolar line defines the position of the corresponding point of the binocular image, and the projection point p_2 (coordinate $\boldsymbol{p}_2$) of the right image plane corresponding to the projection point p_1 (coordinate $\boldsymbol{p}_1$) of the space point W on the left image plane must be on the epipolar line L_2. On the contrary, the projection point of the left image plane corresponding to the projection point of the spatial point W on the right image plane must be on the epipolar line L_1.

There is a corresponding relation between epipoles and epipolar lines. Consider a stereo vision system with two sets of monocular optical systems as shown in Figure 3.11. In Figure 3.11, there are a set of points (p_1, p_2, ...) on the Imaging plane 1, and each point corresponds to a light in 3-D space. Each ray projects a line (L_1, L_2, ...) on the Imaging plane 2. Because all light rays converge to the optical center of the first camera, these lines must intersect at a point on the Imaging plane 2. This point is the image of the optical center of the first camera in the second camera, which is called the epipole. Similarly, the image of the optical center of the second camera in the first camera is also an epipole. These projection lines are just epipolar lines.

The epipolar line defines the position of the corresponding points on the binocular image. The projection point of the right image plane corresponding to the projection point of the spatial point W on the left image plane must be on the epipolar line L_2; the projection point on the left image plane corresponding to the projection point must be on the epipolar line L_1. This is the **epipolar line constraint**.

In binocular vision, when an ideal parallel optical axis model is used (that is, the lines of sight of two cameras are parallel), the epipolar line coincides with the image scan line. At this time, the stereo vision system is called the parallel stereo vision system. In a parallel stereo vision system, epipolar constraints can also be used to reduce the search range of stereo matching. In an ideal situation, the use of epipolar constraints can change the search for the entire image into a search for one line of the image.

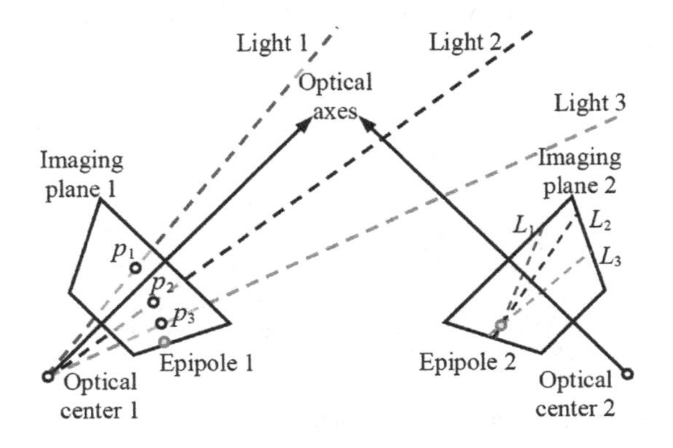

FIGURE 3.11 Correspondence between epipoles and epipolar lines.

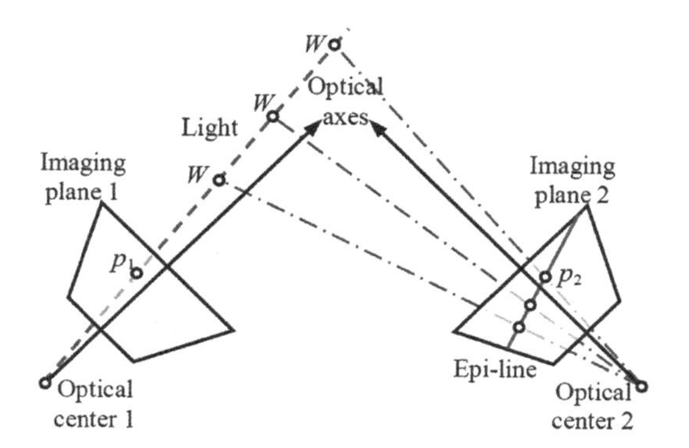

FIGURE 3.12 Epipolar line constraint diagram.

However, it should be noted that the epipolar constraint is only a local constraint. For a spatial point, there may be more than one projection point on the epipolar line. For example, in Figure 3.12, a camera (on the left) is used to observe a spatial point W, and the imaged point p_1 should be on the line connecting the optical center of the camera and point W. However, all the points on the line will be imaged at the point p_1, so the position/distance of the specific point W cannot be completely determined from the point p_1. Now use the second camera to observe the same spatial point W, and the imaged point p_2 should also be on the line connecting the optical center of the camera and point W. All points W on this line are projected onto a straight line on the Imaging plane 2, and this straight line is called an **epipolar line.**

From the geometric relationship in Figure 3.12, we can see that for any point p_1 on the Imaging plane 1, the Imaging plane 2 and all its corresponding points are (constrained) on the same straight line. This is the epipolar constraint mentioned earlier.

3.3.2.2 Essential Matrix and Fundamental Matrix

The relationship between the projected coordinate points of the space point W on the two images can be described by an **essential matrix** (also called the **eigen-matrix**) $\boldsymbol{E}$ with five degrees of freedom (Davies 2005); $\boldsymbol{E}$ can be decomposed into an orthogonal rotation matrix $\boldsymbol{R}$ followed by a translation matrix $\boldsymbol{T}$ ($\boldsymbol{E} = \boldsymbol{RT}$). If the coordinates of the projection point in the left image are represented by $\boldsymbol{p}_1$, and the coordinates of the projection point in the right image are represented by $\boldsymbol{p}_2$, then

$$p_2^{\mathrm{T}} \boldsymbol{E} p_1 = 0 \tag{3.33}$$

The epipolar lines passing through $\boldsymbol{p}_1$ and $\boldsymbol{p}_2$ on the corresponding image satisfy $\boldsymbol{L}_2 = \boldsymbol{E}\boldsymbol{p}_1$ and $\boldsymbol{L}_1 = \boldsymbol{E}^{\mathrm{T}}\boldsymbol{p}_2$, respectively. On the corresponding image, the poles passing through $\boldsymbol{p}_1$ and $\boldsymbol{p}_2$ satisfy $\boldsymbol{E}\boldsymbol{e}_1 = 0$ and $\boldsymbol{E}^{\mathrm{T}}\boldsymbol{e}_2 = 0$, respectively.

The essential matrix indicates the relationship between the projection point coordinates of the same spatial point W (coordinates $\boldsymbol{W}$) on the two images, which can be derived with the help of Figure 3.13. Assuming that the projection positions p_1 and p_2 (coordinates $\boldsymbol{p}_1$ and $\boldsymbol{p}_2$) of the point W on the image can be observed, and the rotation matrix $\boldsymbol{R}$ and the

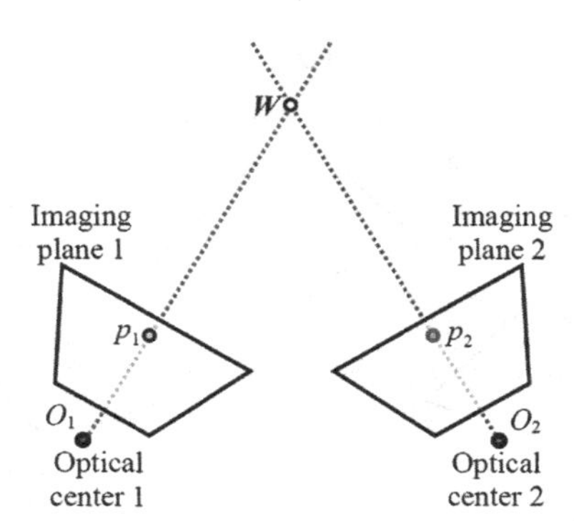

FIGURE 3.13 Derivation of the essential matrix.

translation matrix $\boldsymbol{T}$ between the two cameras are also known, then three 3-D vectors: $\boldsymbol{O_1O_2}$, $\boldsymbol{O_1W}$ and $\boldsymbol{O_2W}$ can be obtained. The three 3-D vectors must be coplanar, because in mathematics, the criterion that three 3-D vectors $\boldsymbol{a}$, $\boldsymbol{b}$, and $\boldsymbol{c}$ are coplanar can be written as $\boldsymbol{a} \cdot (\boldsymbol{b} \times \boldsymbol{c}) = 0$, so this criterion can be used to derive the essential matrix.

According to the perspective relationship of the second camera: vector $\boldsymbol{O_1W} \propto \boldsymbol{Rp}_1$, vector $\boldsymbol{O_1O_2} \propto \boldsymbol{T}$, and vector $\boldsymbol{O_2W} = \boldsymbol{p}_2$. Combining these relationships with the coplanar condition, the desired result is obtained:

$$\boldsymbol{p}_2^{\mathrm{T}}\left(\boldsymbol{T} \times \boldsymbol{R}\boldsymbol{p}_1\right) = \boldsymbol{p}_2^{\mathrm{T}}\boldsymbol{E}\boldsymbol{p}_1 = 0 \tag{3.34}$$

The epipolar lines passing through the points p_1 and p_2 on the corresponding image satisfy $\boldsymbol{L}_2 = \boldsymbol{E}\boldsymbol{p}_1$ and $\boldsymbol{L}_1 = \boldsymbol{E}^{\mathrm{T}}\boldsymbol{p}_2$, respectively. On the corresponding image, the epipoles e_1 and e_2 passing through the points p_1 and p_2 satisfy $\boldsymbol{E}\boldsymbol{e}_1 = 0$ and $\boldsymbol{E}^{\mathrm{T}}\boldsymbol{e}_2 = 0$, respectively.

In the above discussion, it is assumed that $\boldsymbol{p}_1$ and $\boldsymbol{p}_2$ are the pixel coordinates after the camera has been calibrated. If the camera has not been calibrated, the original pixel coordinates $\boldsymbol{q}_1$ and $\boldsymbol{q}_2$ need to be used. Suppose the internal parameter matrix of the camera is $\boldsymbol{G}_1$ and $\boldsymbol{G}_2$, then

$$\boldsymbol{p}_1 = \boldsymbol{G}_1^{-1}\boldsymbol{q}_1 \tag{3.35}$$

$$\boldsymbol{p}_2 = \boldsymbol{G}_2^{-1}\boldsymbol{q}_2 \tag{3.36}$$

Substituting the above two equations into Equation (3.33), we get $\boldsymbol{q}_2^{\mathrm{T}}(\boldsymbol{G}_2^{-1})^{\mathrm{T}}\boldsymbol{E}\boldsymbol{G}_1^{-1}\boldsymbol{q}_1 = 0$, which can be written as

$$\boldsymbol{q}_2^{\mathrm{T}}\boldsymbol{F}\boldsymbol{q}_1 = 0 \tag{3.37}$$

where

$$\boldsymbol{F} = \left(\boldsymbol{G}_2^{-1}\right)^{\mathrm{T}}\boldsymbol{E}\boldsymbol{G}_1^{-1} \tag{3.38}$$

which is called the **fundamental matrix**, because it contains all the information used for camera calibration. The basic matrix has seven degrees of freedom (each epipole requires two parameters, plus three parameters to map three epipolar lines from one image to another, because the projection transformation in the two 1-D projection spaces has three degrees of freedom), the essential matrix has five degrees of freedom, so the basic matrix has two more free parameters than the essential matrix, but comparing Equation (3.33) and Equation (3.37) shows that the functions of these two matrices are similar.

The essential matrix and the fundamental matrix are related to the internal and external parameters of the camera. If the internal and external parameters of the camera are given, it can be known from the epipolar line constraint that for any point on the Imaging plane 1, only a 1-D search needs to be performed on the Imaging plane 2 to determine the position of the corresponding point. Further, the correspondence constraint is a function of the internal and external parameters of the camera. Given the internal parameters, the external parameters can be determined by the observed pattern of the corresponding points, and the geometric relationship between the two cameras can be established.

3.3.2.3 Calculation of Optical Properties

Using the grayscale information from the binocular image, it is possible to further calculate some optical properties of the surface of the object. Two factors relevant to the reflection characteristics of the surface are the scattering caused by the surface roughness and the specular reflection caused by the surface compactness. These two factors are combined as follows: let $\boldsymbol{N}$ be the unit vector in the normal direction of the surface panel, $\boldsymbol{S}$ is the unit vector in the direction of the point light source, and $\boldsymbol{V}$ is the unit vector in the direction of the observer's line of sight. The reflected brightness $I(x, y)$ obtained on the surface element is the product of the synthetic reflectance $\rho\ (x, y)$ and the synthetic reflectance $R[\boldsymbol{N}(x, y)]$, namely

$$I(x,y) = \rho(x,y)R\big[N(x,y)\big] \tag{3.39}$$

where

$$R\big[N(x,y)\big] = (1-\alpha)\boldsymbol{N}\cdot\boldsymbol{S} + \alpha(\boldsymbol{N}\cdot\boldsymbol{H})^k \tag{3.40}$$

where ρ, α, k, are coefficients related to the optical properties of the surface, which can be calculated from the image data.

The first term on the right side of Equation (3.40) considers the scattering effect, which does not vary with the line-of-sight angle; the second term considers the specular reflection effect. Let $\boldsymbol{H}$ be the unit vector in the direction of the specular reflection angle:

$$\boldsymbol{H} = (\boldsymbol{S}+\boldsymbol{V}) / \sqrt{2\big[1+(\boldsymbol{S}\cdot\boldsymbol{V})\big]} \tag{3.41}$$

The second term on the right side of Equation (3.40) reflects the change of the sight vector $\boldsymbol{V}$ through the vector $\boldsymbol{H}$. In the coordinate system used in Figure 3.10:

$$V' = \{0,\ 0,\ -1\} \qquad V'' = \{-\sin\theta,\ 0,\ \cos\theta\} \tag{3.42}$$

3.4 BINOCULAR STEREO MATCHING BASED ON FEATURES

The disadvantage of the region-based matching method is that it relies on the statistical characteristics of the image gray level, so it is more sensitive to the surface structure of the scenery and the light reflection, therefore it lacks enough texture details on the scenery surface in space and/or it has certain difficulties in the case where the imaging distortion is large (such as where the baseline is too long). Taking into account the characteristics of the actual image, some salient feature points (also called control points, key points or matching points) in the image can be determined first, and then these feature points can be used for matching. Feature points are not very sensitive to changes in ambient lighting during matching, and their performance is relatively stable.

3.4.1 Basic Steps

The main steps of feature-based matching are as follows:

(1) Select feature points for matching in the image; the most commonly used feature points are some special points in the image, such as edge points, corner points, inflection points, landmark points, etc.

(2) Match the feature point pairs in the stereo image pair.

(3) Calculate the disparity of the matching point pair and obtain the depth at the matching point (similar to the previous region-based method).

(4) Interpolate the sparse depth value result to obtain a dense depth map (because the feature points are discrete, the dense disparity field cannot be directly obtained after matching).

3.4.1.1 Matching Using Edge Points

For an image $f(x, y)$, the feature point image can be obtained by calculating the edge points:

$$t(x,y) = \max\{H, V, L, R\} \tag{3.43}$$

Among them, H, V, L, R are all calculated with the help of grayscale gradient

$$H = [f(x,y) - f(x-1,y)]^2 + [f(x,y) - f(x+1,y)]^2 \tag{3.44}$$

$$V = [f(x,y) - f(x,y-1)]^2 + [f(x,y) - f(x,y+1)]^2 \tag{3.45}$$

$$L = [f(x,y) - f(x-1,y+1)]^2 + [f(x,y) - f(x+1,y-1)]^2 \tag{3.46}$$

$$R = [f(x,y) - f(x+1,y+1)]^2 + [f(x,y) - f(x-1,y-1)]^2 \tag{3.47}$$

Then divide $t(x, y)$ into small regions W that do not overlap each other, and select the point with the largest calculated value as the feature point in each small region.

Now consider matching the image pair formed by the left image and the right image. For each feature point of the left image, all possible matching points in the right image can be formed into a possible matching point set. In this way, a label set can be obtained for each feature point of the left image, where the label l is either the disparity between the feature point of the left image and its possible matching points, or a special label representing no matching point. For each possible matching point, calculate the following formula to set the initial matching probability $P^{(0)}(l)$:

$$A(l) = \sum_{(x,y)\in W} \left[f_L(x,y) - f_R(x+l_x, y+l_y)\right]^2 \tag{3.48}$$

where $l = (l_x, l_y)$ is the possible parallax. $A(l)$ represents the grayscale fit between the two regions, which is inversely proportional to the initial matching probability $P^{(0)}(l)$. In other words, $P^{(0)}(l)$ is related to the similarity in the neighborhood of possible matching points. Accordingly, with the aid of the relaxation iteration method, the points with close disparity in the neighborhood of possible matching points are given a positive increment, and the points with farther disparity in the neighborhood of the possible matching points are adjusted to $P^{(0)}(l)$ for performing iterative update. As the iteration progresses, the iterative matching probability $P^{(k)}(l)$ of the correct matching point will gradually increase, while the matching probability $P^{(k)}(l)$ of other points will gradually decrease. After a certain number of iterations, the point with the largest matching probability $P^{(k)}(l)$ is determined as the matching point.

3.4.1.2 Matching Using Zero-Crossing Points

When matching the feature points, the **zero-crossing mode** can also be used to obtain the matching primitives (Kim and Aggarwal 1987). Use the Laplacian (of Gaussian function) to perform convolution to get the zero-crossing point. Considering the connectivity of the zero-crossing point, 16 different zero-crossing modes can be determined, as shown by the shadow in Figure 3.14.

For each zero-crossing mode of the left image, all possible matching points in the right image form a possible matching point set. In stereo matching, all non-horizontal zero-crossing modes in the left image can be formed into a point set with the help of horizontal

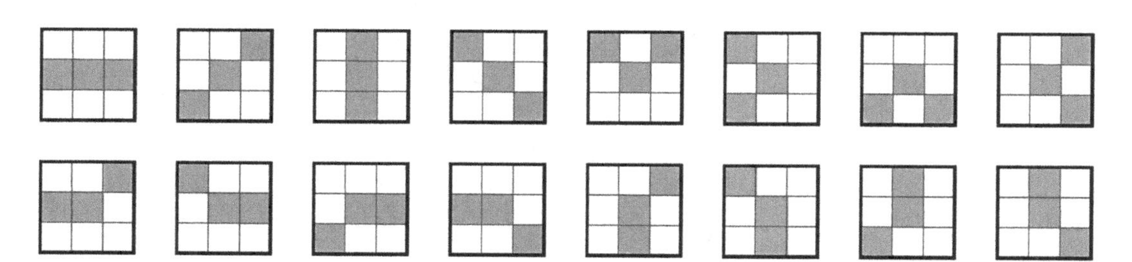

FIGURE 3.14 Diagram of 16 different zero-crossing modes.

epipolar line constraints, a label set is assigned to each point, and an initial matching probability is determined. The final matching point can also be obtained through relaxation iteration by adopting a similar method to the matching method using edge points.

3.4.1.3 Feature Point Depth

Let's use Figure 3.15 (it is to remove the epipolar line in Figure 3.10, and then move the baseline to the X-axis to facilitate the description, where the meaning of each letter is the same as Figure 3.10) to explain the correspondence between the feature points.

In 3-D space coordinates, a feature point $W(x, y, -z)$ is orthogonally projected on the left and right images, respectively:

$$(u',v') = (x,y) \tag{3.49}$$

$$(u'',v'') = \left[(x-B)\cos\theta - z\sin\theta, y\right] \tag{3.50}$$

The calculation of u'' here is based on the coordinate transformation of first translation and then rotation. Equation (3.50) can also be derived with the help of Figure 3.16 (a schematic diagram parallel to the XZ plane in Figure 3.15 is given here):

$$u'' = \overline{OS} = \overline{ST} - \overline{TO} = \left(\overline{QE} + \overline{ET}\right)\sin\theta - \frac{B-x}{\cos\theta} \tag{3.51}$$

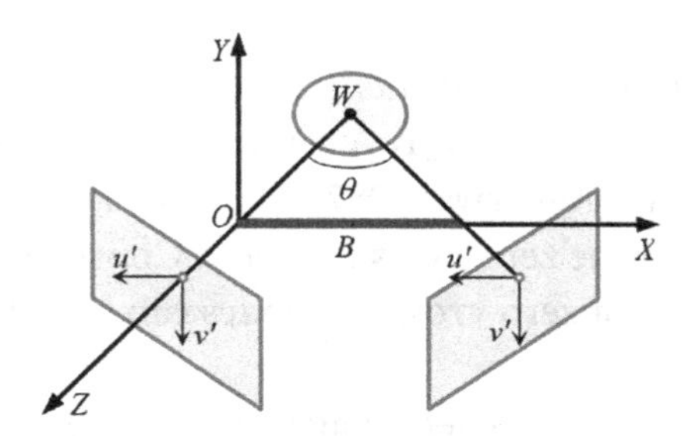

FIGURE 3.15 Schematic diagram of the coordinate system of binocular vision.

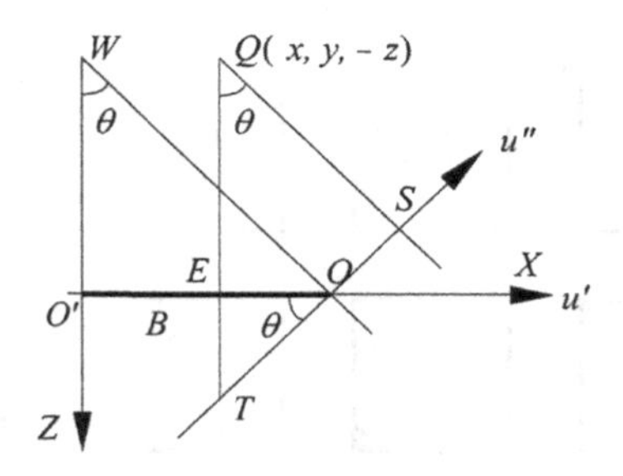

FIGURE 3.16 Calculate the coordinate arrangement of binocular stereo matching disparity.

Note that W is on the $-Z$ axis, so there is

$$u'' = -z\sin\theta + (B-x)\tan\theta\sin\theta - \frac{B-x}{\cos\theta} = (x-B)\cos\theta - z\sin\theta \tag{3.52}$$

If u'' has been determined by u' (that is, the matching between the feature points has been established), the depth of the feature points projected to u' and u'' can be inversely solved from Equation (3.50) as

$$-z = u''\csc\theta + (B-u')\cot\theta \tag{3.53}$$

3.4.1.4 Sparse Matching Points

From the above discussion, it can be seen that the characteristic points are only some specific points on the object, and there is a certain interval between them. Only the sparse matching points cannot directly obtain the dense parallax field, so the shape of the object may not be restored accurately. For example, Figure 3.17(a) shows four points that are coplanar in space (equal distance from another space plane). These points are sparse matching points obtained by disparity calculation. Suppose these points are located on the outer surface of the object, but there can be infinitely many curved surfaces passing these four points, as shown in Figures 3.17(b)–3.17(d), which give several possible examples. It can be seen that only the sparse matching points cannot uniquely restore the shape of the object, and some other conditions or interpolation of the sparse matching points need to be combined to obtain a dense disparity map such as region matching.

3.4.2 Scale-Invariant Feature Transformation

Scale-invariant feature transformation (SIFT) can be regarded as a method of detecting **salient features** in an image (Nixon and Aguado 2008). It can not only determine the position of a point with salient features in the image, but also give a description vector of the point. Also known as SIFT operator or descriptor, it is a local descriptor that contains three types of information: location, scale, and direction.

The basic ideas and steps of SIFT are as follows. First, the image is represented in multiple scales, which can be convolved with the image using a Gaussian convolution

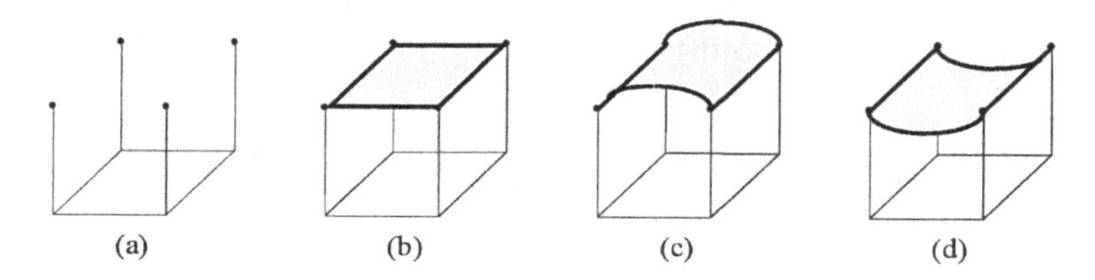

FIGURE 3.17 Only the sparse matching points cannot uniquely restore the shape of the object.

kernel (the only linear kernel). The Gaussian convolution kernel is a Gaussian function with a variable scale:

$$G(x,y,\sigma)=\frac{1}{2\pi\sigma^2}\exp\left[\frac{-(x^2+y^2)}{2\sigma^2}\right] \tag{3.54}$$

where σ is the scale factor. The multi-scale representation of the image after the convolution of Gaussian convolution kernel and the image is represented as

$$L(x,y,\sigma)=G(x,y,\sigma)\otimes f(x,y) \tag{3.55}$$

The Gaussian function is a low-pass function, and the image will be smoothed after its convolution with the image. The size of the scale factor is related to the degree of smoothness. Large σ corresponds to a large scale, which mainly gives an overview of the image after convolution; a small σ corresponds to a small scale, and the details of the image are retained after convolution. In order to make full use of image information of different scales, a series of Gaussian convolution kernels and image convolutions with different scale factors are used to construct a Gaussian pyramid. Generally, the scale factor coefficient between two adjacent layers of the Gaussian pyramid is k. If the scale factor of the first layer is σ, the scale factor of the second layer is $k\sigma$, the scale factor of the third layer is $k^2\sigma$, and so on.

SIFT then searches for **salient feature points** in the multi-scale representation of the image, using the **difference of Gaussian** (DoG) operator for this purpose. DoG is the difference between the convolution results of two Gaussian kernels of different scales, which is similar to the **Laplacian of Gaussian** (LoG) operator. If h and k are used to represent the coefficients of different scale factors, the DoG pyramid can be expressed as

$$D(x,y,\sigma)=\left[G(x,y,k\sigma)-G(x,y,h\sigma)\right]\otimes f(x,y)=L(x,y,k\sigma)-L(x,y,h\sigma) \tag{3.56}$$

The multi-scale representation space of the DoG pyramid of the image is a 3-D space (image plane and scale axis). In order to search for extreme values in such a 3-D space, it is necessary to compare the value of a point in the space with the values of its 26 neighboring voxels. The result of this search determines the location and scale of the salient feature points.

Next, the gradient distribution of pixels in the neighborhood of the salient feature point is used to determine the direction parameter of each point. The modulus (amplitude) and direction of the gradient at (x, y) in the image are, respectively (the scale used for each L is the scale of each salient feature point):

$$m(x,y)=\sqrt{\left[L(x+1,y)-L(x-1,y)\right]^2+\left[L(x,y+1)-L(x,y-1)\right]^2} \tag{3.57}$$

$$\theta(x,y)=\arctan\left\{\left[L(x,y+1)-L(x,y-1)\right]/\left[L(x+1,y)-L(x-1,y)\right]\right\} \tag{3.58}$$

After obtaining the direction of each point, the direction of the pixels in the neighborhood can be combined to obtain the direction of the salient feature point. For details, please refer to Figure 3.18. First (on the basis of determining the location and scale of the salient feature point) a 16 × 16 window centered on the salient feature point, as shown in Figure 3.18(a), is taken. Divide the window into 16 times of 4 × 4 groups, as shown in Figure 3.18(b). Calculate the gradient of each pixel in each group to obtain the gradient of the pixels in the group (the direction of the arrow indicates the direction of the gradient, and the length of the arrow is proportional to the size of the gradient), as shown in Figure 3.18(c). Use an 8-direction (interval 45°) histogram to count the gradient direction of the pixels in each group, and take the peak direction as the gradient direction of the group, as shown in Figure 3.18(d). In this way, for 16 groups, each group can get an 8-D direction vector, which can be joined together to get a 16 × 8=128-D vector. This vector is normalized, and finally used as the description vector of each salient feature point, that is, the SIFT descriptor. In practice, the coverage region of the SIFT descriptor can be square or round; it is also called a **salient patch**.

The SIFT descriptor is invariant to image scale scaling, rotation, and illumination changes, and it is also relatively stable to affine transformation, viewing angle changes, local shape distortion, noise interference, and so on. This is because in the process of obtaining the SIFT descriptor, the influence of rotation is eliminated by calculation and adjustment of the gradient direction, the influence of illumination changes is eliminated with the aid of vector normalization, and the combination of pixel direction information in the neighborhood is used to enhance the robustness. In addition, the SIFT descriptor is rich in information and has good uniqueness (compared to edge points or corner points that only contain position and extreme value information, the SIFT descriptor has a 128-D description vector). Also due to its uniqueness or particularity, a large number of salient patches can often be identified in an image for different applications to choose from. Of course, due to the high dimension of the description vector, the calculation amount of the SIFT descriptor is often relatively large. There are also many improvements to SIFT, including replacing the gradient histogram with PCA (for effective dimensionality reduction), limiting the amplitude of each direction of the histogram (some non-linear illumination changes mainly affect the amplitude), and so on.

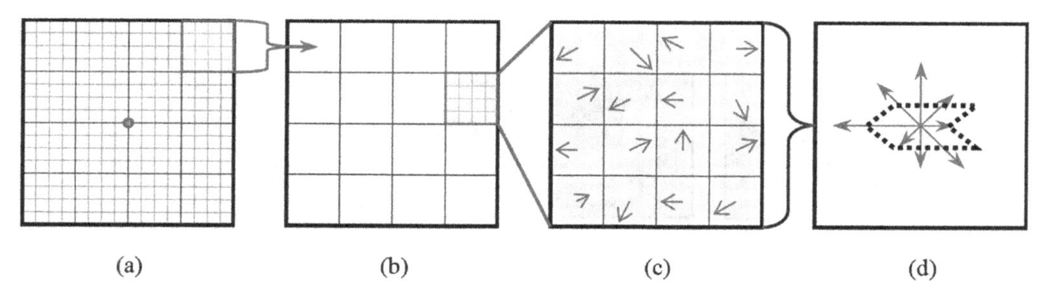

FIGURE 3.18 Calculation steps for SIFT description vector.

3.4.3 Speed-Up Robust Feature

Speed-up robust feature (SURF) can also be regarded as a method to detect **salient feature points** in images. The basic idea is to accelerate SIFT. Therefore, in addition to the stability of SIFT method, it also reduces the computational complexity and has good real-time ability for detection and matching.

3.4.3.1 Determine the Point of Interest Based on the Hessian Matrix

The SURF algorithm determines the position and scale information of the point of interest by calculating the determinant of the second-order **Hessian matrix** of the image. The Hessian matrix of image $f(x, y)$ at position (x, y) and scale σ is defined as follows:

$$H[x,y,\sigma]=\begin{bmatrix} h_{xx}(x,y,\sigma) & h_{xy}(x,y,\sigma) \\ h_{xy}(x,y,\sigma) & h_{yy}(x,y,\sigma) \end{bmatrix} \tag{3.59}$$

Among them, $h_{xx}(x, y, \sigma)$, $h_{xy}(x, y, \sigma)$, and $h_{yy}(x, y, \sigma)$ are the result of convolution of Gaussian second-order differentials $[\partial^2 G(\sigma)]/\partial x^2$, $[\partial^2 G(\sigma)]/\partial x \partial y$, and $[\partial^2 G(\sigma)]/\partial y^2$ with image $f(x, y)$ at (x, y), respectively.

The determinant of the Hessian matrix is

$$\det(H)=\frac{\partial^2 f}{\partial x^2}\frac{\partial^2 f}{\partial y^2}-\frac{\partial^2 f}{\partial xy}\frac{\partial^2 f}{\partial xy} \tag{3.60}$$

Its maximum point in scale space and image space is called the **point of interest**. The value of the Hessian matrix determinant is the eigenvalue of the Hessian matrix, and it can be judged whether the point is an extreme point according to the positive or negative value of the determinant at the image point.

The Gaussian filter is optimal in the analysis of the scale space, but in practice, after discretization and quantization, it will lose repeatability when the image is rotated by an odd multiple of 45° angle (because the template is square and anisotropic). For example, Figures 3.19(a) and 3.19(b) show discretized and quantized Gaussian second-order partial differential responses along the X direction as well as along the center line of X and Y bisector direction, respectively. There is a big difference between them.

In practical applications, a box filter can be used to approximate the Hessian matrix, so as to obtain a faster calculation speed (and has nothing to do with the size of the filter) with

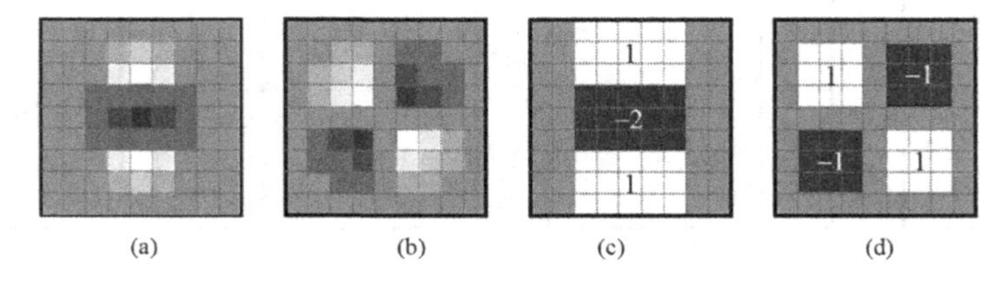

FIGURE 3.19 Gaussian second-order partial differential response and its approximation (light color represents positive value, dark color represents negative value, middle gray represents 0).

the help of the integral image (Zhang 2017). For example, Figures 3.19(c) and 3.19(d) are, respectively, the approximations of the Gaussian second-order partial differential response of Figures 3.19(a) and 3.19(b), in which the 9 × 9 box filter has a scale of 1.2 and is the approximation of the Gaussian filter that represents the lowest scale of the calculated response (i.e., the highest spatial resolution). The approximate values of $h_{xx}(x, y, \sigma)$, $h_{xy}(x, y, \sigma)$, and $h_{yy}(x, y, \sigma)$ are denoted as A_{xx}, A_{xy}, and A_{yy} respectively. The determinant of the approximate Hessian matrix is

$$\det(H_A) = A_{xx}A_{yy} - (wA_{xy})^2 \tag{3.61}$$

where w is the relative weight of the balanced filter response (that is, the approximate compensation used for the Gaussian convolution kernel is not used), which is used to maintain the energy between the Gaussian kernel and the approximate Gaussian kernel, which can be calculated as follows:

$$w = \frac{\|h_{xy}(1.2)\|_F \|A_{yy}(9)\|_F}{\|h_{yy}(1.2)\|_F \|A_{xy}(9)\|_F} = 0.912 \approx 0.9 \tag{3.62}$$

where $\|\cdot\|_F$ stands for Frobenius norm.

In theory, the weight depends on the scale, but in practice it can be kept as a constant, because its change has little effect on the result. Furthermore, the filter response should be normalized in respect to the size, so that a constant Frobenius norm can be guaranteed for any filter size. Experiments show that the performance of approximate calculation is equivalent to that of Gaussian filter after discretization and quantization.

3.4.3.2 Scale Space Representation

The detection of points of interest needs to be carried out on different scales. The scale space is generally represented by a pyramid structure. However, due to the use of box filters and integral images, it is not necessary to use the same filter for each layer of the pyramid, but box filters of different sizes are directly used for the original image (the calculation speed is the same). Therefore, the (Gaussian kernel) filter can be up-sampled without iteratively reducing the image size. The output of the previous 9 × 9 box filter is used as the initial scale layer, and the following scale layers can be obtained by filtering the image with larger and larger box sizes. Since the image is not down-sampled and high-frequency information is retained, no **aliasing effects** will occur.

The scale space is divided into several groups, and each group represents a series of filter response maps obtained by convolving the same input image with a filter of increasing size. There is a double relationship between the groups, as shown in Table 3.2.

Each group is divided into constant scale layers. Due to the discrete nature of the integral image, the smallest scale difference between two adjacent scales depends on the length l_0 of the positive or negative lobe in the corresponding direction of the second-order partial differential (this length is 1/3 of the filter size). For the 9 × 9 filter, $l_0 = 3$. For two adjacent

TABLE 3.2 Scale Space Grouping Situation

Group			1					2			...
Interval	1	2	3	4	...	1	2	3	4	...	...
Box Filter Side Length	9	15	21	27	...	15	27	39	51	...	...
σ = Side Length ×1.2/9	1.2	2	2.8	3.6	...	2	3.6	5.2	6.8	...	...

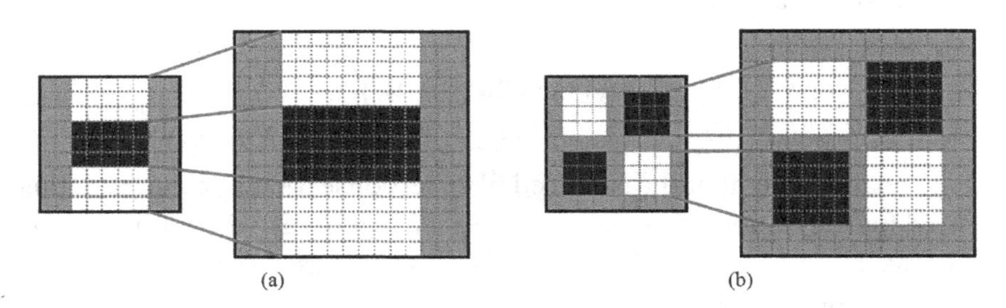

FIGURE 3.20 The filter between two adjacent scale layers (9 × 9 and 15 × 15).

layers, the size in either direction must be increased by at least two pixels to ensure that the final size is odd (so that the filter has a center pixel), which results in a total increase of the mask (side) size by six pixels. The construction of the scale space starts with the use of 9 ××9 box filters, followed by filters with sizes of 15 × 15, 21 × 21, and 27 × 27. Figures 3.20(a) and 3.20(b) respectively show the filters A_{yy} and A_{xy} between two adjacent scale layers (9 × 9 and 15 × 15). The length of the black lobe can only be increased by an even number of pixels. Note that for directions different from l_0, such as the width of the center band of the vertical filter, the scaling mask will introduce rounding errors. However, since these errors are much smaller than l_0, this approximation is acceptable.

The same considerations apply to other groups. For each new group, the filter size increases exponentially. At the same time, the sampling interval used to extract points of interest is increased exponentially for each new group, which can reduce the calculation time, and the loss in accuracy is comparable to the traditional method for image subsampling. The filter sizes for the second group are 15, 27, 39, 51. The filter sizes for the third group are 27, 51, 75, 99. If the size of the original image is still larger than the size of the corresponding filter, then the fourth group of calculations can be performed, using filters with sizes 51, 99, 147, and 195. Figure 3.21 gives a complete picture of the filters used.

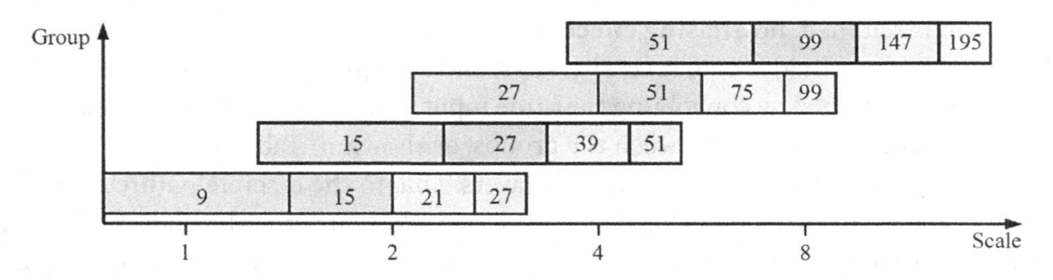

FIGURE 3.21 Graphical representation of the filter side lengths in different groups (logarithmic horizontal axis).

The groups overlap each other to ensure smooth coverage of all possible scales. In a typical scale space analysis, the number of points of interest that can be detected in each group decreases very quickly.

The large-scale change, especially the change between the first filters of these groups (the change from 9 to 15 is 1.7), makes the sampling of the scale quite rough. For this purpose, a scale space with a finer sampling scale can also be used. At this time, the image is first scaled by 2 times, and then the first group is started with a filter of size 15. The following filter sizes are 21, 27, 33, 39. Then the second group starts, and its size increases in steps of 12 pixels. The following groups can be deduced by analogy. So the scale change between the first two filters is only 1.4 (21/15). The smallest scale that can be detected by quadratic interpolation is $\sigma = (1.2\times18/9)/2 = 1.2$.

Since the Frobenius norm remains constant for filters of any size, it can be considered that it has been normalized on the scale, and it is no longer necessary to weight the response of the filter.

3.4.3.3 Description and Matching of Points of Interest

The SURF descriptor describes the brightness distribution in the neighborhood of the point of interest, similar to the gradient information extracted by SIFT. The difference is that SURF is based on the response of the first-order Haar wavelet in the X and Y directions instead of the gradient, which can make full use of the integral image to increase the calculation speed; and the descriptor length is only 64, which can improve the robustness while reducing feature calculation and matching time.

Matching with the help of the SURF descriptor includes three steps: (i) determining an orientation around the point of interest; (ii) constructing a square region aligned with the selected orientation and extracting the SURF description from it; and (iii) matching the descriptive characteristics between the two regions.

(1) Determine the orientation

In order to obtain the invariance to the image rotation, an orientation must be determined for the point of interest. First, calculate the response of the Haar wavelet along the X and Y directions in a circular neighborhood with a radius of 6σ around the point of interest, where σ is the scale at which the point of interest is detected. The sampling step size depends on the scale and is set as σ. In order to be consistent with other parts, the size of the wavelet also depends on the scale and is set to a side length of 4σ. In this way, the integral image can be used again for fast filtering. According to the characteristics of the Haar wavelet mask, only six operations are needed to calculate the response in the X and Y directions at any scale.

Once the wavelet response is calculated and weighted with the Gaussian distribution centered at the point of interest, the response can be expressed as a point in the coordinate space, with the horizontal coordinate corresponding to the horizontal response intensity and the vertical coordinate corresponding to the vertical response intensity. The orientation can be obtained by calculating the sum of the responses in a fan-shaped sliding window with a radian size of $\pi/3$ (step $\pi/18$), as shown in Figure 3.22.

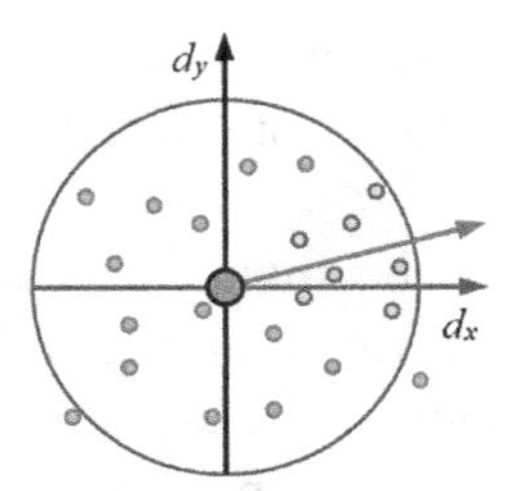

FIGURE 3.22 Determining the direction of response.

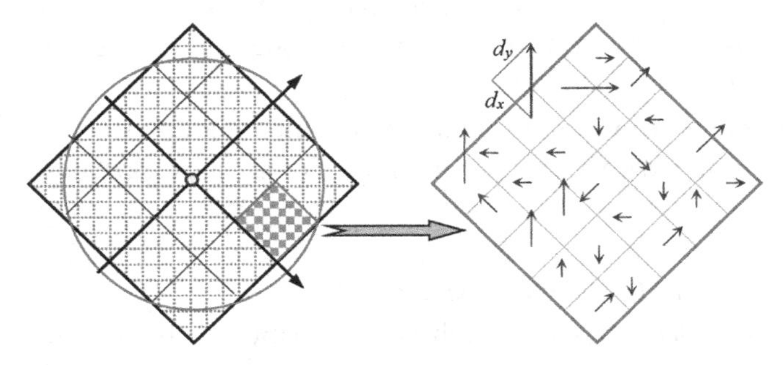

FIGURE 3.23 The square region surrounding the point of interest.

The horizontal and vertical responses in the window are summed separately, and these two sums can form a local orientation vector. The longest vector in all windows defines the direction of the point of interest. The size of the sliding window is a parameter that needs to be carefully selected. A small size mainly reflects the gradient of a single advantageous gradient, while a large size tends to produce an insignificant maximum value in the vector.

(2) Descriptor based on the sum of Haar wavelet responses

To extract the descriptor, the first step is to construct a square region around the point of interest to have the orientation determined above (to ensure rotation invariance). The size of the window is 20σ. These square regions are regularly split into smaller $4 \times 4 = 16$ sub-regions, so that important spatial information can be preserved. For each sub-region, calculate the Haar wavelet response in a regular 5×5 grid. For simplicity, let d_x represent the Haar wavelet response along the horizontal direction, and d_y represent the Haar wavelet response along the vertical direction. Here "horizontal" and "vertical" are relative to the selected point of interest, as shown in Figure 3.23.

Next, the wavelet responses d_x and d_y are summed, respectively; and in order to use the polarization information about the brightness change, the absolute values of the wavelet responses d_x and d_y, that is, $|d_x|$ and $|d_y|$ are also summed, respectively. In this way, a 4-D description vector V can be obtained from each sub-region, $V = (\Sigma d_x, \Sigma d_y, \Sigma|d_x|, \Sigma|d_y|)$. For all 16 sub-regions, their description vectors are linked to obtain a new description vector with a length of 64-D. The wavelet response obtained in this way is not sensitive to changes in illumination. The invariance to the contrast (scalar) is obtained by converting the descriptor into a unit vector.

Figure 3.24 is a schematic diagram of three different brightness patterns and the descriptors obtained from the corresponding sub-regions. On the left is a uniform pattern, and each component of the descriptor is very small; in the middle is an alternating pattern along the X direction, with only $\Sigma|d_x|$ being large, and the remaining terms small; on the right is a pattern in which the brightness gradually increases in the horizontal direction, and the values of Σd_x and $\Sigma|d_x|$ are both large. It can be seen that there are obvious differences in the descriptors for different brightness patterns. It is also conceivable that if these three local brightness patterns are combined, a specific descriptor can be obtained.

The principle of SURF is similar to the principle of SIFT to some extent, and they are all based on the spatial distribution of gradient information. However, SURF often performs better than SIFT in practice. The reason here is that SURF aggregates all the gradient information in the sub-region, while SIFT only depends on the orientation of each independent gradient. This difference makes SURF more resistant to noise. An example is shown in Figure 3.25. When there is no noise, SIFT has only one gradient direction; when there is noise (the edge is no longer smooth), SIFT also has certain gradient components in other directions in addition to the main gradient direction. However, the response of SURF is basically the same in both cases (the noise is smoothed).

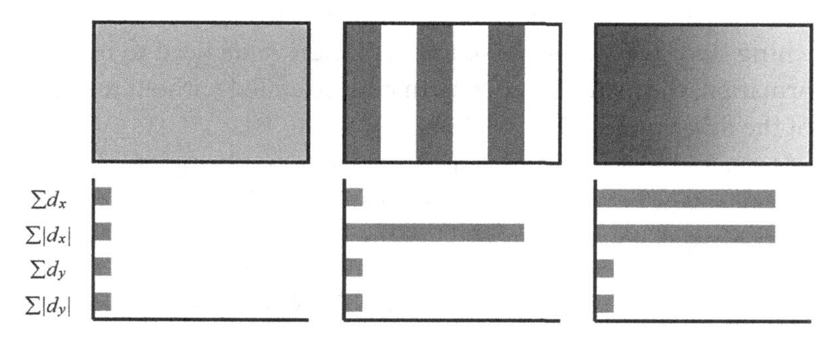

FIGURE 3.24 Different brightness patterns and their descriptors.

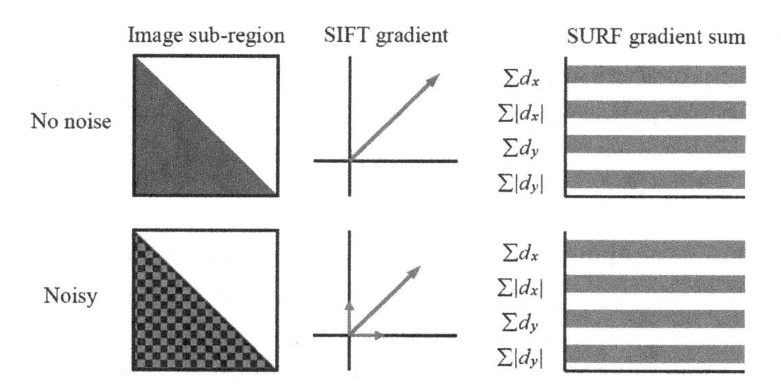

FIGURE 3.25 Comparison of SIFT and SURF.

The evaluation experiments on the number of sampling points and the number of sub-regions show that the square sub-regions divided by 4 × 4 can give the best results. Further subdivision will result in poor robustness and a large increase in matching time. On the other hand, short descriptors obtained using 3 × 3 sub-regions (SURF-36, i.e., 3 × 3 = 9 sub-regions, 4 responses per sub-region) will slightly reduce the performance (it is acceptable by comparing to other descriptors), but the calculation is much faster.

In addition, there is a variant of SURF descriptor, namely SURF-128. It also uses the previous summation, but divides the values into more detail. The summation of d_x and $|d_x|$ is divided according to $d_y < 0$ and $d_y \geq 0$. Similarly, the summation of d_y and $|d_y|$ is also calculated according to $d_x < 0$ and $d_x \geq 0$. In this way, the number of features is doubled, and the robustness and reliability of the descriptor are improved. However, although the descriptor itself is faster to calculate, it will still increase the amount of calculation due to the high dimensionality during matching.

(3) Quick index to match

In order to quickly index during matching, the sign of the Laplacian value of the point of interest (that is, the rank of the Hessian matrix) can be considered. In general, points of interest are detected and processed in **blob**-like structures. The sign of the Laplace value can distinguish bright patches in a dark background from dark patches in a bright background. No additional calculation is needed here, because the sign of the Laplacian value has already been calculated in the detection step. In the matching step, only the signs of the Laplacian value need to be compared. With this information, the matching speed can be accelerated without reducing the performance of the descriptor.

The advantages of the SURF algorithm include not being affected by image rotation and scale changes, and anti-blur; while the disadvantages include being greatly affected by changes in viewpoint and lighting.

3.5 SOME RECENT DEVELOPMENTS AND FURTHER RESEARCH

In the following sections, some technical developments and promising research directions from the last few years are briefly overviewed.

3.5.1 Biocular and Stereopsis

Binocular vision enables **human stereopsis**, that is, the ability to perceive depth. This is based on the difference in position of an image between the left and right retinas due to the slightly different perspective of each eye (binocular disparity).

3.5.1.1 Biocular and Binocular

In the introduction of visual displays used in **helmet-mounted displays** (HMDs), three types of visual displays are presented (Posselt and Winterbottom 2021): (i) **Monocular** (using one eye to view one image); (ii) **biocular** (using two eyes to view two same images); and (iii) **binocular** (using two eyes to view two different images). A summary of these different systems is provided in Figure 3.26.

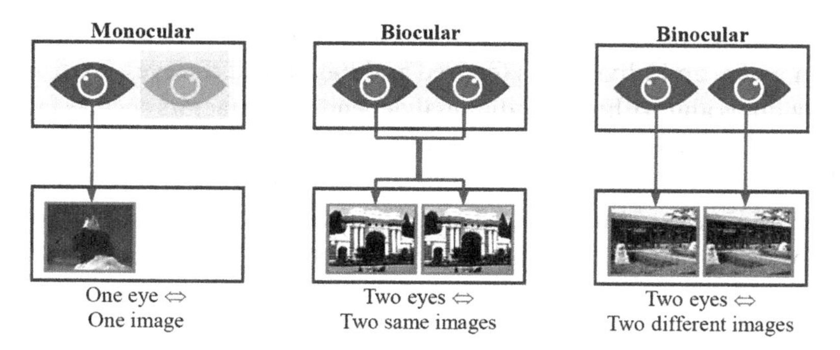

FIGURE 3.26 Comparison of the three types of visual displays used in HMDs.

There is a tendency from monocular to binocular. The initial display was monocular, the current display is biocular, but has two separate optical pathways, thus have the potential ability to display information using disparate images to create stereo depth as in binocular mode (to display stereoscopic or 3-D imagery).

3.5.1.2 From Monocular to Binocular

In the real world, when one focuses on an object, both eyes will converge so that the object will fall on each fovea with almost zero disparity. This situation can be illustrated in Figure 3.27.

In Figure 3.27, the target F is the object of focus; the arc passing through this fixation point is called the **horopter**. This sets a baseline from which relative depths are judged. On either side of the horopter exists a range over which images can be fused to perceive a single object appearing to be at a different depth to the focus point object. This spatial area is termed **Panum's zone of fusion** or the **zone of clear single binocular vision** (ZCSBV). Objects that fall in front of (crossed disparity) or behind (uncrossed disparity) Panum's zone will appear **diplopic**, which, although they can still contribute to depth perception in the form of qualitative stereopsis, are less reliable or accurate. For example, the object A is within Panum's zone of fusion, thus will be seen as a single image, whereas the object B falls outside the fusional area and will appear diplopic.

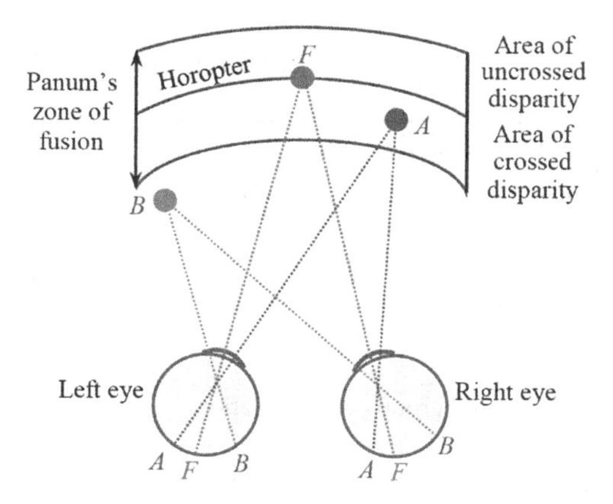

FIGURE 3.27 Horopter and Panum's zone of fusion.

Spatial positioning of the horopter and consequently Panum's zone alters constantly depending on where an individual is focused and has verged the eyes. Its size varies both between individuals and within an individual depending on factors such as fatigue, as well as luminance and pupil size, but disparities of roughly 10~15 arcmin from the focal point can give clear patent depth cues when viewed at the macula. However, only the middle third of the ZCSBV (approximately 0.5 dioptres crossed and uncrossed) is thought to offer comfortable binocular viewing without adverse symptoms.

3.5.2 Stereo Matching Methods Based on Deep Learning

Unlike the traditional matching algorithm based on artificial features, the **stereo matching** algorithm based on deep learning can extract more image features for cost calculation by performing non-linear transformation on the image through operations such as convolution, pooling, and full connection. Compared with artificial features, deep learning can obtain more contextual information, make more use of the global information of images, and obtain model parameters through pre-training, which improves the robustness of the algorithm. At the same time, the use of GPU acceleration technology can also obtain faster processing speed and meet real-time requirements in many application fields.

At present, the image networks used for stereo matching mainly include image pyramid network, Siamese network and generative adversarial network (Chen et al. 2020).

3.5.2.1 Methods Using Image Pyramid Networks

A spatial pyramid pooling layer is set between the convolutional layer and the fully connected layer to convert image features of different sizes into fixed-length representations (He et al. 2015). This can circumvent the repeated calculation of convolution and ensure the consistency of input image size.

Some typical methods using **image pyramid networks**, their characteristics and principles, as well as effects, are shown in Table 3.3.

TABLE 3.3 Several Typical Methods Using Image Pyramid Networks, their Characteristics and Principles, and Effects

Methods	Characteristics and Principles	Effects
Žbontar and Lecun (2015)	Using convolutional neural networks to extract image features for cost computation.	Deep learning features replace artificial features.
Chang, and Chen (2018)	Introducing pyramid pooling module into feature extraction, using multi-scale analysis and 3D-CNN structure.	Solving the problem of gradient disappearance and gradient explosion, suitable for weak texture, occlusion, uneven lighting, etc.
Guo et al. (2019a, 2019b)	Making constructed group cost calculation.	Replacing 3D convolutional layers to improve computational efficiency.
Zhang et al. (2019)	A semi-global aggregation layer and a local bootstrap aggregation layer are designed.	Replacing 3D convolutional layers to improve computational efficiency.

3.5.2.2 Methods Using Siamese Networks

The basic structure of the **Siamese network** is shown in Figure 3.28 (Bromley et al. 1993). The two input images to be matched are converted into two feature vectors with the help of two weight-sharing **convolutional neural networks** (CNNs), and the similarity of the two images can be determined according to the L_1 distance between the two feature vectors.

Some improvements have been made on the basic structure of Siamese networks. Their characteristics and principles, as well as effects, are shown in Table 3.4.

3.5.2.3 Methods Using Generative Adversarial Networks

Generative adversarial networks (GANs) consist of a generative model and a discriminative model. The generative model learns the sample features to make the generated image similar to the original image, while the discriminative model is used to distinguish the "generated" image from the real image (Luo et al. 2017). This process runs iteratively, and the final discrimination result reaches the Nash equilibrium, that is, the true and false concepts are both 0.5.

Some modifications have been made to the fundamental method using GAN. Their characteristics and principles, as well as effects, are shown in Table 3.5.

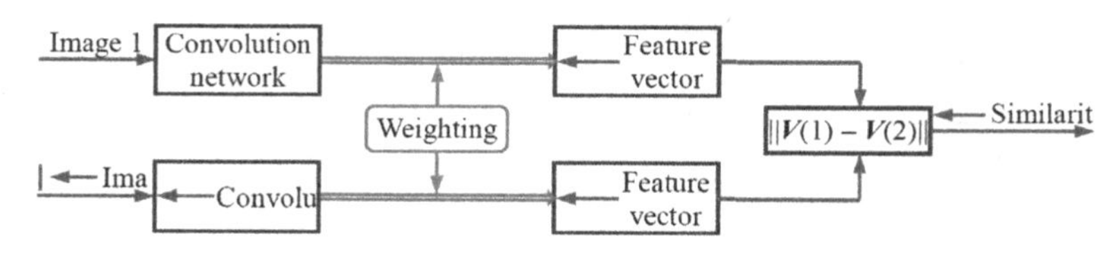

FIGURE 3.28 The basic structure of the Siamese network.

TABLE 3.4 Several Improvements Using Siamese Networks, their Characteristics and Principles, and Effects

Methods	Characteristics and Principles	Effects
Zagoruyko and Komodakis (2015)	Deepen convolutional layers using ReLU function and small convolution kernel.	Matching accuracy is improved.
Khamis et al., (2018)	When extracting features, first calculate the disparity map in the low-resolution cost convolution, and then use the hierarchical refinement network to introduce high-frequency details.	Uses color input as a guide, the high-quality boundaries can be generated.
Liu et al. (2019)	Pyramid pooling is used to connect two sub-networks. The first sub-network is composed of a Siamese network and a 3D convolutional network, which can generate low-precision disparity maps; the second sub-network is a fully convolutional network, which restores the initial disparity map to the original resolution.	Multi-scale features can be obtained.
Guo et al. (2019a, 2019b)	Depth discontinuity is processed on the low-resolution disparity map, and restored to the original resolution in the disparity refinement stage.	The continuity of depth discontinuities is improved.

TABLE 3.5 Several Improvements Using GAN, their Characteristics and Principles, and Effects

Methods	Characteristics and Principles	Effects
Pilzer et al. (2018)	Using a binocular vision-based GAN framework, including two generative sub-networks and one discriminative network. The two generative networks are used to train and reconstruct the disparity map respectively in the adversarial learning. Through mutual restriction and supervision, two disparity maps from different perspectives are generated, and the final data is output after fusion.	This unsupervised model works well under uneven lighting conditions.
Matias et al. (2019)	Use generative models to process occluded regions.	Recoverable to get a good parallax effect.
Lore et al. (2018)	Generative adversarial models using deep convolutions to obtain multiple depth maps with adjacent frames.	The visual effect of depth map for occluded regions is improved.
Liang et al. (2019)	Use two images from the left and right cameras to generate a brand new image for improving the poorly matched part of the disparity map.	The disparity map for regions with poor lighting is improved.

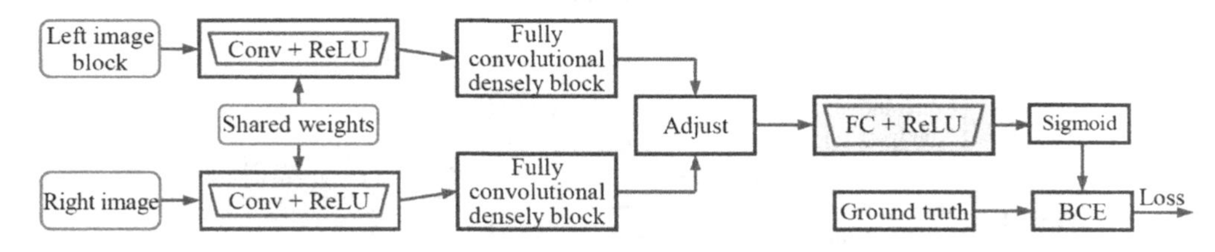

FIGURE 3.29 Schematic of feature cascade CNN.

3.5.3 Matching Based on Feature Cascade CNN

In order to improve the accuracy of disparity estimation in difficult scenes such as complex scenes, illumination changes, and weak textures and robustness, a binocular stereo matching method based on **feature cascade convolutional neural network** (FCCNN) is proposed (Wu et al. 2021).

The schematic of FCCNN for binocular matching is shown in Figure 3.29. It uses the image block as input to overcome the false matching problem that was encountered when relying on a single piece of gray-level information in the weak texture region. For feature extraction, a cascade (the trapezoids in Figure 3.29) of Conv + ReLU is used to produce the initial feature map. A fully convolutional densely block module (see below) is followed to enhance high-frequency information, and to produce a feature tensor. The dimension of the feature tensor is adjusted. These feature tensors are classified and re-organized by a stacked layer of fully connected (FC) and ReLU. Finally, Sigmoid function is used to predict similarities. The performance of network model can be evaluated by **binary cross entropy** (BCE) loss function values.

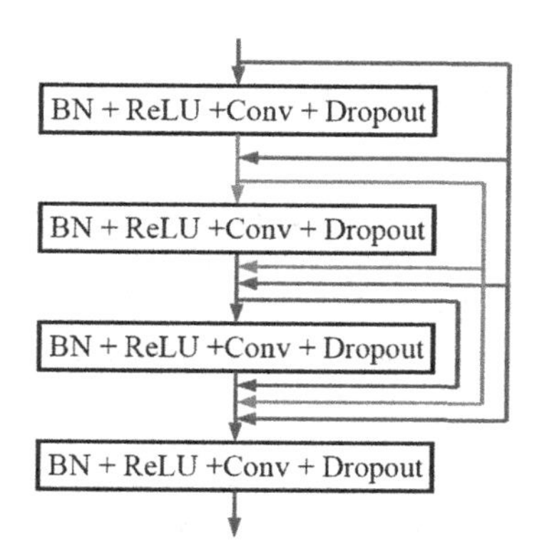

FIGURE 3.30 Detailed structure of fully convolutional densely block.

The detailed structure of fully convolutional dense blocks for feature reuse is shown in Figure 3.30. Compared with the standard CNN model, the dense connection mechanism iteratively concatenates the feature maps of all previous layers in a feedforward manner (Huang et al. 2017). Its output is the result of four consecutive operations. These four operations are batch normalization (BN), linear rectification function (ReLU), convolution (Conv), and drop-out with a certain random loss rate. The feature map extracted from the shallow layer is cascaded to the subsequent sub-layers by using the "skip connection" mechanism, and the local feature information lost by deep convolution is compensated. Using such a dense connection method to build a stereo matching model can effectively reduce the space complexity of the network model and enhance the image texture details.

REFERENCES

Ballard, D.H. and C.M. Brown. 1982. *Computer Vision*. London: Prentice-Hall.

Bromley, J., J.W. Bentz, L. Bottou, et al. 1993. Signature verification using a "Siamese" time delay neural network. *International Journal of Pattern Recognition and Artificial Intelligence*, 7(4): 669–688.

Chang, J. and Y. Chen. 2018. Pyramid stereo matching network. *IEEE Conference on Computer Vision and Pattern Recognition (CVPR)*, 5410–5418.

Chen, Y., L.L. Yang and Z.P. Wang. 2020. Literature survey on stereo vision matching algorithms. *Journal of Graphics*, 41(5): 702–708.

Davies. E.R. 2005. *Machine Vision: Theory, Algorithms, Practicalities*, 3rd Ed. The Netherlands, Amsterdam: Elsevier.

Forsyth, D. and J. Ponce. 2012. *Computer Vision: A Modern Approach*, 2nd Ed. London: Prentice Hall.

Goshtasby, A.A. 2005. *2-D and 3-D Image Registration – for Medical, Remote Sensing, and Industrial Applications*. Hoboken: Wiley-Interscience.

Guo, C.G., D.Y. Chen and Z.Q. Huang. 2019a. Learning efficient stereo matching network with depth discontinuity aware super-resolution. *IEEE Access*, 7: 159712–159723.

Guo, X.Y., K. Yang, W.K. Yang, et al. 2019b. Group-wise correlation stereo network. *IEEE Conference on Computer Vision and Pattern Recognition (CVPR)*, 3268–3277.

He, K.M., X.Y. Zhang, S.Q. Ren, et al. 2015. Spatial pyramid pooling in deep convolutional networks for visual recognition. *IEEE Transactions on Pattern Analysis and Machine Intelligence*, 37(9): 1904–1916.

Huang, G., Z. Liu, L. Van Der Maaten, et al. 2017. Densely connected convolutional networks. *IEEE Conference on Computer Vision and Pattern Recognition*, 4700–4708.

Kanade T., A. Yoshida, K. Oda, et al. 1996. A stereo machine for video-rate dense depth mapping and its new applications. *Proceedings of the 15CVPR*, 196–202.

Khamis, S., S. Fanello, C. Rhemann, et al. 2018. StereoNet: Guided hierarchical refinement for real-time edge-aware depth prediction. *Proceedings of the ECCV*, 596–613.

Kim, Y.C. and J.K. Aggarwal. 1987. Positioning three-dimensional objects using stereo images. *IEEE Journal on Robotics and Automation*, 1: 361–373.

Lew, M.S., T.S. Huang and K. Wong. 1994. Learning and feature selection in stereo matching. *IEEE Transactions on Pattern Analysis and Machine Intelligence*, 16(9): 869–881.

Liang, H., L. Qi, S.T. Wang, et al. 2019. Photometric stereo with only two images: a generative approach. *IEEE 2nd International Conference on Information Communication and Signal Processing (ICICSP)*, 363–368.

Liu, G.D., G.L. Jiang, R. Xiong, et al. 2019. Binocular depth estimation using convolutional neural network with Siamese branches. *IEEE International Conference on Robotics and Biomimetics (ROBIO)*, 1717–1722.

Lore, K.G., K. Reddy, M. Giering, et al. 2018. Generative adversarial networks for depth map estimation from RGB video. *IEEE Conference on Computer Vision and Pattern Recognition Workshops (CVPRW)*, 1177–1185.

Luo, J.Y., Y. Xu, C.W. Tang, et al. 2017. Learning inverse mapping by AutoEncoder based generative adversarial nets. *Proceedings of International Conference on Neural Information Processing*, 207–216.

Matias, L.P.N., M. Sons, J.R. Souza, et al. 2019. VeIGAN: Vectorial inpainting generative adversarial network for depth maps object removal. *IEEE Intelligent Vehicles Symposium (IV)*, 310–316.

Nixon, M.S. and A.S. Aguado. 2008. *Feature Extraction and Image Processing*, 2nd Ed. Maryland: Academic Press.

Pilzer, A., D. Xu, M. Puscas, et al. 2018. Unsupervised adversarial depth estimation using cycled generative networks. *International Conference on 3D Vision (3DV)*, 587–595.

Posselt, B.N. and M. Winterbottom. 2021. Are new vision standards and tests needed for military aircrew using 3D stereo helmet-mounted displays? *BMJ Military Health*, 167: 442–445.

Roberts, L.G. 1965. Machine perception of three-dimensional solids. In: *Optical and Electro-Optical Information Processing*, PhD thesis, MIT, 159–197.

Wu, J.J., Z. Chen and C.X. Zhang. 2021. Binocular stereo matching based on feature cascade convolutional network. *Acta Electronica Sinica*, 49(4): 690–695.

Zagoruyko, S. and N. Komodakis. 2015. Learning to compare image patches via convolutional neural networks. *IEEE Conference on Computer Vision and Pattern Recognition (CVPR)*, 4353–4361.

Žbontar, J. and Y. Lecun. 2015. Computing the stereo matching cost with a convolutional neural network. *IEEE Conference on Computer Vision and Pattern Recognition (CVPR)*, 1592–1599.

Zhang, F.H., V. Prisacariu, R.G. Yang, et al. 2019. GA-net: Guided aggregation net for end-to-end stereo matching. *IEEE Conference on Computer Vision and Pattern Recognition (CVPR)*, 185–194.

Zhang, Y.-J. 2017. *Image Engineering, Vol.2: Image Analysis*. Germany: De Gruyter.

Zhang, Y.-J. 2021. *Handbook of Image Engineering*. Singapore: Springer Nature.

Zhang, Y.-J. and H.B. Lu. 2002. A hierarchical organization scheme for video data. *Pattern Recognition*, 35(11): 2381–2387.

CHAPTER 4

Generalized Matching

THE TASK OF UNDERSTANDING images and understanding scenes is complex, including processes such as perception/observation, scene recovery, matching cognition, scene interpretation, and so on. Matching cognition attempts to connect the unknown with the known through matching, and then use the known to explain the unknown. For example, scene matching technology uses the data of the scene reference map for autonomous navigation and positioning and uses real-time matching to obtain accurate navigation and positioning information.

Multiple forms of image input and other forms of knowledge often coexist in a complex image understanding system. **Matching** uses the existing representations and models stored in the system to perceive the information in the image input, and finally establish the correspondence with the external world to realize the interpretation of the scene. For this purpose, matching can be understood as a technique or process of combining various representations and knowledge to interpret a scene.

Commonly used image-related matching methods and techniques can be classified into two categories: one is more specific, and the set of low-level pixels or pixel groups corresponding to the image is collectively referred to as **image matching**; the other is more abstract, mainly related to the image object, to its properties and connections, and even to the description and interpretation of the scene, and is collectively referred to as **generalized matching**. A number of image-matching techniques were introduced in Chapter 3; this chapter focuses on generalized matching methods and techniques.

The chapter is organized as follows. Section 4.1 introduces general matching strategies and classification methods, discusses the similarities and differences between matching and registration, and analyzes commonly used image-matching evaluation criteria. Section 4.2 discusses the principles and metrics of general object matching, and introduces several basic techniques, including correspondence point matching, string matching, inertia-equivalent ellipse matching, and shape matrix matching. Section 4.3 introduces a dynamic pattern-matching technique, in which the pattern representation to be matched is dynamically established during the matching process. The absolute model and relative model used

DOI: 10.1201/9781003362388-4

in it are analyzed, and specific matching experimental results are given. Section 4.4 discusses matching interrelationships between objects. Relationships can express different attributes of object sets, as well as more abstract concepts. Here, objects and their relational representations are introduced in detail, as well as examples of matching connection relations. Section 4.5 introduces some basic definitions of graph theory, including the geometric representation of graphs and subgraphs, and the concepts of graph identity and isomorphism, and discusses how to use several judgments of graph isomorphism to match. Section 4.6 provides a brief introduction to technique developments and promising research directions of the last year.

4.1 MATCHING OVERVIEW

In the understanding of images, matching techniques play an important role. From a visual point of view, "seeing" should be a purposeful "seeing", that is, according to certain knowledge (including the description of the object and the interpretation of the scene); the required scenery should be found in the scene with the help of the image; and "feeling" should be a cognitive "sense", that is, to extract the characteristics of the scene from the input image, and then match with the existing scene model, so as to achieve the purpose of understanding the meaning of the scene. Matching is intrinsically linked to knowledge, and matching and interpretation are inseparable.

4.1.1 Matching Strategies and Categories

Matching can be performed at different (abstract) levels because knowledge has different levels and can be applied at different levels. For each specific match, it can be thought of as finding a correspondence between two representations. If the types of the two representations are comparable, matching can be done in a similar sense. For example, when both representations are image structures, it is called **image matching**; if both representations represent the object in the image, it is called **object matching**; if both representations represent the description of the scene, it is called **scene matching**; if both representations are relational structures, it is called **relational matching**; if the types of the two representations are different (for example, one is an image structure and the other is a relational structure), it can also be matched in an extended sense, sometimes referred to as **fitting**.

Matching is to establish the connection between the two things; this needs to be done through mapping. When reconstructing the scene, the image-matching strategy can be divided into two cases according to the different mapping functions used; see Figure 4.1 (Kropatsch and Bischof 2001).

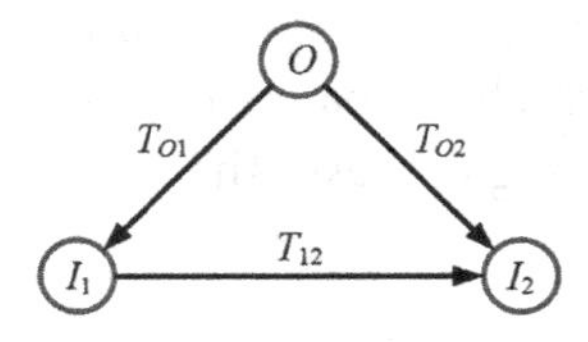

FIGURE 4.1 Matching and mapping.

(1) Matching in the object space

In this case, the object O is directly reconstructed by inverting the perspective transformations T_{O1} and T_{O2}. Here, an explicit representation model for the object O is needed to solve the problem by establishing a correspondence between the image features and the object model features. The advantage of object space matching techniques is that they fit well with the physical world, so even occlusions can be handled if more complex models are used.

(2) Matching of image space

The matching in the image space directly associates the images I_1 and I_2 with the mapping function T_{12}. In this case, the object model is implicitly included in the establishment of T_{12}. This process is generally quite complicated, but if the object surface is relatively smooth, an affine transformation can be used for local approximation, and the computational complexity can be reduced to be comparable to the matching in the object space. In the case of occlusion, the smoothness assumption will be affected and the image-matching algorithm will encounter difficulties.

Image-matching algorithms can be further classified according to the image representation and description model they use.

(3) Raster-based matching

Raster-based matching uses a raster representation of the image, i.e., it attempts to find a mapping function between image regions by directly comparing grayscale or grayscale functions. This class of methods can be highly accurate, but sensitive to occlusion.

(4) Feature-based matching

In feature-based matching, the symbolic description of an image is first decomposed by salient features extracted from the image using feature extraction operators, and then the corresponding features of different images are searched based on assumptions about the local geometric properties of the object to be described, and geometric mapping is performed. These methods are more suitable for situations with surface discontinuities and data approximations than those based on raster-based matching.

(5) Relationship-based matching

Relationship-based matching, also known as structural matching, is based on the similarity of topological relationships between features (topological properties do not change under perspective transformation), and these similarities exist in **feature adjacency graphs** rather than similarity in grayscale or point distributions. Matching of relational descriptions can be applied in many situations, but it may generate a very complex search tree, so its computational complexity may be very large.

The **template matching** theory introduced in Subsection 3.3.1 believes that in order to recognize the content of an image, it is necessary to have its "memory trace" or basic model in past experience, which is also called a template. If the current stimulus matches the template in the brain, you can tell what the stimulus is. However, template matching theory says that the matching is that the external stimulus must exactly match the template. In practice, people in real life can not only recognize images that are consistent with the basic pattern, but also recognize images that do not completely conform to the basic pattern.

Gestalt psychologists came up with the **prototype matching** theory. This theory holds that for a currently observed image of the letter A, no matter what shape it is or where it is placed, it bears a resemblance to an A known to have been perceived in the past. Humans do not store countless templates of different shapes in long-term memory, but use the similarities abstracted from various images as prototypes to test the images to be recognized. If a prototype resemblance can be found from the image to be recognized, then the recognition of the image is achieved. This model of image cognition is more suitable than template matching from the perspective of neurology and memory search process, and it can also explain the cognitive process of some irregular images that have certain aspects similar to the prototype. According to this model, an idealized prototype of the letter A can be formed, which summarizes the common characteristics of various images similar to this prototype. On this basis, the cognitions of all other As that are not identical with the archetype but only similar will be made possible by matching.

Although prototype matching theory can more reasonably explain some phenomena in image cognition, it does not explain how humans discriminate and process similar stimuli. Prototype matching theory does not give a clear image recognition model or mechanism, and it is difficult to realize it in computer programs.

4.1.2 Matching and Registration

Matching and registration are two closely related concepts with many technical similarities. However, upon closer inspection, there are some differences between the two.

4.1.2.1 Registration Technology

The meaning of **registration** is generally narrow, mainly referring to the establishment of correspondence between images obtained in different time or space, especially the geometric correspondence (such as in geometric correction). The final effect is often reflected at the pixel level. Matching can consider both the geometric properties and the grayscale properties of the image, and even its other abstract properties and attributes. From this point of view, registration can be seen as matching of lower-level representations, and **generalized matching** can include registration. The main difference between image registration and stereo matching is that the former not only needs to establish the relationship between point pairs, but also needs to calculate the coordinate transformation parameters between the two images from this correspondence; while the latter only needs to establish the correspondence between point pairs relationship, and then calculate the disparity separately for each pair of points.

In terms of specific implementation techniques, registration can often be facilitated by various coordinate transformations and affine transformations. Most registration algorithms consist of three steps: (i) feature selection; (ii) feature matching; and (iii) computation of transformation functions. The performance of registration techniques is often determined by the following four factors (Lohmann 1998):

(1) The feature space of the features used for registration

(2) A search space that makes it possible for the search process to have a solution

(3) The search strategy for scanning the search space

(4) A similarity measure used to determine whether the registration correspondence holds.

Registration techniques in the image space can be classified into two categories similar to stereo matching techniques (see Sections 3.3 and 3.4). The registration technology in the frequency domain is mainly carried out by the correlation calculation. Here, the image needs to be converted into the frequency domain through Fourier transform, and then the phase information or amplitude information of the spectrum is used in the frequency domain to establish the corresponding relationship and to achieve registration. They can be called the phase correlation method and amplitude correlation method, respectively.

4.1.2.2 Inertia-Equivalent Ellipse Matching

Matching between objects can also be done by means of their inertia-equivalent ellipses, which have been used in registration work for 3-D object reconstruction from sequence images (Zhang 1991a). Unlike matching based on the object contour, matching based on the inertia-equivalent ellipse is performed based on the entire object region. For any object region, the corresponding inertia ellipse can be calculated (Zhang 2017b). With the help of the inertia ellipse corresponding to the object, an inertia-equivalent ellipse can be further calculated for each object. From the point of view of object matching, since each object in the image pair to be matched can be represented by its inertia-equivalent ellipse, the matching problem of the objects can be transformed into the matching of their inertia-equivalent ellipse (see Figure 4.2).

In general object matching, the main consideration is the deviation caused by translation, rotation, and scale transformation, and the corresponding geometric parameters need to be obtained. The parameters required can be calculated by the center coordinates of the equivalent ellipse, the orientation angle (defined as the angle between the major axis of the ellipse and the positive X-axis), and the length of the major axis.

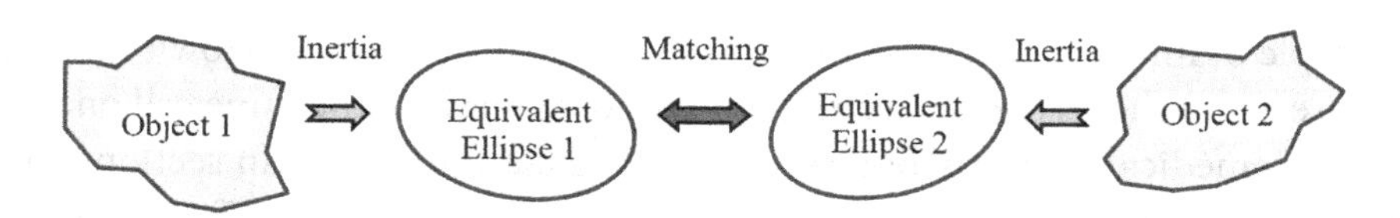

FIGURE 4.2 Matching with the help of inertia-equivalent ellipses.

First consider the center coordinates (x_c, y_c) of the equivalent ellipse, that is, the barycentric coordinates of the object. Assuming that the object region contains N pixels in total, then

$$x_c = \frac{1}{N}\sum_{i=1}^{N} x_i \tag{4.1}$$

$$y_c = \frac{1}{N}\sum_{i=1}^{N} y_i \tag{4.2}$$

The translation parameter can be calculated according to the difference of the center coordinates of the two equivalent ellipses. Secondly, the orientation angle ϕ of the equivalent ellipse can be obtained by using the slopes k and l of the two main axes of the corresponding inertia ellipse (let A be the moment of inertia of the object rotating around the X axis, and B is the moment of inertia of the object rotating around the Y axis)

$$\varphi = \begin{cases} \arctan(k) & A < B \\ \arctan(l) & A > B \end{cases} \tag{4.3}$$

The rotation parameter can be calculated from the difference in the orientation angles of the two ellipses. Finally, the two semi-principal lengths (a and b) of the equivalent ellipse reflect information about the object size. If the object itself is an ellipse, it is identical to its equivalent ellipse. In general, the equivalent ellipse of the object is the approximation of the object in terms of moment of inertia and area (but not equal at the same time). Here, the axis length needs to be normalized by the object area M. After normalization, when $A < B$, the length a of the semi-major principal axis of the equivalent ellipse can be calculated by the following formula (let H represent the product of inertia):

$$a = \sqrt{\frac{2\left[(A+B) - \sqrt{(A-B)^2 + 4H^2}\right]}{M}} \tag{4.4}$$

The scale transformation parameter can be calculated according to the length ratio of the major axes of the two ellipses. The three transformation parameters of the geometric correction required for the above two object matchings can be calculated independently, so the transformations in the equivalent ellipse matching can be performed sequentially (Zhang 1997).

An example of image registration of cells in a biomedical study is shown in Figure 4.3, where Figure 4.3(a) shows two adjacent cross-sectional views of the same cell on two serial sections. In biomedical research, body tissue is often cut into very thin sections so that its structure can be observed under an optical microscope. Due to the effects of translation and rotation when making slices, although the two slices are adjacent, their orientations

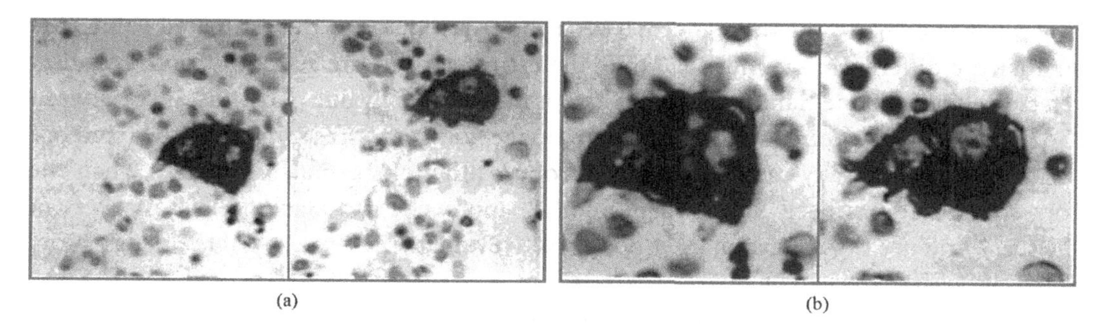

FIGURE 4.3 Registration of cells on two adjacent slices.

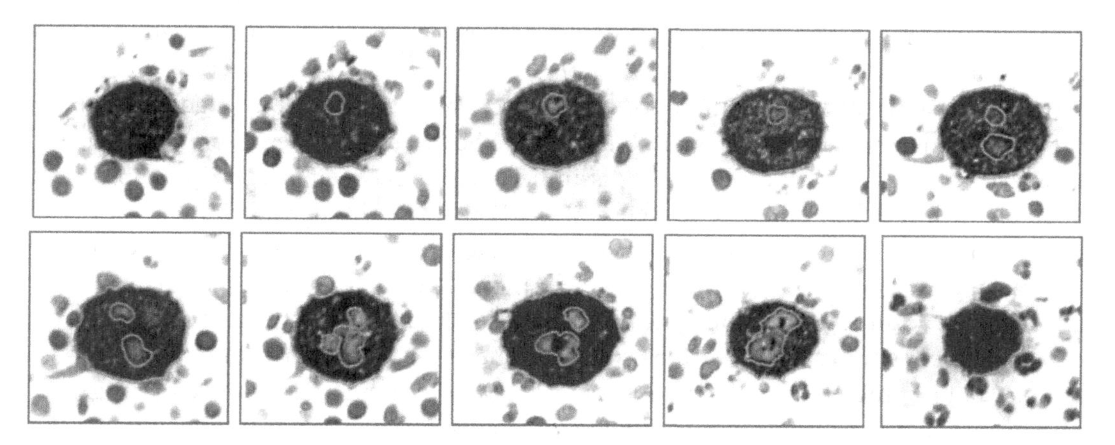

FIGURE 4.4 Results of registering a cell profile spanning 10 serial sections.

are not identical and require registration. Considering that the changes in the internal and surrounding structures of the cell are large, the registration only considering the contour is not effective. The matching result obtained by the above-mentioned equivalent ellipse matching method is shown in Figure 4.3(b). Here, a higher resolution display is used to obtain a clearer result. It can be seen from the registration diagram that the effects of translation and rotation have been eliminated.

When a cell spans many slices, to avoid the cumulative error of pairwise registration of tracking profiles in slice order, each cell profile can be directly registered with a predetermined common reference. Specifically, translational registration can be based on the center of the image. Considering that the above method can determine the absolute orientation of cells, each cell can be rotated and registered according to its long major axis parallel to the X axis. The scaling criterion takes into account the size of the image and ensures that the cell profile does not intersect the edge of the image. Figure 4.4 shows the results of registering a cell profile spanning 10 serial sections using the method described above.

4.1.3 Matching Evaluation

Commonly used image-matching evaluation criteria include accuracy, reliability, robustness and computational complexity (Goshtasby 2005).

Accuracy refers to the difference between the true value and the estimated value. The smaller the difference, the more accurate the estimation. In image registration, accuracy refers to the mean, median, maximum, or root mean square value of the distance between the reference image point and the registered image point (after resampling to the reference image space). Accuracy can be measured from synthetic or simulated images when the correspondence has been determined; another approach is to place fiducial markers in the scene and use the fiducial marker positions to evaluate the accuracy of the registration. The unit of accuracy can be pixels or voxels.

Reliability refers to how many times an algorithm has achieved satisfactory results in the total number of tests performed. If N pairs of images are tested, where M tests give satisfactory results, when N is large enough and N pairs of images are representative, then M/N indicates reliability. The closer M/N is to 1, the more reliable it is. The reliability of an algorithm is predictable.

Robustness refers to the stability of accuracy or the reliability of an algorithm under different changes in its parameters. Robustness can be measured in terms of noise, density, geometric differences, or percentage of dissimilar regions between images. The robustness of an algorithm can be obtained by determining how stable the algorithm's accuracy is or how reliable it is when the input parameters change (if their variance is used, the smaller the variance the more robust the algorithm). If there are many input parameters, each affecting the accuracy or reliability of the algorithm, then the robustness of the algorithm can be defined with respect to each parameter. For example, an algorithm might be robust to noise but not robust to geometric distortions. Saying that an algorithm is robust generally means that the performance of the algorithm does not change significantly as the parameters involved change.

Computational complexity determines the speed of an algorithm, indicating its usefulness in a specific application. For example, in image-guided neurosurgery, the images used to plan the surgery need to be registered within seconds with images that reflect the conditions of the surgery at a particular time. However, matching the aerial imagery acquired by an aircraft often needs to be completed in the order of milliseconds. Computational complexity can be expressed as a function of image size (considering the number of additions or multiplications required for each unit). It is generally expected that the computational complexity of a good matching algorithm is a linear function of image size.

4.2 OBJECT MATCHING

Image matching takes pixels as the unit, the amount of calculation is generally large, and the matching efficiency is low. In practice, objects of interest are often detected and extracted first, and then objects are matched. If a concise object representation is used, the matching effort can be greatly reduced. Since the object can be represented in different ways, the object matching can also take a variety of forms.

4.2.1 Matching Metrics

The effect of object matching should be judged by certain metrics, the core of which is the similarity of the object.

4.2.1.1 Hausdorff Distance

In the image, the object is composed of points (pixels), and the matching of two objects is, in a certain sense, the matching between two sets of points. The method of using Hausdorff distance (HD) to describe the similarity between point sets and matching them through feature point sets is widely used. Given two finite point sets $A = \{a_1, a_2, \ldots, a_m\}$ and $B = \{b_1, b_2, \ldots, b_n\}$, the Hausdorff distance between them is defined as:

$$H(A,B) = \max\left[h(A,B), h(B,A)\right] \tag{4.5}$$

where

$$h(A,B) = \max_{a \in A} \min_{b \in B} \|a - b\| \tag{4.6}$$

$$h(B,A) = \max_{b \in B} \min_{a \in A} \|b - a\| \tag{4.7}$$

The norm $||\bullet||$ in Equations (4.6) and (4.7) can take different forms. The function $h(A, B)$ is called the directed Hausdorff distance from the set A to B, which describes the longest distance from a point $a \in A$ to any point in the point set B; similarly, the function $h(B, A)$ is called the directed Hausdorff distance from set B to A, describing the longest distance from point $b \in B$ to any point in point set A. Since $h(A, B)$ and $h(B, A)$ are not symmetrical, the maximum value between them is generally taken as the Hausdorff distance between the two point sets. The geometric meaning of Hausdorff distance can be explained as follows: if the Hausdorff distance between two point sets A and B is d, then for any point in each point set, in a circle centered on this point with radius d, at least one point in another set of points can be found. If the Hausdorff distance between two point sets is 0, it means that the two point sets are coincident. In the schematic diagram of Figure 4.5: $h(A, B) = d_{21}$, $h(B, A) = d_{22} = H(A, B)$.

The Hausdorff distance as defined above is sensitive to noise points or the outline of the point set. A commonly used improvement method adopts the concept of statistical averaging and replaces the maximum value with the average value, which is called the modified Hausdorff distance (MHD) (Dubuisson and Jain 1994), that is, Equations (4.6) and (4.7) are changed to

$$h_{\text{MHD}}(A,B) = \frac{1}{N_A} \sum_{a \in A} \min_{b \in B} \|a - b\| \tag{4.8}$$

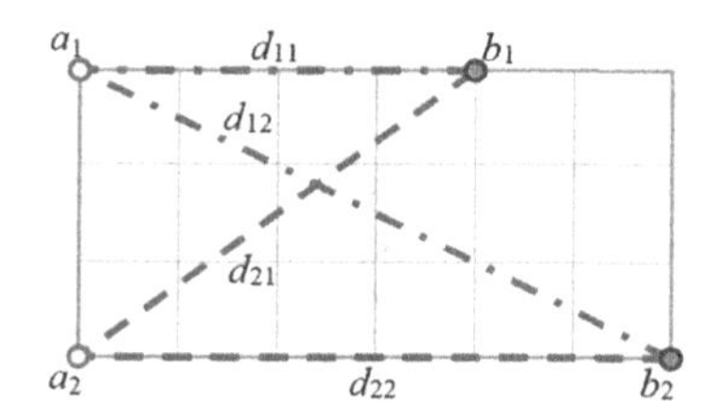

FIGURE 4.5 Hausdorff distance diagram.

$$h_{\text{MHD}}(B,A) = \frac{1}{N_B}\sum_{b \in B} \min_{a \in A} \|b - a\| \tag{4.9}$$

where N_A represents the number of points in point set A, and N_B represents the number of points in point set B. Substituting them into Equation (4.5) gives

$$H_{\text{MHD}}(A,B) = \max\left[h_{\text{MHD}}(A,B), h_{\text{MHD}}(B,A)\right] \tag{4.10}$$

When using the Hausdorff distance to calculate the correlation matching between the template and the image, a clear point relationship between the template and the image is not required. In other words, it is not necessary to establish a one-to-one relationship of point correspondence between the two point sets, which is one of its important advantages.

4.2.1.2 Structural Matching Metrics

Objects are often decomposable, that is, into their individual components. Different objects can have the same components but different structures. For structural matching, most matching metrics can be explained by the "template and spring" physical analogy model (Ballard and Brown 1982). Considering that structure matching is a match between the reference structure and the structure to be matched, if the reference structure is viewed as a structure depicted on a transparency, the matching can be seen as moving the transparency on the structure to be matched and deforming it to obtain a fit of the two structures.

Matching often involves similarities that can be quantitatively described. A match is not a simple correspondence, but a correspondence that is quantitatively described according to a certain goodness index, and this goodness corresponds to the matching metric. For example, the goodness of fit of two structures depends both on how well the components of the two structures match each other and on the amount of work required to deform the transparencies.

In practice, to achieve the deformation is to consider the model as a set of rigid templates connected by springs, such as a face template and spring model (Figure 4.6). Here the

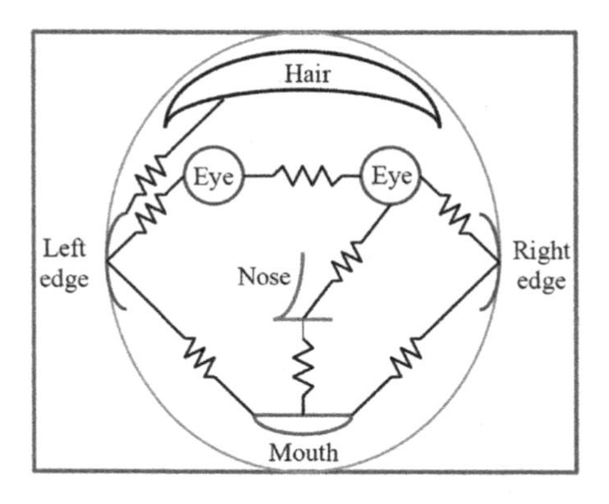

FIGURE 4.6 A template and a spring model of a face.

templates are connected by springs, and the spring functions describe the relationship between the templates. The relationship between templates generally has certain constraints. For example, in a face image, the two eyes are generally on the same horizontal line, and the distance is always within a certain range. The quality of the match is a function of the goodness of the local fit of the template and the energy required to elongate the spring to fit the structure to be matched to the reference structure.

The matching metric of template and spring can be expressed in general form as follows:

$$C=\sum_{d\in Y} C_T\left[d,F(d)\right]+\sum_{(d,e)\in(Y\times E)} C_S\left[F(d),F(e)\right]+\sum_{c\in(N\cup M)} C_M(c) \tag{4.11}$$

where C_T represents the dissimilarity between the template d and the structure to be matched, C_S represents the dissimilarity between the structure to be matched and the object part e, C_M represents the penalty for missing parts, and $F(\bullet)$ is the mapping to convert the reference structure template to the structural components to be matched. F divides reference structures into two categories: structures that can be found in the structures to be matched (belonging to set Y), and structures that are not found in the structures to be matched (belonging to set N). Similarly, components can also be divided into components that exist in the structure to be matched (belonging to set E) and components that do not exist in the structure to be matched (belonging to set M).

Normalization issues need to be considered in structural matching metrics because the number of matched parts may affect the value of the final matching metric. For example, if a "spring" always has a finite cost, such that the more elements matched, the greater the total energy, that doesn't mean that having more parts matched is worse than having fewer parts. Conversely, delicate matching of a part of the structure to be matched with a specific reference object often makes the remaining part unmatched, and this "sub-match" is not as good as making most of the parts to be matched closely matched. In Equation (4.11), this is avoided by a penalty for missing parts.

4.2.2 Corresponding Point Matching

The matching between two objects (or a model and an object) can be carried out by means of the correspondence between them when there are specific landmark points or feature points on the object. If these landmark points or feature points are different from each other (they have different properties), there are two pairs of points to match. If these landmark points or feature points are the same as each other (they have the same attributes), at least three non-collinear corresponding points (three points must be coplanar) need to be determined on each of the two objects.

In 3-D space, if perspective projection is used, since any set of three points can match any other set of three points, the correspondence between the two sets of points cannot be determined at this point, whereas if a weak perspective projection is used, the matching is much less ambiguous.

Consider a simple case. Suppose a group of three points P_1, P_2 and P_3 on the object are on the same circumference, as shown in Figure 4.7(a). Let the center of gravity of the triangle be C, and the straight line connecting C with P_1, P_2 and P_3 intersects with the circumference at Q_1, Q_2 and Q_3 respectively. Under the condition of weak perspective projection, the distance ratio $P_iC : CQ_i$ remains unchanged after projection. In this way, the circumference will become an ellipse after projection (but the straight line is still a straight line after projection, and the distance ratio remains unchanged), as shown in Figure 4.7(b). When P'_1, P'_2 and P'_3 are observed in the image, C' can be calculated, and then the positions of points Q'_1, Q'_2 and Q'_3 can be determined. In this way, there are six points to determine the position and parameters of the ellipse (in fact, at least five points are required). Once the ellipse is known, the matching becomes ellipse matching.

If the distance ratio is calculated incorrectly, Q_i will not fall on the circumference, as shown in Figure 4.7(c). In this way, the ellipses passing through P'_1, P'_2, P'_3 and Q'_1, Q'_2, Q'_3 cannot be obtained after projection, and the above calculation is impossible.

More general ambiguity cases can be found in Table 4.1, which gives the number of solutions obtained when matching objects with corresponding points in the image in each case. When the number of solutions ≥ 2, it indicates that there is ambiguity. All 2s occur when coplanar, corresponding to perspective inversion. Any non-coplanar point (when there are more than three points in the corresponding plane) provides enough information to disambiguate. In Table 4.1, two cases of coplanar point and non-coplanar point respectively are considered, and the perspective projection and weak perspective projection are also compared.

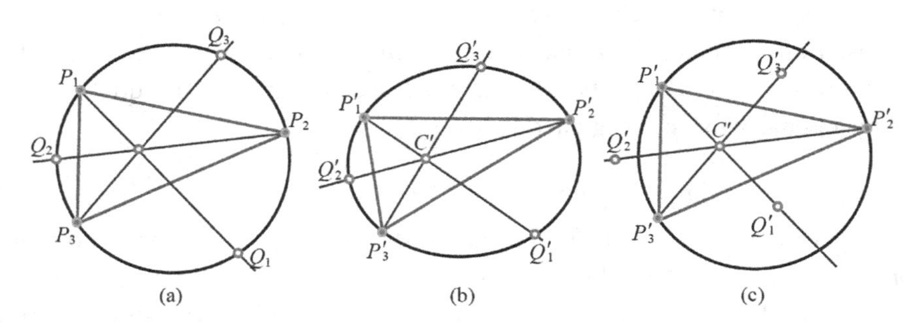

FIGURE 4.7 Three points matching under weak perspective projection.

TABLE 4.1 Ambiguity When Using Corresponding Point Matching

Distribution of Points			**Coplanar**					**Non-coplanar**		
The Number of Corresponding Point Pairs	≤ 2	3	4	5	≥ 6	≤ 2	3	4	5	≥ 6
Perspective Projection	∞	4	1	1	1	∞	4	2	1	1
Weak Perspective Projection	∞	2	2	2	2	∞	2	1	1	1

4.2.3 String Matching

String matching can be used to match the contours of two object regions (essentially it can be seen as matching two feature point sequences, so it is easy to generalize to other structures or relationships). Suppose the two region contours A and B have been encoded as strings $a_1a_2...a_n$ and $b_1b_2...b_m$ (Zhang 2017b), respectively. Starting from a_1 and b_1, if there is $a_k = b_k$ at the k-th position, the two contours are said to have a match. If M is used to represent the total number of matches between two strings, the number of unmatched symbols is

$$Q = \max(\|A\|, \|B\|) - M \tag{4.12}$$

where $||\text{arg}||$ represents the length (number of symbols) of the string representation of arg. It can be shown that $Q = 0$ if and only if A and B are congruent.

A simple similarity metric between A and B is

$$R = \frac{M}{Q} = \frac{M}{\max(\|A\|, \|B\|) - M} \tag{4.13}$$

It can be seen from Equation (4.13) that a larger R value indicates a better match. When A and B match exactly, the value of R is infinity; when no symbols in A and B match ($M = 0$), the value of R is zero.

Because string matching is done symbol by symbol, the determination of the starting point is important to reduce the amount of computation. If the calculation starts from any point, and then moves the position of one symbol to calculate again, according to Equation (4.13), the whole calculation will be very time-consuming (proportional to $||A|| \times ||B||$), so in practice, the string representation often needs to be normalized first.

The similarity between two strings can also be described by the Levenshtein distance (edit distance). The distance is defined as the (minimum) number of operations required to convert one string into another. The operations here mainly include editing operations on strings, such as deletion, insertion, replacement, etc. For these operations, weights can also be defined, allowing a finer measure of the similarity between two strings.

4.2.4 Shape Matrix Matching

The object regions to be matched in the two images often have differences in translation, rotation and scale. Considering the local characteristics of the images, if the images do not represent the deformed scene, the local nonlinear geometric differences between the images can be ignored. In order to determine the correspondence between the objects to be matched in two images, it is necessary to seek the similarity between objects that does not depend on the difference of translation, rotation and scale. The **shape matrix** is a representation of the object shape quantized in polar coordinates (Goshtasby 2005). As shown in Figure 4.8(a), the coordinate origin is placed at the center of gravity of the object, and the object is resampled along the radial and circumferential directions. These sampled data are independent of the position and orientation of the object. Let the radial increment be

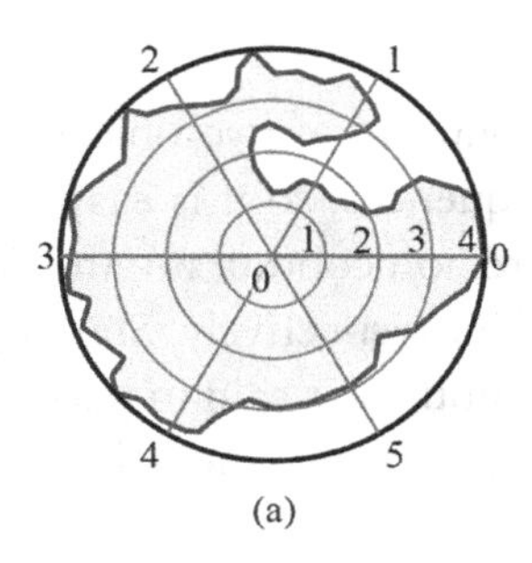

(a)

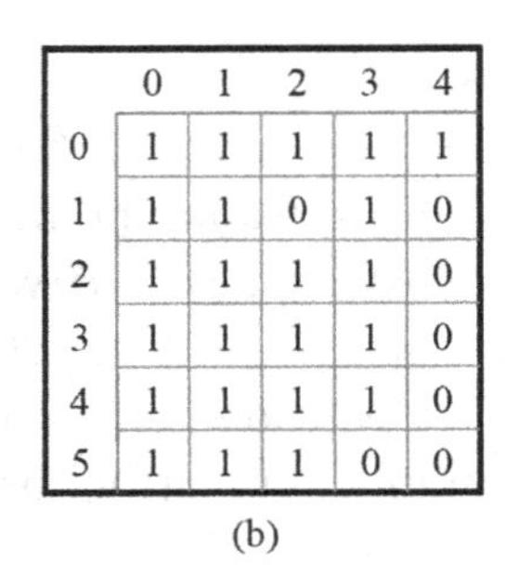

	0	1	2	3	4
0	1	1	1	1	1
1	1	1	0	1	0
2	1	1	1	1	0
3	1	1	1	1	0
4	1	1	1	1	0
5	1	1	1	0	0

(b)

FIGURE 4.8 Object and its shape matrix.

a function of the maximum radius, that is, always quantify the maximum radius into the same number of intervals, and the resulting representation is called the shape matrix, as shown in Figure 4.8(b). The shape matrix is scale independent.

The shape matrix contains both the object boundary and interior information, so it can also represent objects with holes (instead of only outer contours). The shape matrix can provide a standardized manner for representing the projection, orientation and scale of the object. Given two shape matrices M_1 and M_2 of size $m \times n$, the similarity between them is (note that the matrix is a binary matrix)

$$S = \sum_{i=0}^{m-1}\sum_{j=0}^{n-1}\frac{1}{mn}\left\{\left[M_1(i,j) \wedge M_2(i,j)\right] \vee \left[\bar{M}_1(i,j) \wedge \bar{M}_2(i,j)\right]\right\} \tag{4.14}$$

where the upper horizontal line represents the logical NOT operation. When $S = 1$, it means that the two objects are exactly the same, and as S gradually decreases and tends to 0, the two objects become more and more dissimilar. If the sampling is dense enough when building the shape matrix, the original object region can be reconstructed from the shape matrix.

If a logarithmic scale is used for sampling radially when constructing the shape matrix, the difference in scale between the two objects will translate to a difference in position along the horizontal axis in a logarithmic coordinate system. If you start at any point in the object region (rather than at the largest radius) when quantizing the region circumference, you will get values along the vertical axis in a logarithmic coordinate system. The log-polar mapping can convert both the rotation and scale differences between two regions into translation differences, which simplifies the task of object matching.

4.3 DYNAMIC PATTERN MATCHING

In the previous discussion of various matching techniques, it is considered that the representations that need to be matched have been established in advance, or the matching is performed on predetermined features or representations. But in fact, sometimes the representations to be matched are created dynamically during the matching process, or the different representations need to be adaptively created for matching according to the data to be matched in the matching process. A specific method, called **dynamic pattern matching** (Zhang 1990), is introduced below in combination with a practical application.

4.3.1 Matching Process

In the process of reconstructing 3-D cells from sequential medical slice images, determining the correspondence between the individual profiles of the same cell in adjacent slices is a critical step (which is the basis for the cell registration mentioned earlier). Due to the complex slicing process, thin slices and easy deformation, the number of cell profiles on adjacent slices may be different, and their distribution and arrangement may also be different. In order to reconstruct 3-D cells, it is necessary to determine the corresponding relationship between each profile of each cell, that is, to find the corresponding profile of the same cell on each slice image. The overall process of completing this work is shown in Figure 4.9. Here, the two slices to be matched are called matched slices and matching slices, respectively. The matched piece is a reference piece. After each profile on a piece to be matched is registered with the corresponding matched profile on the matched slice, the slice to be matched (matching) becomes a matched slice and can be used as the reference slice of the next slice to be matched. By continuing to match in this way, all profiles in a serial of slices can be registered (Figure 4.9 only takes one profile as an example).

Referring to the flow chart in Figure 4.9, this matching mainly has six steps.

(1) Select a matched profile from the matched slice.

(2) Construct a model representation of the selected matched profile.

(3) Determine the candidate region on the slice to be matched (with the help of prior knowledge to reduce the amount of calculation and ambiguity).

(4) Select the profile to be matched in the candidate region.

(5) Construct the pattern representation of each selected profile to be matched.

(6) Use the similarity between profile modes to check to determine the correspondence between profiles.

4.3.2 Absolute Pattern and Relative Pattern

Due to the large number of cell profiles on each slice, matching at too high a resolution is not suitable. In fact, each profile is treated as a point when matching, and the profile size and shape information cannot be used at this time. If only the position information of the

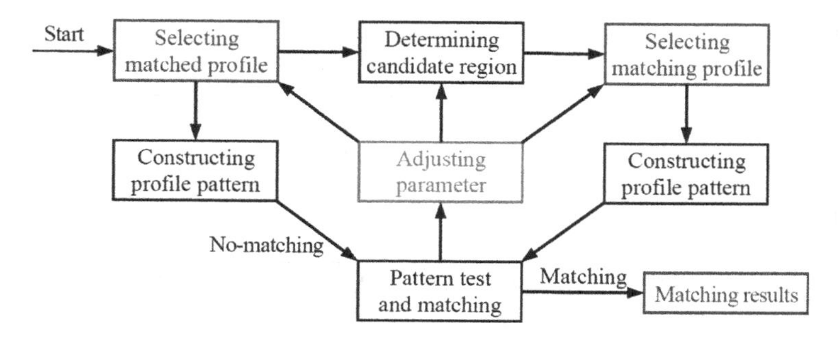

FIGURE 4.9 Flowchart of matching for dynamic patterns.

profile is used, the matching may not be possible due to the different placement of slices and the deformation caused by sectioning.

Considering that the distribution of cell profiles on slices is not uniform, in order to complete the above matching steps, it is necessary to dynamically establish a pattern representation for each profile that can be used for matching. Here, it can be considered to construct the unique pattern for each profile by using the relative positional relationship of each profile to its several adjacent profiles. This contains more information than a single profile point. The pattern thus constructed can be represented by a vector. If the relationship used is the length and orientation of the line between each profile point and its adjacent profile point (or the angle between the adjacent lines), then the two profile patterns $\boldsymbol{P}_{\mathrm{l}}$ and $\boldsymbol{P}_{\mathrm{r}}$ (both represented by a vector) can be written as:

$$P_{\mathrm{l}} = P\left(x_{\mathrm{l}0}, \quad y_{\mathrm{l}0}, \quad d_{\mathrm{l}1}, \quad \theta_{\mathrm{l}1}, \quad \cdots, \quad d_{\mathrm{l}m}, \quad \theta_{\mathrm{l}m}\right)^{\mathrm{T}} \tag{4.15}$$

$$P_{\mathrm{r}} = P\left(x_{\mathrm{r}0}, \quad y_{\mathrm{r}0}, \quad d_{\mathrm{r}1}, \quad \theta_{\mathrm{r}1}, \quad \cdots, \quad d_{\mathrm{r}n}, \quad \theta_{\mathrm{r}n}\right)^{\mathrm{T}} \tag{4.16}$$

In the formula, $x_{\mathrm{l}0}$, $y_{\mathrm{l}0}$ and $x_{\mathrm{r}0}$, $y_{\mathrm{r}0}$ are the center coordinates of the two profile points, respectively; each d represents the length of the line connecting other profile points and the matching profile point on the same slice; and each θ represents the angle between the adjacent lines from the matching profile point to the surrounding two adjacent profile points. So the pattern can be seen as contained in a circle with a definite radius of action. Note that m and n can be different here. When m is different from n, a part of the point construction patterns can also be selected for matching. In addition, the selection of m and n should be the result of the balance between the amount of computation and the uniqueness of the pattern. Larger m and n will increase the amount of computation, but at the same time, it will increase the probability of pattern uniqueness. Small m and n will reduce the amount of calculation, but it is possible to make the pattern composed by these points not unique. The specific value can be adjusted by determining the pattern radius.

In order to match the profiles, the corresponding patterns need to be translated and rotated. The pattern constructed above can be called an **absolute pattern** because it contains the absolute coordinates of the central profile point. Figure 4.10(a) shows an example

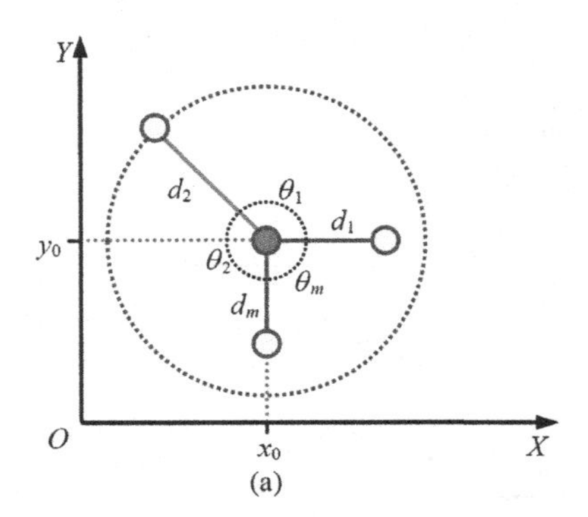

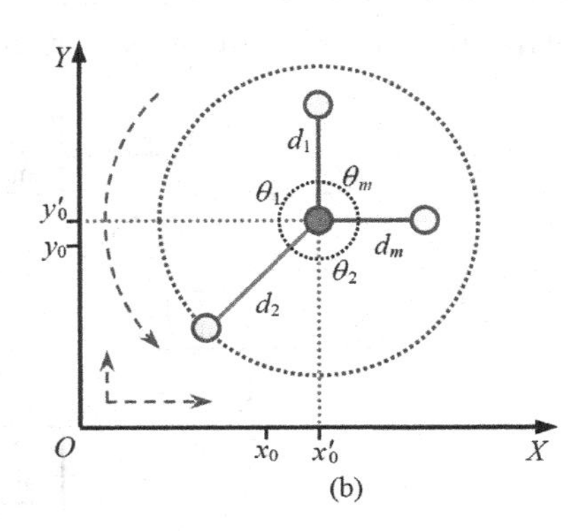

FIGURE 4.10 Illustration of absolute pattern.

of $\boldsymbol{P}_l$. The absolute pattern has rotation invariance to the origin (central profile point), that is, after the entire pattern is rotated, each d and θ are not changed. However, it can be seen from the results in Figure 4.10(b), which is the rotation and translation of Figure 4.10(a), that the vectors are not translation invariant because x_0 and y_0 change when the entire pattern is translated.

In order to obtain translation invariance, the coordinates of the center point in the absolute pattern can be removed, and the following **relative pattern** can be constructed:

$$Q_l = Q(d_{l1},\ \theta_{l1},\ \cdots,\ d_{lm},\ \theta_{lm})^{\mathrm{T}} \tag{4.17}$$

$$Q_r = Q(d_{r1},\ \theta_{r1},\ \cdots,\ d_{rn},\ \theta_{rn})^{\mathrm{T}} \tag{4.18}$$

The relative pattern corresponding to the absolute pattern shown in Figure 4.10(a) is shown in Figure 4.11(a). From the results of Figure 4.11(b) that are the rotation translation of Figure 4.11(a), it can be seen that the relative pattern is not only rotationally invariant, but also has translational invariance. In this way, two relative pattern representations can be matched by rotation and translation, and their similarity can be calculated, so as to achieve the purpose of matching profiles.

It can be seen from the above analysis that the main characteristics of dynamic pattern matching are: the pattern is established dynamically, and the matching is completely automatic. This method is quite general and flexible, and its basic idea can be applied to a variety of situations (Zhang 1991a).

Figure 4.12 shows the actual distribution of cell profiles on two adjacent medical slices (Zhang 1991b), where each cell profile point is represented by a dot. Since the diameter of the cells is much larger than the thickness of the slices, many cells span multiple slices. In other words, there should be many corresponding cell profiles on adjacent slices. However, it can be seen from Figure 4.12 that the distribution of points on each slice is very different, and the number of points is also very different. Figure 4.12(a) has 112 profile points, while Figure 4.12(b) has 137 profile points. The reasons include that some cell profile points in Figure 4.12(a) are the last profiles of cells and do not continue to extend to Figure 4.12(b); and some cell profiles in Figure 4.12(b) are new beginnings, not continued from Figure 4.12(a).

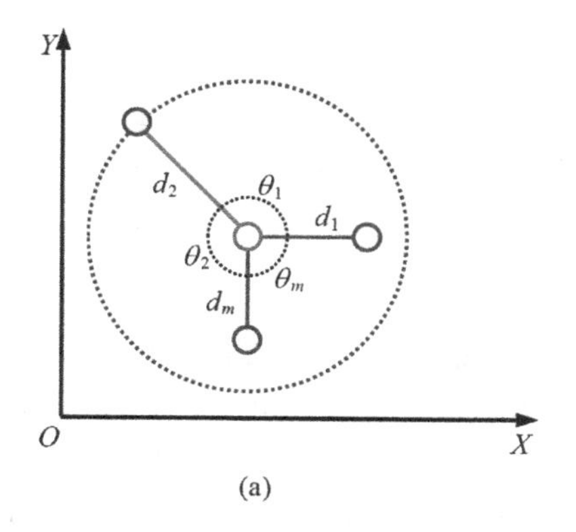

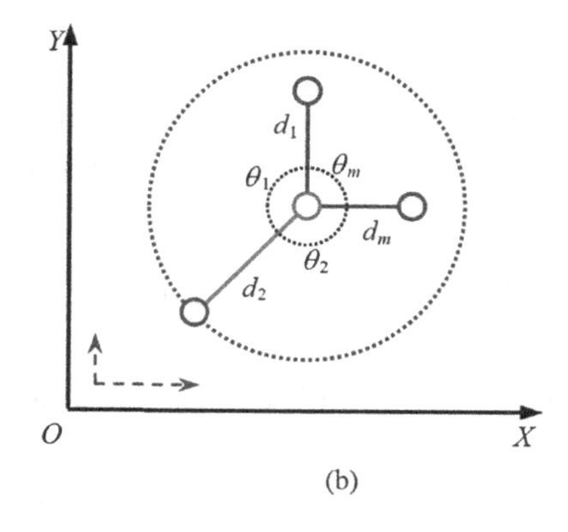

FIGURE 4.11 Illustration of relative pattern.

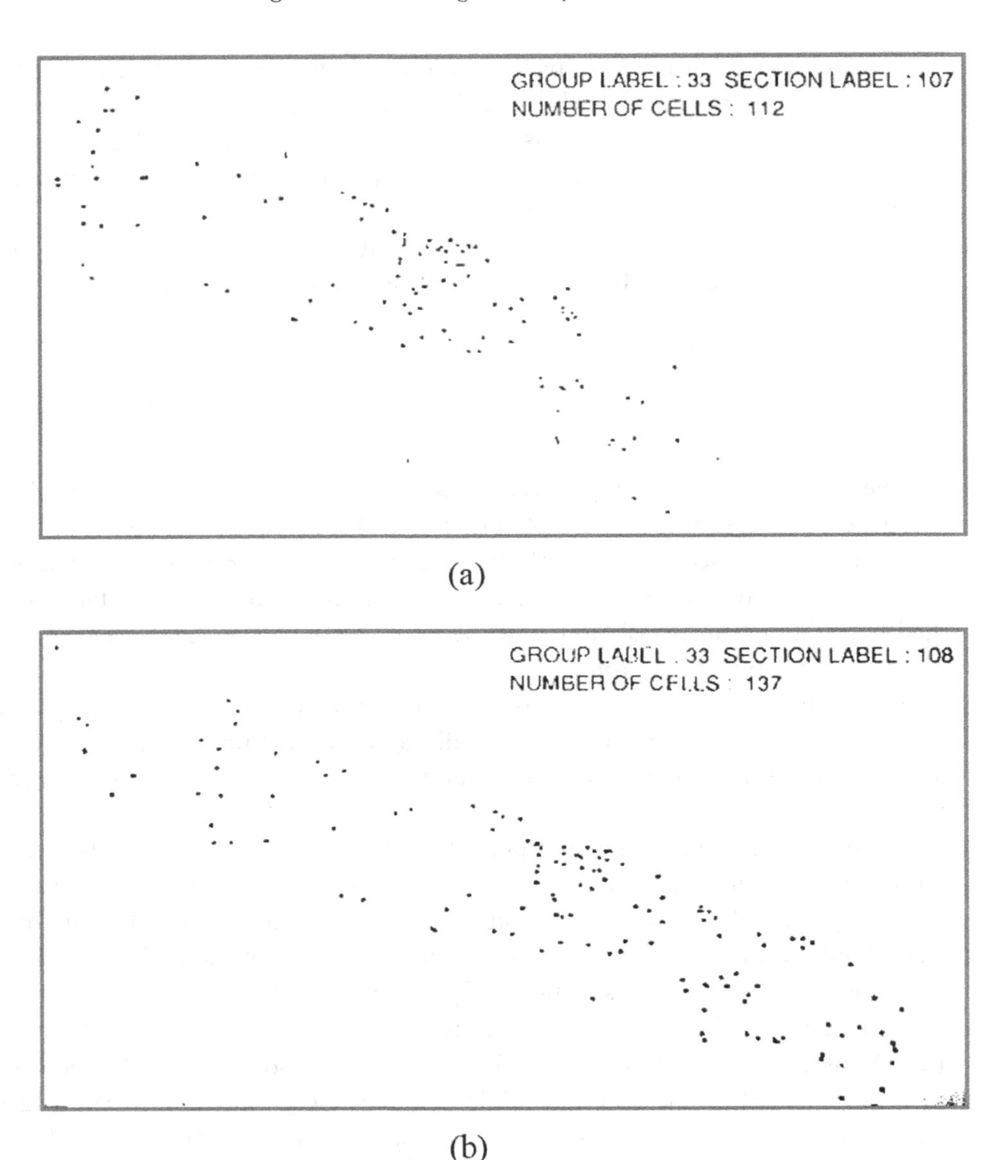

FIGURE 4.12 Distribution of cell profiles on two adjacent medical slices.

Using the dynamic pattern-matching method to match the cell profiles in these two figures resulted in 104 profiles in 4.12(a) finding the correct corresponding profiles (92.86%) in Figure 4.12(b).

4.4 RELATIONSHIP MATCHING

The objective scene can be decomposed into multiple objects, and each object can be decomposed into multiple components/parts, and there are different relationships between them. The images collected from the objective scene can be represented by the collection of various interrelationships between objects, so relationship matching is an important step in image understanding. Similarly, the object in the image can be represented by the set of interrelationships among the various elements of the object, so the object can also be identified by using relational matching. The two representations to be matched in relationship matching are relations; one of them is usually called the object to be matched, and the other is called the model.

4.4.1 Objects and Relational Representations

The main steps of relationship matching are described below with an example. Here we consider a given object to be matched, and find a model that matches it. There are two relation sets: X_l and X_r, where X_l belongs to the object to be matched and X_r belongs to the model. They are respectively represented as

$$X_l = \{R_{l1}, \quad R_{l2}, \quad \cdots, \quad R_{lm}\} \tag{4.19}$$

$$X_r = \{R_{r1}, \quad R_{r2}, \quad \cdots, \quad R_{rm}\} \tag{4.20}$$

where $R_{l1}, R_{l2}, \ldots, R_{lm}$ and $R_{r1}, R_{r2}, \ldots, R_{rn}$ represent the representations of different relationships between the parts of objects to be matched and components in the model, respectively.

Figure 4.13(a) gives a schematic representation of objects in an image (think of a front view of a table). It has three elements, which can be represented as $Q_l = \{A, B, C\}$, and the set of relations between these elements can be expressed as $X_l = \{R_1, R_2, R_3\}$. R_1 represents the connection relationship, $R_1 = \{(A, B)\ (A, C)\}$; R_2 represents the upper-lower relationship, $R_2 = \{(A, B)\ (A, C)\}$; and R_3 represents the left-right relationship, $R_3 = \{(B, C)\}$. Figure 4.13(b) gives a schematic diagram of another object (which can be seen as a front view of a table with a middle drawer), which has four elements, and can be represented as $Q_r = \{1, 2, 3, 4\}$; the set of relations between these elements can be represented as $X_r = (R_1, R_2, R_3)$. R_1 represents the connection relationship, $R_1 = \{(1, 2)\ (1, 3)\ (1, 4)\ (2, 4)\ (3, 4)\}$; R_2 represents the upper and lower relationship, $R_2 = \{(1, 2)\ (1, 3)\ (1, 4)\}$; and R_3 represents the left-right relationship, $R_3 = \{(2, 3)\ (2, 4)\ (4, 3)\}$.

Now consider the distance between X_l and X_r, denoted as dis(X_l, X_r). dis(X_l, X_r) is composed of the difference of the corresponding items represented by each pair of corresponding relations in X_l and X_r, namely each dis(R_l, R_r). The matching of X_l and X_r is the matching of each pair of corresponding relations in the two sets. The following first considers one of the relations, and uses R_l and R_r to represent the corresponding relations, respectively:

$$R_l \subseteq S^M = S(1) \times S(2) \times \cdots \times S(M) \tag{4.21}$$

$$R_r \subseteq T^N = T(1) \times T(2) \times \cdots \times T(N) \tag{4.22}$$

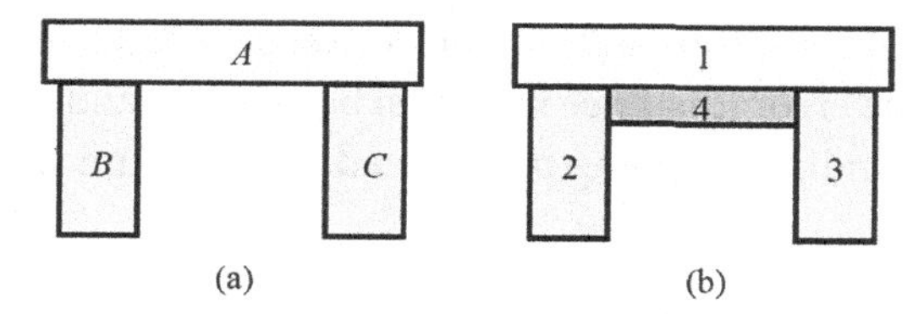

FIGURE 4.13 Schematic representation of objects and their relationships.

Define p as the corresponding transformation (mapping) of S to T, and p^{-1} as the corresponding transformation (inverse mapping) of T to S. It is further defined that the operation symbol $\oplus$ represents a **composite operation**, $R_l \oplus p$ means transforming R_l with transformation p, that is, mapping S^M to T^N, and $R_r \oplus p^{-1}$ means transforming R_r with inverse transformation p^{-1}, that is, mapping T^N to S^M:

$$R_l \oplus p = f\left[T(1), T(2), \cdots, T(N)\right] \in T^N \tag{4.23}$$

$$R_r \oplus p^{-1} = g\left[S(1), S(2), \cdots, S(M)\right] \in S^M \tag{4.24}$$

Here f and g respectively represent the combined function of a certain relational representation.

Now consider dis(R_l, R_r) in detail. If the corresponding items in these two relational representations are not equal, then for any corresponding relation p, there may be the following four errors:

$$\begin{aligned} E_1 &= \left\{R_l \oplus p - (R_l \oplus p) \cap R_r\right\} \\ E_2 &= \left\{R_r - (R_l \oplus p) \cap R_r\right\} \\ E_3 &= \left\{R_r \oplus p^{-1} - \left(R_r \oplus p^{-1}\right) \cap R_l\right\} \\ E_4 &= \left\{R_l - \left(R_r \oplus p^{-1}\right) \cap R_l\right\} \end{aligned} \tag{4.25}$$

The distance between the two relational representations R_l and R_r is the weighted sum of the various errors in Equation (4.25). The weight is W, and here is the weighting of the influence of each error:

$$\text{dis}(R_l, R_r) = \sum_i W_i E_i \tag{4.26}$$

If the corresponding terms in the two relational representations are equal, a corresponding mapping p can always be found. According to the composite operation, there are $R_r = R_l \oplus p$ and $R_r \oplus p^{-1} = R_l$, that is, the distance calculated by Equation (4.26) is zero. At this point it can be said that R_l and R_r have an exact match.

In fact, the error in terms of E can be represented by $C(E)$, and Equation (4.29) can be rewritten as

$$\text{dis}^C(R_l, R_r) = \sum_i W_i C(E_i) \tag{4.27}$$

From the previous analysis, to match R_l and R_r, we should try to find a corresponding mapping to minimize the error (distance in terms) between R_l and R_r. Note that E is a function of p, so the corresponding mapping p that needs to be found should satisfy

$$\text{dis}^C(R_l, R_r) = \inf_p \left\{\sum_i W_i C\left[E_i(p)\right]\right\} \tag{4.28}$$

Going further back to Equations (4.19) and (4.20), to match two relation sets Xl and Xr, a series of corresponding mappings p^j should be found such that

$$\text{dis}^C(X_l, X_r) = \inf_p \left\{ \sum_j^m V_j \sum_i W_{ij} C\left[E_{ij}(p_j) \right] \right\} \tag{4.29}$$

Here $n > m$, and V_j is the weighting of the importance of various relationships.

4.4.2 Connection Relationship Matching

Now consider only the connection relationship, to match the two objects in Figure 4.13. From Equations (4.21) and (4.22), we have

$$R_l = \{(A,B)\ (A,C)\} = S(1) \times S(2) \subseteq S^M \tag{4.30}$$

$$R_r = \{(1,2)\ (1,3)\ (1,4)\ (2,4)\ (3,4)\} = T(1) \times T(2) \times T(3) \times T(4) \times T(5) \subseteq T^N \tag{4.31}$$

When there is no Element 4 in Q_r, R_r = [(1, 2) (1, 3)], which gives p = {(A, 1) (B, 2) (C, 3)}, p^{-1} = {(1, A) (2, B) (3, C)}, $R_l \oplus p$ = {(1, 2) (1, 3)}, $R_r \oplus p^{-1}$ = {(A, B) (A, C)}. At this time, the four errors in Equation (4.25) are, respectively

$$\begin{aligned}
E_1 &= \{R_l \oplus p - (R_l \oplus p) \cap R_r\} = \{(1,2)\ (1,3)\} - \{(1,2)\ (1,3)\} = 0 \\
E_2 &= \{R_r - (R_l \oplus p) \cap R_r\} = \{(1,2)\ (1,3)\} - \{(1,2)\ (1,3)\} = 0 \\
E_3 &= \{R_r \oplus p^{-1} - (R_r \oplus p^{-1}) \cap R_l\} = \{(A,B)\ (A,C)\} - \{(A,B)\ (A,C)\} = 0 \\
E_4 &= \{R_l - (R_r \oplus p^{-1}) \cap R_l\} = \{(A,B)\ (A,C)\} - \{(A,B)\ (A,C)\} = 0
\end{aligned} \tag{4.32}$$

So there is dis(R_l, R_r) = 0.

If Q_r has Element 4 and R_r = [(1, 2) (1, 3) (1, 4) (2, 4) (3, 4)], then p = {(A, 4) (B, 2) (C, 3)}, p^{-1} = {(4, A) (2, B) (3, C)}, $R_l \oplus p$ = {(4, 2) (4, 3)}, $R_r \oplus p^{-1}$ = {(B, A) (C, A)}. At this time, the four errors in Equation (4.25) are respectively

$$\begin{aligned}
E_1 &= \{(4,2)\ (4,3)\} - \{(4,2)\ (4,3)\} = 0 \\
E_2 &= \{(1,2)\ (1,3)\ (1,4)\ (2,4)\ (3,4)\} - \{(2,4)\ (3,4)\} = \{(1,2)\ (1,3)\ (1,4)\} \\
E_3 &= \{(B,A)\ (C,A)\} - \{(A,B)\ (A,C)\} = 0 \\
E_4 &= \{(A,B)\ (A,C)\} - \{(A,B)\ (A,C)\} = 0
\end{aligned} \tag{4.33}$$

If the connection relationship alone is considered, the order of each element can be changed. From the above results, dis(R_l, R_r) = {(1, 2) (1, 3) (1, 4)}. In terms of error terms, $C(E_1) = 0$, $C(E_2) = 3$, $C(E_3) = 0$, $C(E_4) = 0$, so dis^C(R_l, R_r) = 3.

4.4.3 Matching Process

Matching means using the model stored in the computer to identify unknown patterns in the object to be matched, so after finding a series of corresponding mappings p_j, it is necessary to determine their corresponding models. Assume that the object to be identified, X, defined by Equation (4.19), can find a correspondence that conforms to Equation (4.29) for each of the multiple models $Y_1, Y_2, \ldots, Y_L$ (they can be represented by Equation (4.20)), Suppose they are respectively $p_1, p_2, \ldots, p_L$, that is to say, the distance $\text{dis}^C(X, Y_q)$ after matching X and multiple models with their respective corresponding relationships can be obtained. If, for model Y_q, its distance from X satisfies

$$\text{dis}^C(X, Y_q) = \min\{\text{dis}^C(X, Y_i)\} \quad i = 1, 2, \cdots, L \tag{4.34}$$

Then if $q \leq L$, $X \in Y_q$ holds, the object to be matched X is considered to match the model Y_q. It is clear that the matching process can be summarized into four steps.

(1) Determine the same relationship (relationship between components), that is, for a given relationship in X_l to determine the same relationship in X_r. This requires $m \times n$ comparisons:

$$X_l = \begin{bmatrix} R_{l1} \\ R_{l2} \\ \vdots \\ R_{lm} \end{bmatrix} \longrightarrow \begin{bmatrix} R_{r1} \\ R_{r2} \\ \vdots \\ R_{rn} \end{bmatrix} = X_r \tag{4.35}$$

(2) Determine the corresponding mapping of the matching relationship (relationship representation correspondence), that is, determine p that can satisfy Equation (4.28). Suppose p has K possible forms, then it is to find the p that minimizes the weighted sum of errors among the K transformations:

$$R_l \left\{ \begin{array}{c} \xrightarrow{p_1:\ \text{dis}^C(R_l, R_r)} \\ \xrightarrow{p_2:\ \text{dis}^C(R_l, R_r)} \\ \cdots\cdots \\ \xrightarrow{p_K:\ \text{dis}^C(R_l, R_r)} \end{array} \right\} R_r \tag{4.36}$$

(3) Determine the corresponding mapping series of the matching relationship set, that is, weight again according to the K dis function values:

$$\text{dis}^C(X_l, X_r) \Leftarrow \begin{cases} \text{dis}^C(R_{l1}, R_{r1}) \\ \text{dis}^C(R_{l2}, R_{r2}) \\ \cdots\cdots \\ \text{dis}^C(R_{lm}, R_{rm}) \end{cases} \tag{4.37}$$

Note that $m \leq n$ is set in the above equation, that is, only m pairs of relations can find correspondence, and $n - m$ relations only exist in the relation set X_r.

(4) Determine the model to which it belongs (find the minimum value in L $\text{dis}^C(X_l, X_r)$):

$$X \begin{cases} \xrightarrow{p_1} Y_1 \rightarrow \text{dis}^C(X, Y_1) \\ \xrightarrow{p_2} Y_2 \rightarrow \text{dis}^C(X, Y_2) \\ \quad \cdots \quad \cdots \\ \xrightarrow{p_L} Y_L \rightarrow \text{dis}^C(X, Y_L) \end{cases} \tag{4.38}$$

4.5 GRAPH ISOMORPHISM MATCHING

Seeking correspondence is key in relation matching. Because there are many different combinations of correspondences, if the search method is not appropriate, the workload will be too large to be carried out. Graph isomorphism is one way to solve this problem.

4.5.1 Introduction to Graph Theory

Some basic concepts, definitions, and representations of graph theory are given first.

4.5.1.1 Basic Definitions

In graph theory, a **graph** G is defined by a limit and non-empty **vertex set** $V(G)$ and a limit **edge set** $E(G)$, which can be denoted as:

$$G = \left[V(G),\ E(G)\right] = \left[V, E\right] \tag{4.39}$$

where each element of $E(G)$ corresponds to an unordered pair of vertices in $V(G)$, called an edge of G. A graph is also a relational data structure.

Below, the elements in set V are represented by uppercase letters, and the elements in set E are represented by lowercase letters. Generally, the edge e formed by the disordered pair of vertices A and B is recorded as $e \leftrightarrow AB$ or $e \leftrightarrow BA$, and A and B are called the endpoints of e; the edge e is described as **connecting** A and B. In this case, vertices A and B are associated or incident with edge e, which is associated with vertices A and B. Two vertices associated with the same edge are **adjacent**, as are two edges that share a common vertex. Two edges are called **multiple edges** or **parallel edges** if they have the same two endpoints. If the two endpoints of an edge are the same, it is called a **loop**, otherwise it is called a **link**.

In the definition of a graph, the two elements (i.e., two vertices) of each unordered pair can be the same or different, and any two unordered pairs (i.e., two edges) can be the same or different. Different elements can be represented by vertices of different colors, which is called the chromaticity of vertices (meaning that vertices are labeled with different colors). Different relationships between elements can be represented by edges of different colors, which is called edge chromaticity (meaning that edges are marked with different colors). So a generalized **colored graph** G can be represented as:

$$G = [(V,C), (E,S)] \tag{4.40}$$

where V is the vertex set, C is the vertex chromaticity set, E is the edge set, and S is the edge chromaticity set. They are

$$V = \{V_1, V_2, \cdots, V_N\} \tag{4.41}$$

$$C = \{C_{V_1}, C_{V_2}, \cdots, C_{V_N}\} \tag{4.42}$$

$$E = \{e_{V_iV_j} | V_i, V_j \in V\} \tag{4.43}$$

$$S = \{s_{V_iV_j} | V_i, V_{,j} \in V\} \tag{4.44}$$

where each vertex can have a specific color, and each edge can also have a specific color.

4.5.1.2 Geometric Representation of Graphs

The vertices of the graph are represented by dots, and the edges are represented by straight lines or curves connecting the vertices, and a **geometric representation** or **geometric realization** of the graph can be obtained. Graphs with edges greater than or equal to 1 can have an infinite number of geometric representations.

An example of the geometric representation of a graph is shown in Figure 4.14. Consider a graph G, where $V(G) = \{A, B, C\}$, $E(G) = \{a, b, c, d\}$, $a \leftrightarrow AB$, $b \leftrightarrow AB$, $c \leftrightarrow BC$, $d \leftrightarrow CC$. In Figure 4.14, edges a, b, and c are adjacent to each other, and edges c and d are adjacent to each other, but edges a and b are not adjacent to edge d. Likewise, vertices A and B are adjacent, and vertices B and C are adjacent, but vertices A and C are not adjacent. In terms of edge types, edges a and b are multiple edges, edge d is a loop, and edges a, b, and c are links.

An example of the geometric representation of two colored graphs (representing two objects in Figure 4.13) is shown in Figure 4.15. Figure 4.15 gives a geometric representation of two colored graphs, where vertex chromaticity is distinguished by vertex shapes, and link chromaticity is distinguished by line types. The information reflected by the colored map is more comprehensive and intuitive.

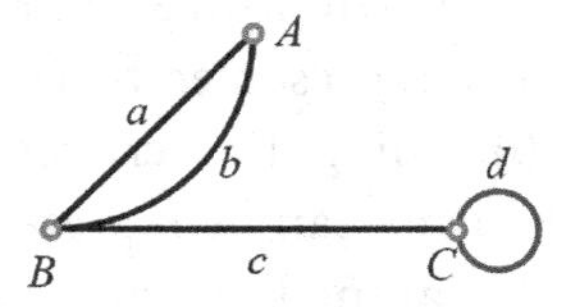

FIGURE 4.14 Geometric representation of a graph.

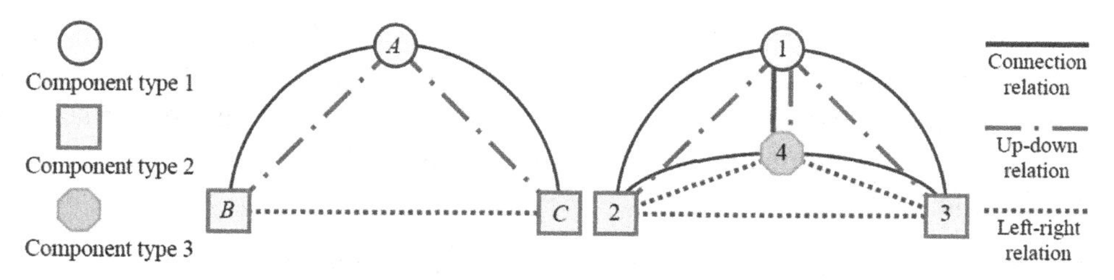

FIGURE 4.15 Geometric representation of color graphs.

4.5.1.3 Subgraphs

For two graphs G and H, if $V(H) \subseteq V(G)$, $E(H) \subseteq E(G)$, then graph H is called a **subgraph** of graph G, denoted as $H \subseteq G$. Conversely, graph G is called the **supergraph** of graph H. If graph H is a subgraph of graph G, but $H \neq G$, then graph H is called the **proper subgraph** of graph G, and graph G is called the **proper supergraph** of graph H (Sun 2004).

If $H \subseteq G$ and $V(H) = V(G)$, then graph H is called the **spanning subgraph** of graph G, and graph G is called the **spanning supergraph** of graph H. For example, in Figure 4.16, Figure 4.16(a) gives a graph G, while Figures 4.16(b), 4.16(c), and 4.16(d) each give a spanning subgraph of graph G (they are all spanning subgraphs of G but distinct from each other). If all multiple edges and loops are removed from the graph G, the resulting simple spanning subgraph is called the **underlying simple graph** of the graph G. Among the three spanning subgraphs given in Figures 4.16(b), 4.16(c), and 4.16(d), only Figure 4.16(d) is the underlying simple graph.

The four operations to obtain the underlying simple graph are described below with the help of the graph G given in Figure 4.17(a).

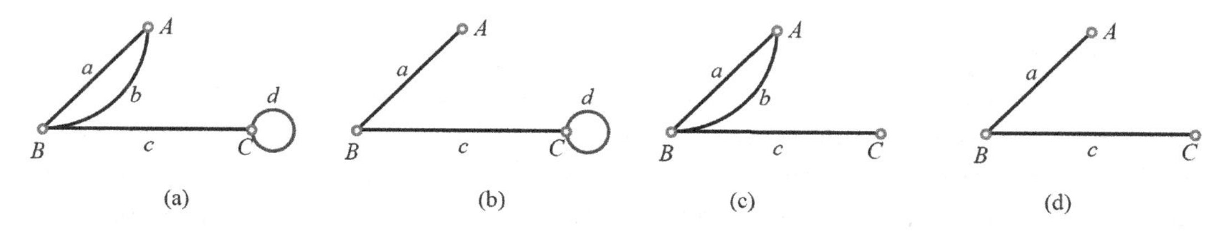

FIGURE 4.16 Graph and spanning subgraph examples.

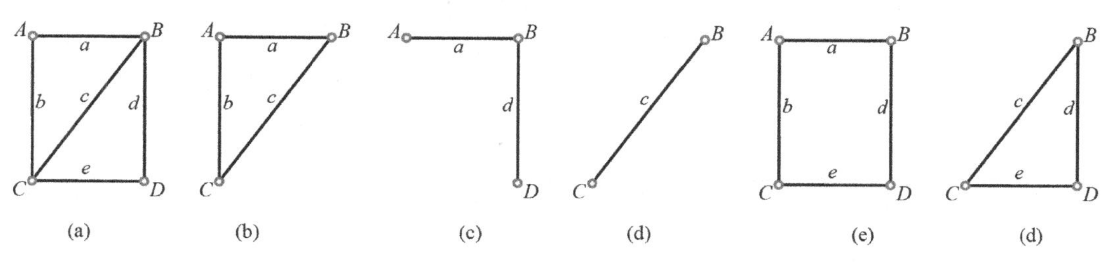

FIGURE 4.17 Several operations to obtain subgraphs.

(1) For the non-empty vertex subset $V'(G) \subseteq V(G)$ of graph G, if there is a subgraph of graph G with $V'(G)$ as the vertex set, and with all edges (in $V'(G)$) of the two endpoints in graph G as the edge sets, then the subgraph is called the **induced subgraph** of graph G, and is denoted as $G[V'(G)]$ or $G[V']$. Figure 4.17(b) gives the graph of $G[A, B, C] = G[a, b, c]$.

(2) Similarly, for the non-empty edge subset $E'(G) \subseteq E(G)$ of graph G, if there is a subgraph of graph G with $E'(G)$ as the edge set, and with all endpoint of the edge as the vertex set, then the subgraph is called the **edge-induced subgraph** of the graph G, denoted as $G[E'(G)]$ or $G[E']$. Figure 4.17(c) gives the graph of $G[a, d] = G[A, B, D]$.

(3) For the non-empty vertex proper subset $V'(G) \subseteq V(G)$ of graph G, if there is a subgraph of graph G with the vertices after removing $V'(G) \subset V(G)$ as the vertex set, and with the edges after removing all edges associated with $V'(G)$ in the graph G as the edge set, then the subgraph is the remaining subgraph of the graph G, denoted as $G-V'$. Here $G-V' = G[V \setminus V']$. Figure 4.17(d) gives the graph $G-\{A, D\} = G[B, C] = G[\{A, B, C, D\} - \{A, D\}]$.

(4) For the proper subset $E'(G) \subseteq E(G)$ of non-empty edges of graph G, if there is a subgraph of graph G with the edges after removing $E'(G) \subset E(G)$ as the edge set, then the subgraph is the spanning subgraph of the graph G, denoted as $G-E'$. Note here that $G-E'$ and $G[E \setminus E']$ have the same set of edges, but they are not necessarily identical. Among them, the former are always induced subgraphs, while the latter may or may not be. Figure 4.17(e) gives an example of the former, $G-\{c\} = G[a, b, d, e]$. An example of the latter is given in Figure 4.17(f), $G[\{a, b, c, d, e\} -\{a, b\}] = G-A \neq G-[\{a, b\}]$.

4.5.2 Graph Isomorphism and Matching

The matching of graphs is achieved by means of graph isomorphism.

4.5.2.1 Identity and Isomorphism of Graphs

According to the definition of graph, for two graphs G and H, if and only if $V(G) = V(H)$, $E(G) = E(H)$, the graphs G and H are said to be **identical**, and the two graphs can be represented by the same geometric representation. For example, graphs G and H in Figure 4.18 are identical. However, if two graphs can be represented by the same geometric representation, they are not necessarily identical. For example, the graphs G and I in Figure 4.18 are not identical (the vertices and edges are labeled differently), although they can be represented by two geometric representations of the same shape.

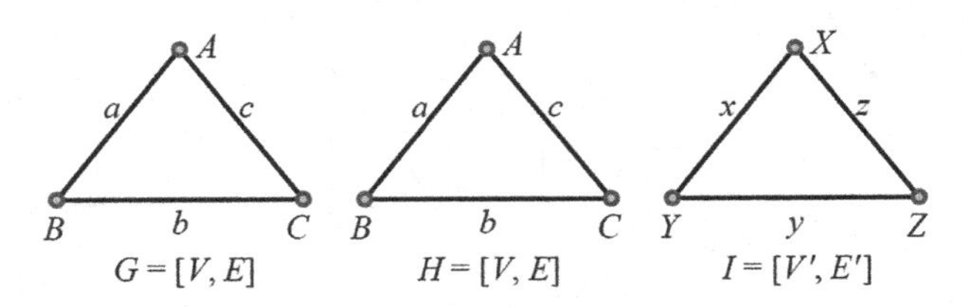

FIGURE 4.18 Identity of graphs.

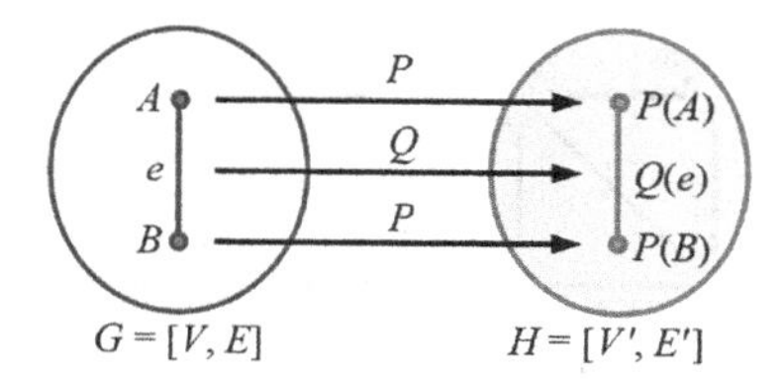

FIGURE 4.19 Isomorphism of graphs.

For two graphs that have the same geometric representation but are not identical, as long as the labels of the vertices and edges of one graph are appropriately renamed, a graph that is identical to the other graph can be obtained, which can be called **isomorphism** for such two graphs. In other words, a two-graph isomorphism indicates that there is a one-to-one correspondence between the vertices and edges of the two graphs. The isomorphism of two graphs G and H can be written as $G \cong H$, and the necessary and sufficient conditions are that the following mappings exist between $V(G)$ and $V(H)$, as well as $E(G)$ and $E(H)$:

$$P:\ V(G) \to V(H) \tag{4.45}$$

$$Q:\ E(G) \to E(H) \tag{4.46}$$

and the mappings P and Q maintain an incident relationship, that is, $Q(e) = P(A)P(B)$, $\forall e \leftrightarrow AB \in E(G)$, as shown in Figure 4.19.

4.5.2.2 Judgment for Isomorphism

It can be seen from the previous definition that isomorphic graphs have the same structure, and the only difference is that the labels of vertices or edges are not exactly the same. Graph isomorphism is more focused on describing mutual relationships, so graph isomorphism can have no geometric requirements, that is, it is more abstract (of course, it can also have geometric, that is, more specific, requirements). **Graph isomorphism matching** is essentially a tree search problem, where different branches represent heuristics on different combinations of correspondences.

Now consider several cases of graph-to-graph isomorphism. For simplicity, all graph vertices and edges are not labeled here, that is, all vertices are considered to have the same color, and all edges also have the same color. For clarity, a monochromatic line graph (which is a special case of G) is used for explanation:

$$B = \left[(V),\ (E)\right] = \left[V, E\right] \tag{4.47}$$

V and E in Equation (4.47) are still given by Equations (4.38) and (4.39), respectively, but here all elements in each set are the same. In other words, there is only one type of vertex and one type of edge each. Referring to Figure 4.20, given two graphs $B_1 = [V_1, E_1]$ and $B_2 = [V_2, E_2]$, the isomorphism between them can be classified into the following types (Ballard and Brown 1982):

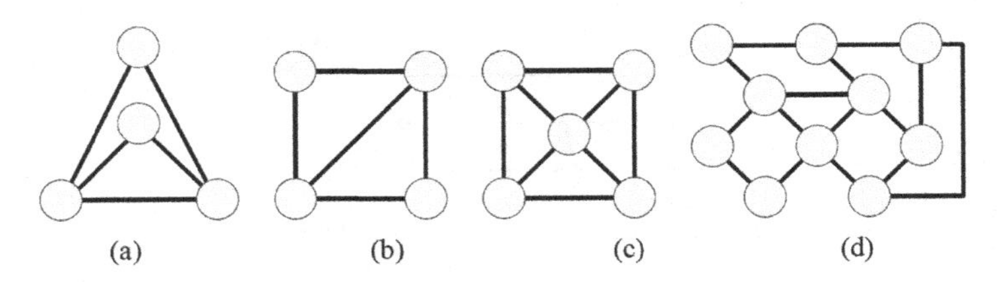

FIGURE 4.20 Several cases of graph isomorphism.

(1) Graph isomorphism

Graph isomorphism refers to the one-to-one mapping between B_1 and B_2. For example, Figure 4.20(a) and Figure 4.20(b) satisfy graph isomorphism. Generally speaking, if the mapping is represented by f, then for $e_1 \in E_1$ and $e_2 \in E_2$, there must be $f(e_1) = e_2$, and for each connection in E_1 to any pair of vertices e_1 and e'_1 ($e_1, e'_1 \in E_1$), there must be a connection between $f(e_1)$ and $f(e'_1)$ in E_2. When identifying the object, it is necessary to establish a graph isomorphism between the graph representing the object and the graph of the object model.

(2) Subgraph isomorphism

Subgraph isomorphism refers to the isomorphism between a part (subgraph) of B_1 and the whole graph of B_2. For example, four subgraphs (as indicated by respective triangles in Figure 4.21) can be obtained from Figure 4.20(c), which are isomorphic to Figure 4.20(a). When detecting objects in a scene, the object model needs to be used to search for isomorphic subgraphs in the scene graph.

(3) Double subgraph isomorphism

Double subgraph isomorphism refers to all isomorphism between the subgraphs of B_1 and the subgraphs of B_2. For example, the subgraph of Figure 4.20(a) indicated by the triangles in Figure 4.22(a) and the subgraph of Figure 4.20(d) indicated by the

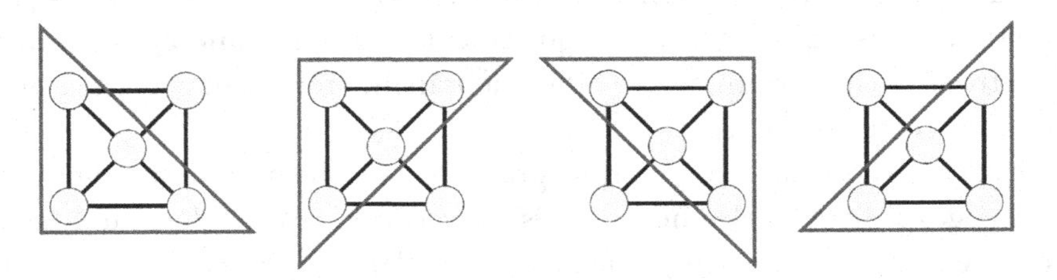

FIGURE 4.21 Four subgraphs obtained from Figure 4.20(c).

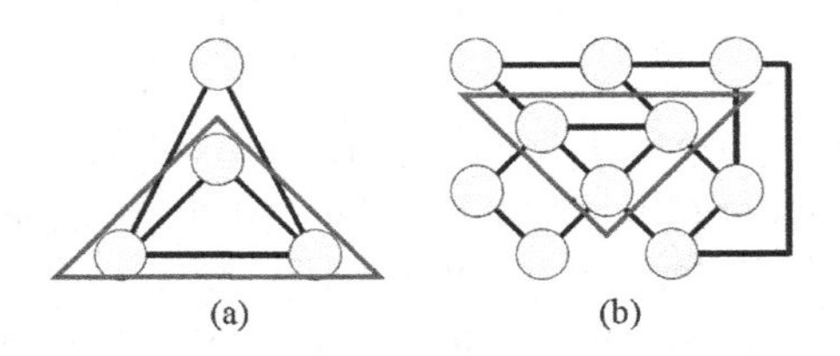

FIGURE 4.22 Two subgraphs obtained from Figure 4.20(a) and (b).

triangles in Figure 4.22(b) satisfy the double subgraph isomorphism. In fact, there are several double subgraphs of Figure 4.20(a) and Figure 4.20(d) that are isomorphic. When it is necessary to find a common object in two scenarios, the task can be transformed into the problem of double subgraph isomorphism.

There are many algorithms for finding graph isomorphism. For example, each graph to be determined can be converted into a certain standard form, so that isomorphism can be easily determined. In addition, it is also possible to perform an exhaustive search on the tree of possible matches between corresponding vertices in the graph, but this method requires a large amount of computation when the number of vertices in the graph is large.

A method that is less restrictive and converges faster than isomorphic methods is **association graph matching** (Snyder and Qi 2004). In association graph matching, the graph is defined as $G = [V, P, R]$, where V represents the set of nodes, P represents the set of unit predicates used for the nodes, and R represents the set of binary relations between nodes. Here the predicate represents a statement that takes only one of the two values TRUE or FALSE, and the binary relation describes the properties that a pair of nodes has. Given two graphs, an association graph can be constructed. Association graph matching is the matching between nodes and nodes in two graphs, as well as binary relationships and binary relationships.

4.6 SOME RECENT DEVELOPMENTS AND FURTHER RESEARCH

In the following sections, some technical developments and promising research directions from the last few years are briefly reviewed.

4.6.1 Image Registration and Matching

Image registration and image matching are closely related. Many image registration tasks can be accomplished with the help of various matching techniques.

4.6.1.1 Heterogeneous Remote Sensing Image Registration Based on Feature Matching

Difficulties in the registration of heterogeneous images are often caused by different imaging modality, resolutions cross-modality, time phases, etc. To treat these problems, a deep learning feature matching method (cross-modality matching network) has been proposed (Lan et al. 2021).

The flow chart of this matching method is shown n Figure 4.23. It consists of two stages. In the feature extraction stage, the **convolutional neural network** (CNN) is first used to

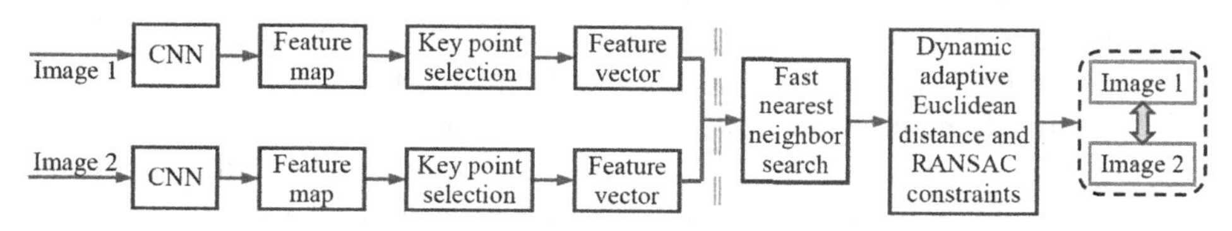

FIGURE 4.23 CNN feature matching method flow chart.

extract the high-dimensional feature map of a pair of heterogeneous remote sensing images. Then, the key points on the feature map are selected according to the two conditions of channel maximum and local maximum. At the end, the 512-dimensional descriptor of the corresponding position is extracted. In the matching stage, the fast nearest neighbor search is first used for feature matching. Next, a purification algorithm based on dynamic adaptive Euclidean distance calculation and a **random sample consensus** (RANSAC) is used to eliminate the mismatched points and to maintain the correct matching points between the pair of images.

To achieve robust feature matching of heterogeneous remote sensing images, an invariant feature representation method to reduce the influence of radiation and geometric differences in heterogeneous images needs to be found. Three aspects are considered.

(1) Because the high-level abstract semantic information is more adaptable to the changes in radiation and geometry than the low-level gradient information, the feature maps from the deeper layers of the CNN are selected and the original input image range (receptive field) corresponding to the extracted features is appropriately expanded.

(2) The CNN network is trained using the data that has been paired with large differences in illumination and shooting angles, so that the CNN feature extractor can learn the invariant features of changing images such as illumination and geometry.

(3) Adopt the "reliability with more" strategy, so that a large number of candidate features are extracted first, and then effectively restricted by improving the screening mechanism of the matching process, so as to obtain more reliable and more uniform matching pairs.

On the basis of feature vectors for matching, a fast nearest neighbor search is performed to find those pairs of matching points that have a large ratio (greater than a threshold value) of the closest Euclidean distance over the second-closest Euclidean distances. Generally, the smaller the distance of the first matching point is than that of the second matching point, the better the matching quality. Further, the average value of all matching points can be calculated and subtracted from each of the closest Euclidean distances. If the result is negative, this pair of matching points is kept and RANSAC is performed to finally select the real matching point pairs.

4.6.1.2 Image Matching Based on Spatial Relation Reasoning

Scene images often contain many objects. The representation of an image space relationship describes the geometric relationship of image objects in Euclidean space. Due to the complexity of the real world and the randomness of scene shooting, the imaging of the same object on different images will change significantly. It is difficult to accurately match images by simply relying on the overall representation of the image to calculate the image similarity. On the other hand, when the same object is imaged in different images, its imaging morphology will change significantly, but its spatial relationship with adjacent objects generally remains stable. To take advantage of this fact, a method that solves the

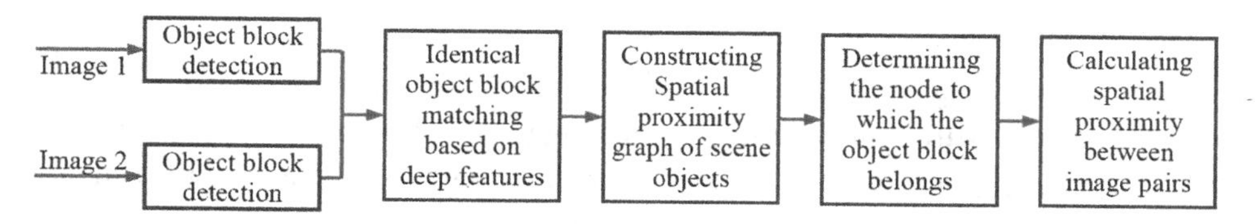

FIGURE 4.24 Spatial relation reasoning flow chart.

whole image-matching problem by analyzing the spatial proximity of objects in the image by reasoning is proposed (Li et al. 2021).

The flow chart of the image-matching algorithm is shown in Figure 4.24. First, for the image pairs from the scene, object blocks (patches) are detected and depth features are extracted for matching, so as to determine the spatial proximity of objects and construct a spatial proximity graph of objects in the scene. Then, based on the constructed spatial proximity graph, the spatial proximity relationship of objects in the image is analyzed, and the proximity degree of image pairs is quantitatively calculated. Finally, the matching images are found.

Detailed information is as follows:

(1) To extract the deep features of object block, an object block feature extraction network based on the contrast mechanism is constructed. The network contains two completely identical channels with shared weights. Each channel is a deep convolutional network containing seven convolutional layers and two fully connected layers. Based on the deep features, the same object blocks in the two images are matched.

(2) To construct the spatial proximity graph of scene objects, the spatial proximity relationship of different objects in the scene is reasoned and analyzed according to the distribution of each object on the previous images. The construction process is an iterative search process with initialization and update steps (Li et al. 2021). The constructed spatial proximity graph summarizes all objects present in the scene and quantitatively represents the proximity between different objects, in which the identical object blocks on different images are aggregated in the same node.

(3) To determine the matching images, search the node of the object in the image in the spatial proximity graph, and determine the proximity relationship between the objects in the image according to the connection weights between the nodes. Each test image may include several object blocks that can be searched for their belonging node in the node set.

(4) To calculate image pair spatial proximity, the object blocks contained in the image are detected, and the nodes to which each object block belongs is determined to form a set of nodes. The connection weights between the belonging nodes represent the proximity relationship between object blocks in the image. The spatial relationship between two images can be represented by the proximity relationship of object blocks in the images, and the spatial relationship matching of images is completed by quantitatively calculating the spatial proximity of the images.

4.6.2 Multimodal Image Matching

General image matching aims to identify corresponding or similar structure/content from two or more images. **Multimodal image matching** (MMIM) can be seen as a specific case. Generally, the images and/or objects to be matched have significant nonlinear appearance differences that are not only caused by different imaging sensors, but also by different imaging conditions (such as day–night, cross-weather, or cross-season), and input data types (such as image–paint–sketch, and image–text).

The problem of multimodal image matching can be formulated as follows. Given a reference image I_R and a matching image I_M, of different modalities, to find their correspondence (or the objects in them) according to similarity between them. The object can be represented by the region they occupied or by the features they have. So, matching techniques can be classified into region based and feature based.

4.6.2.1 Region-Based Techniques

Region-based techniques consider the intensity information of the object. Two groups can be distinguished: traditional group with handcraft framework, and recent group with learning framework.

The flow chart of traditional region-based techniques is shown in Figure 4.25. There are three important modules: (i) Measure metric, (ii) transformation model, (iii) optimization method (Jiang et al. 2021).

(1) Measure metric

The accuracy results of matching are dependent on the metric (matching criteria). Different metrics can be devised depending on the assumptions about the intensity relationship between two images. Frequently used manual metrics can be briefly classified into correlation-like and information theory-based methods.

(2) Transformation model

Transformation models generally explain the geometrical relations between the image pairs whose parameters need to be accurately estimated to guide the image manipulation for matching. Existing transformation models can be briefly classified into linear models and nonlinear models. The latter can be further classified into physical models (derived from physical phenomena and represented by partial differential equations) and interpolation models (derived from interpolation or approximation theory).

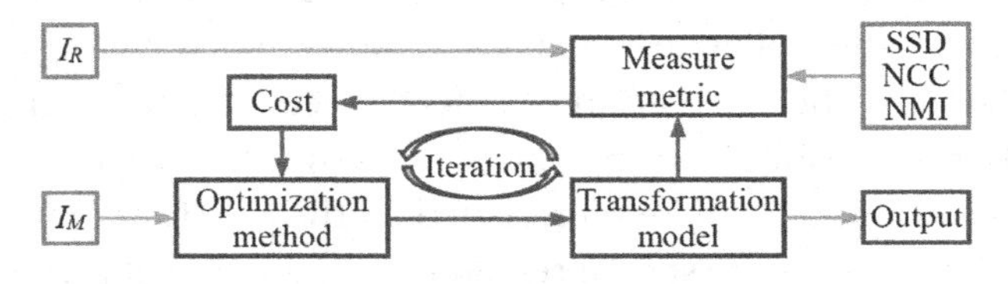

FIGURE 4.25 Traditional region-based techniques flow chart.

TABLE 4.2 Method Groups in Each Module and Some Typical Techniques

Modules	Method Groups	Some Typical Techniques
Measure metric	Correlation-like	Cross correlation (Avants et al. 2008) Normalized correlation coefficient (NCC), (Luo and Konofagou 2010)
	Information theory-based	Mutual information (MI) (Viola, and Wells III 1997) Normalized version of MI (NMI) (Studholme et al. 1999) Conditional MI (cMI), (Loeckx et al. 2009)
Transformation model	Linear models	Rigid, affine, and projective transformation (Zhang 2017a)
	Nonlinear physical models	Diffeomorphism-based (Trouvé 1998) Large deformation diffeomorphic metric mapping (Marsland and Twining 2004)
	Nonlinear interpolation models	Radial basis function (RBF) (Zagorchev and Goshtasby 2006) Thin-plate splines (TPS) (Bookstein 1989) Free-form deformations (FFDs) (Sederberg and Parry 1986)
Optimization method	Continuous methods	Gradient descent (Zhang 2021) Conjugate gradient (Zhang 2021)
	Discrete methods	Graph-based (Ford Jr and Fulkerson 2015) Message passing (Pearl 2014) Linear programming (Komodakis and Tziritas 2007)

(3) Optimization method

Optimization methods are needed to search the optimal transformation based on the given measure metric to achieve the required matching accuracy and efficiency. Considering the nature of variables that optimization methods try to infer, they can be briefly classified into continuous methods and discrete methods. Continuous optimization assumes the variables as real values that require the objective function to be differentiable. A discrete optimization assumes the solution space as a discrete set.

The method groups in each module and some typical methods are listed in Table 4.2.

Recently, deep learning techniques have been used to drive an iterative optimization procedure or directly estimate the geometrical transformation parameters end to end. The methods for the premier class are called deep iterative learning methods and those for the second class are called deep transformation estimation methods. The latter can be broadly classified into supervised methods and unsupervised methods, according to the training strategies.

Typical methods in these three classes are listed in Table 4.3.

4.6.2.2 Feature-Based Techniques

Feature-based techniques generally have three steps: (i) feature detection, (ii) feature description, and (iii) feature matching. The modality difference has been suppressed in the feature detection and description steps, so the matching step can be performed well by general methods. According to the use or non-use of local image descriptors, the matching step can be conducted in an indirect or direct manner. The flow chart of feature-based techniques is shown in Figure 4.26 (Jiang et al. 2021).

TABLE 4.3 Typical Techniques in Three Classes of Deep Learning Region-based Methods

Method classes	Some typical techniques
Deep iterative learning methods	Training superior measure metrics with the stacked autoencoder (Cheng et al. 2018)
	Combining deep similarity metrics and handcrafted ones as an enhanced measurement (Blendowski and Heinrich 2019)
	Using the reinforcement learning (RL) paradigm to iteratively estimate the transformation parameters (Liao et al. 2016)
Supervised transformation estimation methods	Applying statistical appearance models to ensure that the generated data can better simulate the real data that can be used as ground-truth data to define the loss functions in supervised estimation (Uzunova et al. 2017)
	Employing the fully convolutional (FC) layers of U-Net to represent the high-dimensional parametric space and output deformable fields or displacement vector fields to define the loss functions (Hering et al. 2019)
	Using generative adversarial networks (GANs) to learn to estimate the transformations to force the predicted transformations to be realistic or close to the ground truth (Yan et al. 2018)
Unsupervised transformation estimation methods	Taking the ability of spatial transformer network (STN) to predict the geometrical transformations in an end-to-end manner, by using only traditional similarity measurements, together with a regularization term that constrains the complexity or smoothness of the transformation model, to construct loss functions (Sun and Zhang 2018)
	Performing first image binarization then calculating the Dice score between reference images and matching images to cope with multimodal image pairs (Kori and Krishnamurthi 2019)

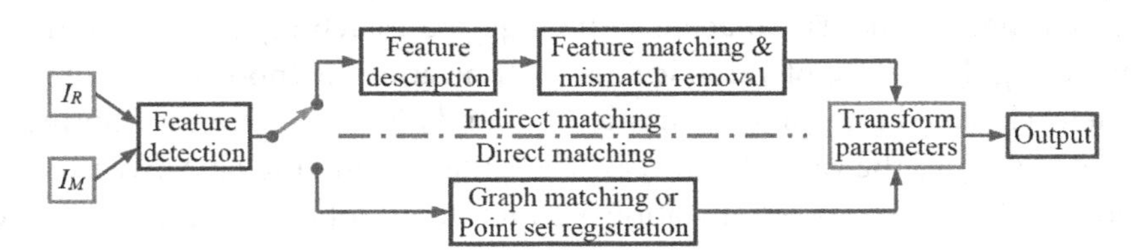

FIGURE 4.26 Feature-based techniques flow chart.

(1) Feature detection

The detected features usually represent specific semantic structures in an image or the real world. Commonly used features can be classified into corner (the crossing point of two straight lines typically located in the texture area or edges), blob (a local closed region, inside which the pixels are considered similar and thus distinct from surrounding neighborhoods), line/edge, and morphological region features. The core idea for feature detection is to construct a response function to distinguish features that differ from one another, along with flat and non-distinctive image regions. Commonly used functions can be subsequently classified into gradient-based, intensity-based, second-order derivative- based, contour curvature-based, region segmentation-based, and learning-based detectors.

Deep learning has shown great potential in key-point detection particularly from two images with a significant difference in appearance, which often occurs in cross-modal image matching. The three groups of CNN-based detectors are:

(i) supervised (Zhang et al. 2017), (ii) self-supervised (DeTone et al. 2018), and (iii) unsupervised (Laguna et al. 2019).

(2) Feature description

Feature description refers to mapping the local intensity around a feature point into a stable and discriminative vector form, enabling the fast and easy matching of the detected features. According to the criteria used (Ma et al. 2020), such as image cues (e.g., gradient, intensity, etc.) and the form of descriptor generation (e.g., comparison, statistic, learning, etc.), the existing descriptors can be classified into float descriptors, binary descriptors, and learnable descriptors. Float descriptors are often generated by statistical methods based on gradient or intensity cues. **SIFT** and **SURF** (Chapter 3) are both typical float descriptors. Binary descriptors are typically based on the comparison strategy of local intensities. **BRIEF** and **ORB** (Rublee et al. 2011) are both typical binary descriptors. Learnable descriptors take the data-driven strategy that can largely enhance discrimination because high-order image cues or semantic information between two images can be extracted in CNNs. Many deep descriptors have been proposed, which are different in terms of training strategy, model structure, and loss design.

(3) Feature matching

Feature matching aims to establish correct feature correspondences from two extracted feature sets.

A direct method is to make these two sets correspond by directly using the spatial geometrical relations and an optimization method. Two representative strategies are: graph matching and point set registration. An indirect method is to treat the feature matching as a two-stage problem. In the first stage, a putative match set is constructed based on the similarity of local feature descriptors. In the second stage, the false matches are rejected by imposing additional local and/or global geometrical constraints.

The **random sample consensus** (RANSAC) is a classic resampling-based method for mismatch removal and parameter estimation. Inspired by the classical RANSAC, a learning technique to eliminate the outliers and/or estimate model parameters through training a deep regressor (Kluger et al. 2020) is proposed to estimate the transformation model. As well as learning with **multilayer perceptron** (MLP), **graph convolutional networks** (GCNs) can also be used (Sarlin et al. 2020).

REFERENCES

Avants, B.B., C.L. Epstein, M. Grossman, et al. 2008. Symmetric diffeomorphic image registration with cross-correlation: evaluating automated labeling of elderly and neurodegenerative brain. *Medical Image Analysis*, 12(1): 26–41.

Ballard, D.H. and C.M. Brown. 1982. *Computer Vision*. London: Prentice-Hall.

Blendowski, M. and M.P. Heinrich. 2019. Combining MRF-based deformable registration and deep binary 3D-CNN descriptors for large lung motion estimation in COPD patients. *International Journal of Computer Assisted Radiology and Surgery*, 14(1): 43–52.

Bookstein, F.L. 1989. Principal warps: Thin-plate splines and the decomposition of deformations. *IEEE Transactions on Pattern Analysis and Machine Intelligence*, 11(6): 567–585.

Cheng, X., L. Zhang and Y. Zheng. 2018. Deep similarity learning for multimodal medical images. *Computer Methods in Biomechanics and Biomedical Engineering: Imaging and Visualization*, 6(3): 248–252.

DeTone, D., T. Malisiewicz and A. Rabinovich, 2018. Superpoint: Self-supervised interest point detection and description. *Proceedings of the IEEE Conference on Computer Vision and Pattern Recognition Workshops*, 224–236.

Dubuisson, M. and A.K. Jain. 1994. A modified Hausdorff distance for object matching. *Proceedings of the 12ICPR*, 566–568.

Ford Jr, L.R. and D.R. Fulkerson. 2015. *Flows in Networks*. Princeton, NJ: Princeton University Press.

Goshtasby, A.A. 2005. *2-D and 3-D Image Registration – for Medical, Remote Sensing, and Industrial Applications*. Hoboken: Wiley-Interscience.

Hering, A., S. Kuckertz, S. Heldmann, et al. 2019. Enhancing label-driven deep deformable image registration with local distance metrics for state-of-the-art cardiac motion tracking. *Bildverarbeitung Für Die Medizin*, Springer Science and Business Media, Germany, 309–314.

Jiang, X.Y., J.Y. Ma, G.B. Xiao, et al. 2021. A review of multimodal image matching: Methods and applications. *Information Fusion*, 73: 22–71.

Kluger, F., E. Brachmann, H. Ackermann, et al. 2020. Consac: Robust multi-model fitting by conditional sample consensus. *Proceedings of the IEEE Conference on Computer Vision and Pattern Recognition*, 4634–4643.

Komodakis, N. and G. Tziritas. 2007. Approximate labeling via graph cuts based on linear programming. *IEEE Transactions on Pattern Analysis and Machine Intelligence*, 29(8): 1436–1453.

Kori, A. and G. Krishnamurthi. 2019. Zero shot learning for multi-modal real time image registration, *arXiv preprint* arXiv:1908.06213.

Kropatsch, W. G. and H. Bischof (eds.). 2001. *Digital Image Analysis – Selected Techniques and Applications*. Heidelberg: Springer.

Laguna, A.B., E. Riba, D. Ponsa, et al. 2019. Key. Net: Keypoint detection by handcrafted and learned CNN filters. *arXiv preprint* arXiv:1904.00889.

Lan, C.Z., W.J. Lu, J. Yu, et al. 2021. Deep learning algorithm for feature matching of cross modality remote sensing images. *Acta Geodactica Cartographica Sinica*, 50(2): 189–202.

Li Q., X. You, K. Li, et al. 2021. Spatial relation reasoning and representation for image matching. *Acta Geodactica Cartographica Sinica*, 50(1): 117–131.

Liao, R., S. Miao, P. de Tournemire, et al. 2016. An artificial agent for robust image registration. *arXiv preprint* arXiv: 1611.10336.

Lohmann, G. 1998. *Volumetric Image Analysis*. Hoboken: John Wiley & Sons and Teubner Publishers.

Loeckx, D., P. Slagmolen, F. Maes, et al. 2009. Nonrigid image registration using conditional mutual information. *IEEE Transaction on Medical Imaging*, 29(1): 19–29.

Luo, J. and E.E. Konofagou. 2010. A fast normalized cross-correlation calculation method for motion estimation, *IEEE Transaction on Ultrasonic, Ferroelectrics and Frequency Control*, 57(6): 1347–1357.

Ma, J., X. Jiang, A. Fan, et al. 2020. Image matching from handcrafted to deep features: A survey. *International Journal of Computer Vision*, 1–57.

Marsland, S. and C.J. Twining. 2004. Constructing diffeomorphic representations for the groupwise analysis of nonrigid registrations of medical images. *IEEE Transaction on Medical Imaging*, 23(8): 1006–1020.

Pearl, J. 2014. *Probabilistic Reasoning in Intelligent Systems: Networks of Plausible Inference*. The Netherlands, Elsevier.

Rublee E., V. Rbaud, K. Konolige, et al. 2011. ORB: An efficient alternative to SIFT or SURF. *Proceedings of the ICCV*, 2564–2571.

Sarlin, P.-E. D. DeTone, T. Malisiewicz, et al. 2020. Superglue: Learning feature matching with graph neural networks. *Proceedings of the IEEE Conference on Computer Vision and Pattern Recognition*, 4938–4947.

Sederberg, T.W. and S.R. Parry. 1986. Free-form deformation of solid geometric models. *Proceedings of the 13th Annual Conference on Computer Graphics and Interactive Techniques*, 151–160.

Snyder, W.E. and H.R. Qi. 2004. *Machine Vision*. UK: Cambridge University Press.

Studholme, C., D.L. G. Hill and D.J. Hawkes. 1999. An overlap invariant entropy measure of 3D medical image alignment. *Pattern Recognition*, 32(1): 71–86.

Sun, H.Q. 2004. *Graph Theory and Applications*. Beijing: Science Publisher.

Sun, L. and S. Zhang. 2018. Deformable MRI-ultrasound registration using 3D convolutional neural network. *Simulation, Image Processing, and Ultrasound Systems for Assisted Diagnosis and Navigation*, Springer Verlag, 152–158.

Trouvé, A. 1998. Diffeomorphisms groups and pattern matching in image analysis. *International Journal of Computer Vision*, 28(3): 213–221.

Uzunova, H., M. Wilms, H. Handels, et al. 2017. Training CNNs for image registration from few samples with model-based data augmentation. *Proceedings of International Conference on Medical Image Computing and Computer-Assisted Intervention*, 223–231.

Viola, P. and W.M. Wells III. 1997. Alignment by maximization of mutual information, *International Journal of Computer Vision*, 24(2): 137–154.

Yan, P., S. Xu, A.R. Rastinehad, et al. 2018. Adversarial image registration with application for MR and TRUS image fusion. *International Workshop on Machine Learning in Medical Imaging*, 197–204.

Zagorchev, L. and A. Goshtasby. 2006. A comparative study of transformation functions for nonrigid image registration. *IEEE Transactions on Image Processing*, 15(3): 529–538.

Zhang, X., F.X. Yu, S. Karaman, et al. 2017. Learning discriminative and transformation covariant local feature detectors. *Proceedings of the IEEE Conference on Computer Vision and Pattern Recognition*, 6818–6826.

Zhang, Y.-J. 1990. Automatic correspondence finding in deformed serial sections. *Scientific Computing and Automation (Europe) 1990*, Chapter 5 (39–54).

Zhang, Y.-J. 1991a. 3-D image analysis system and megakaryocyte quantitation. *Cytometry*, 12: 308–315.

Zhang, Y.-J. 1991b. Analytical comparison of differential edge detectors in 3-D space. *Proceedings of Digital Signal Processing-91*, 311–316.

Zhang, Y.-J. 1997. Ellipse matching and its application to 3D registration of serial cell images. *Journal of Image and Graphics*, 2(8, 9): 574–577.

Zhang, Y.-J. 2017a. *Image Engineering, Vol. 1: Image Processing*. Germany: De Gruyter.

Zhang, Y.-J. 2017b. *Image Engineering, Vol. 3: Image Understanding*. Germany: De Gruyter.

Zhang, Y.-J. 2021. *Handbook of Image Engineering*. Singapore: Springer Nature.

CHAPTER 5

Scene Analysis and Semantic Interpretation

THE PURPOSE OF IMAGE understanding is actually to achieve understanding of the scene. The understanding of the visual scene takes place on the basis of the visual perception of scene environment data, combined with various image technologies, mining features and patterns in the visual data from different perspectives such as computational statistics, behavioral cognition and semantics, in order to efficiently realize, analyze and interpret the scene.

Analysis of the scene should be combined with high-level semantic interpretation, in which both scene labeling and classification play a part. Interpretation of scene semantics requires results of the image data to be analyzed through collecting information, learning, and making decisions based on logic. Many mathematical theories and methods can also be used in scene understanding.

This chapter is organized as follows. Section 5.1 introduces aspects of scene understanding, including scene analysis, perception level and semantic interpretation. Section 5.2 introduces the concepts of fuzzy sets and fuzzy operations, and discusses fuzzy inference methods. In Section 5.3, predicate logic based on formal logic is discussed, together with various predicate calculus rules and the theorem proving inference method. Section 5.4 explores scene object labeling, an important way of improving analysis results to the level of abstract concepts. Commonly used methods include discrete labeling and probabilistic labeling. Definitions, concepts and construction methods of some models around scene classification can be found in Section 5.5, mainly the bag-of-words/features model, pLSA model, and LDA model. Section 5.6 reviews technique developments and promising research directions from the last year.

5.1 OVERVIEW OF SCENE UNDERSTANDING

Scene analysis and semantic interpretation are important contents of scene understanding, but the research is relatively immature, and many issues are still under discussion.

DOI: 10.1201/9781003362388-5

5.1.1 Scenario Analysis

Scene analysis needs to use image analysis technology to obtain information about the scenery in the scene and lay the foundations for further scene interpretation.

Object recognition is an important basic step in scene analysis. When recognizing a single object, it is generally believed that the image region of the object can be decomposed into a limited number of sub-regions (often corresponding to the parts of the object). These sub-regions have a relatively fixed geometric relationship, which together constitute the appearance of the object. However, when analyzing natural scenes, there are often many sceneries, and the relationship between them is complex and difficult to predict. Therefore, in addition to considering the internal relationship of the object itself, scene analysis should also pay attention to the relative distribution and mutual relationship between the objects.

From a cognitive point of view, scene analysis is more concerned with human perception and understanding of the scene. A large number of biological, physiological and psychological scene analysis experiments have shown that the analysis of the global properties of the scene often occurs in the early stage of visual attention. However, in scene analysis, the visual content of the scene (scenery and distribution) can be highly uncertain:

(1) Different lighting conditions, which lead to difficulties in scene detection and tracking;

(2) Different external appearances of the scene (sometimes despite similar structural elements), which can lead to ambiguity in the identification of the scene;

(3) Different observation scales, which often affect the identification and description of scene characteristics;

(4) Different scene positions, orientations and mutual occlusion factors, which will increase the complexity of scene cognition.

5.1.2 Scene Awareness Hierarchy

The analysis and semantic interpretation of the scene can be performed hierarchically. Similar to the case in content-based coding, model-based coding can be done at three levels: (i) object-based coding at the lowest level; (ii) knowledge-based coding at the middle level; and (iii) semantic-based coding at the highest level. The analysis and semantic interpretation of the scene can also be carried out at three levels:

(1) *Local layer*. This layer mainly emphasizes the analysis of the local part of the scene or single scenery, and the identification or labeling of the image region.

(2) *Global layer*. This layer considers the overall situation of the entire scene, and looks at the interrelationship of scenes with similar shapes and similar functions.

(3) *Abstraction layer*. This layer corresponds to the conceptual meaning of the scene, and a description and explanation of the abstraction of the scene should be given.

If the analysis and judgment of the scene in the classroom is taken as an example, the following situations can be considered for the above three levels. In the first layer, the main consideration is to extract the objects in the image, such as teachers, students, desks and chairs, screens, projectors, etc. On the second level, determine the position and relationship of each scene, such as the teacher standing in front of the classroom, the student sitting facing the screen, the projector throwing the picture to the screen, etc.; as well as judge the environment and function, such as indoor or outdoor, what type of indoor (office or classroom). On the third level, describe the content of the activities in the classroom (e.g., in class or recess) and the atmosphere (e.g., more relaxed or serious, more calm or energetic).

Humans have a strong ability to perceive a scene. For a new scene (especially a more conceptual scene), the human eye can interpret the meaning of the scene with just a glance. For example, when a sports meeting is taking place on the playground, people are observing the color of the grass and the track (low-level features), the athletes on the track and their running status (middle-level objects), plus inference based on experience and knowledge (high-level concepts). In such a process, the perception of the middle layer has a certain priority. Studies have shown that humans have a strong and very rapid ability to identify most of the objects in the middle layer (such as sports fields), and they can identify and name them faster than the low and high layers. It has been hypothesized that perception of the mid-level is preferred because it maximizes both (same) intra-class similarity (playground with or without athletes) and (different) inter-class dissimilarity (even classrooms in recess are different from playgrounds). From the visual characteristics of the scene itself, the scenes in the middle layer often have similar spatial structures and are endowed with similar behaviors.

In order to obtain a semantic interpretation of the scene, it is necessary to establish the connection between high-level concepts, low-level visual features and mid-level object characteristics, and to identify the relationships between them. To accomplish this, two approaches can be considered.

(1) Low-level scene modeling

The **low-level scene modeling** method directly represents, describes and classifies the low-level attributes (color, texture, etc.) of the scene, then infers its high-level information. For example, the playground has a large area of grass (light color), the running track (dark red); the classroom has many tables and chairs (horizontal or vertical edges, regular geometry). The method can be further divided into global and blocking methods. The former describes the scene with the help of information statistics (such as color histogram) of the whole image, for example dividing it into indoor images and outdoor images, outdoor images can be divided into natural landscape images and artificial building images, and artificial building images can be divided into building images, sports field images, etc. The later method divides the image into multiple blocks (regular or less regular), uses local methods for each block, and then integrates them for further judgment.

(2) Middle-level semantic modeling

Middle-level semantic modeling improves the performance of low-level feature classification and discrimination through object recognition, and solves the semantic gap between high-level concepts and low-level attributes. At present, the most studied method is to use visual vocabulary modeling (see Subsection 5.5.1) and classify scenes into specific semantic categories according to the semantics and distribution of objects. For example, classrooms can be determined from indoor scenes based on desks and chairs, projectors/projection screens, and the like. It should be noted that the semantics of the scene may not be unique (for instance, an indoor place with a projector may be a classroom, but can also be a conference room). Especially for outdoor scenes, the situation is more complicated, because the scene can have any size, shape, position, orientation, lighting, shadow, occlusion, etc., and individual examples of the same type of scene may also look very different.

5.1.3 Scene Semantic Interpretation

Scene semantic interpretation involves a variety of techniques that are being researched and developed:

(1) Video computing technology

(2) Dynamic control strategy of vision algorithm

(3) Self-learning of scene information

(4) Fast or real-time computing technology

(5) Fusion collaboration of various types of sensors

(6) Visual attention mechanism (human-like)

(7) Scenario interpretation combined with cognitive theory

(8) System integration and optimization.

This list is still growing.

5.2 FUZZY INFERENCE

Fuzziness is a concept often opposed to clarity or precision (crispness). In daily life, we often encounter things without clear quantity or quantitative boundaries, and we need to use vague words to describe them. The fuzzy concept can express a variety of loose, uncertain, imprecise knowledge and information (for example, fuzzy mathematics is based on uncertain things), and can even obtain knowledge from conflicting sources. Similar qualifiers or modifiers in human language can be used here, such as high grayscale, medium grayscale, low grayscale, etc., to form fuzzy sets for representing and describing related image knowledge. Based on the representation of knowledge, further inference can be performed. Fuzzy inference requires the help of fuzzy logic and fuzzy operations.

5.2.1 Fuzzy Sets and Fuzzy Operations

A **fuzzy set** S in a fuzzy space X is a set of ordered pairs:

$$S = \{[x, M_S(x)] \mid x \in X\} \tag{5.1}$$

where the membership function $M_S(x)$ represents the membership degree of x in S.

The value of the membership function is always a non-negative real number, usually limited to [0, 1]. Fuzzy sets can often be uniquely described by their **membership functions**. Figure 5.1 shows several examples of using exact set and fuzzy set to express the gray level as "dark", where the horizontal axis corresponds to the image gray level x, and for the fuzzy set it is the definition domain of its membership function $M_S(x)$. Figure 5.1(a) is described in terms of exact sets, giving results that are binary (less than 127 is completely "dark", greater than 127 is not at all "dark"). Figure 5.1(b) is a typical fuzzy set membership function, ranging from 0 to 255 along the horizontal axis, and its degree of membership along the vertical axis from 1 (corresponding to a gray level of 0, fully belonging to a "dark" fuzzy set) to 0 (the corresponding grayscale is 255, which is not part of the "dark" fuzzy set at all). The gradual transition in between shows that the x between them is partly "dark" and partly not "dark". Figure 5.1(c) gives an example of a nonlinear membership function that is somewhat like a combination of Figures 5.1(a) and 5.2(b), but still represents a fuzzy set.

Operations on fuzzy sets can be performed by means of fuzzy logic operations. **Fuzzy logic** is a science based on multi-valued logic, which uses fuzzy sets to study fuzzy thinking, language forms and their laws. There are some fuzzy logic operations with names similar to general logic operations but with different definitions. Let $M_A(x)$ and $M_B(y)$ denote the membership functions corresponding to fuzzy sets A and B, respectively, whose domains are X and Y, respectively. Fuzzy intersection, fuzzy union, and fuzzy complement can be defined point by point as follows:

$$\begin{aligned} \text{Intersection } A \cap B &: M_{A\cap B}(x,y) = \min[M_A(x), M_B(y)] \\ \text{Union } A \cup B &: M_{A\cup B}(x,y) = \max[M_A(x), M_B(y)] \\ \text{Complement} A^c &: M_{A^c}(x) = 1 - M_A(x) \end{aligned} \tag{5.2}$$

Operations on fuzzy sets can also be performed by changing the shape of the fuzzy membership function point by point by means of general algebraic operations. Assuming

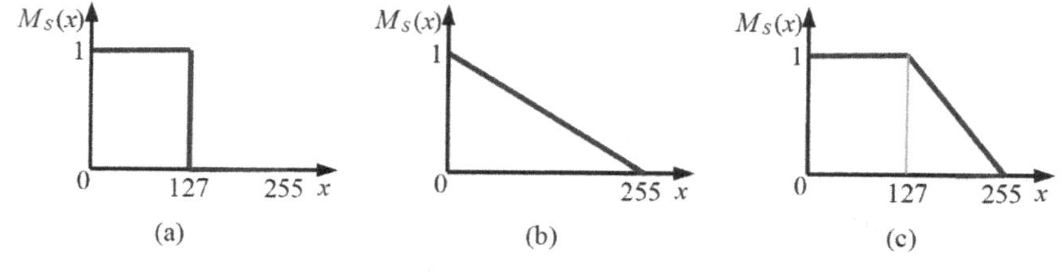

FIGURE 5.1 Representation of exact set and fuzzy set.

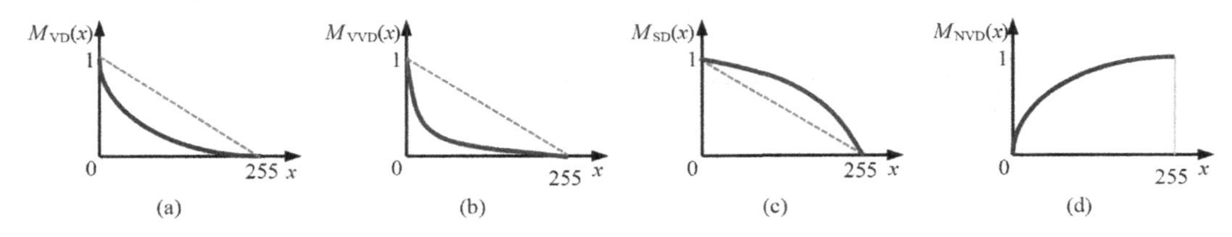

FIGURE 5.2 Some operation results of the original fuzzy set D in Figure 5.1(b).

that the membership function in Figure 5.1(b) represents a fuzzy set D (dark), then the membership function of the enhanced fuzzy set VD (very dark) can be (Figure 5.2(a)):

$$M_{VD}(x) = M_D(x) \cdot M_D(x) = M_D^2(x) \tag{5.3}$$

Such operations can be repeated. For example, the membership function of the fuzzy set VVD (very very dark) is (as shown in Figure 5.2(b)):

$$M_{VVD}(x) = M_D^2(x) \cdot M_D^2(x) = M_D^4(x) \tag{5.4}$$

On the other hand, a weakened fuzzy set SD (somewhat dark) can also be defined, and its membership function is (as shown in Figure 5.2(c)):

$$M_{SD}(x) = \sqrt{M_D(x)} \tag{5.5}$$

Logical operations and algebraic operations can also be combined. For example, to strengthen the negation of the fuzzy set VD, that is, the membership function of the fuzzy set NVD (not very dark) is (as shown in Figure 5.2(d)):

$$M_{NVD}(x) = 1 - M_D^2(x) \tag{5.6}$$

Here NVD can be regarded as N[V(D)], that is, $M_D(x)$ corresponds to D, $M^2{}_D(x)$ corresponds to V(D), and $1-M^2{}_D(x)$ corresponds to N[V(D)].

5.2.2 Fuzzy Inference Methods

In fuzzy inference, the information in each fuzzy set is combined with certain rules to make a decision (Sonka et al. 2014).

5.2.2.1 Basic Model

The basic model and main steps of fuzzy inference are shown in Figure 5.3. Based on **fuzzy rules**, the basic relationship of determining the membership degree in the relevant membership function is called combination. The result of **fuzzy combination** is a **fuzzy solution space**. In order to make a decision based on the solution space, there must be a process of **defuzzification**.

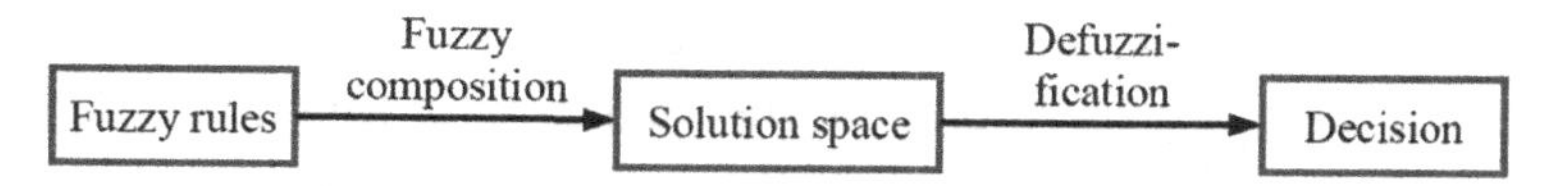

FIGURE 5.3 The model and steps for fuzzy inference.

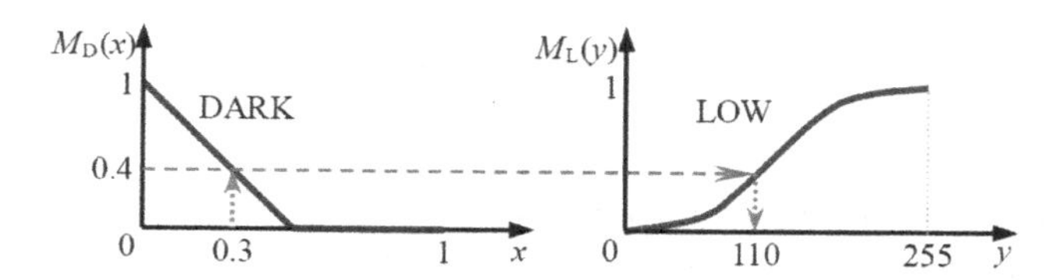

FIGURE 5.4 Monotonic fuzzy inference based on a single fuzzy rule.

Fuzzy rules refer to a series of unconditional and conditional propositions. The form of unconditional fuzzy rules is:

$$x \text{ is } A \tag{5.7}$$

The form of conditional fuzzy rules is:

$$\text{if } x \text{ is } A \quad \text{then } y \text{ is } B \tag{5.8}$$

where A and B are fuzzy sets, and x and y represent scalars in their corresponding domains.

The membership function corresponding to the unconditional fuzzy rule is $M_A(x)$. Unconditional fuzzy rules are used to limit the solution space or define a default solution space. Since these rules are unconditional, they can act directly on the solution space by means of fuzzy set operations.

Now consider conditional fuzzy rules. Among the various existing methods for implementing decision making, the simplest is monotonic **fuzzy inference**, which can directly obtain the solution without using the fuzzy combination and defuzzification described below. For example, let x represent the illuminance value of the outside world, and y represent the grayscale value of the image, then the fuzzy rule that represents the high–low level of the image grayscale is: if x is DARK then y is LOW.

Figure 5.4 shows the principle of monotonic fuzzy inference, assuming that according to the determined external illuminance value (here $x = 0.3$), the membership value $M_D(0.3) = 0.4$ can be obtained. If this value is used to represent the membership value $M_L(y) = M_D(x)$, the expectation for the gray level of the image is $y = 110$ (the range is 0 to 255), which belongs to the low range.

5.2.2.2 Fuzzy Combination

The knowledge related to the decision-making process is often contained in more than one fuzzy rule. But not every fuzzy rule makes the same contribution to decision making. There are different combination mechanisms for combining rules, and the most commonly used is the **min-max rule**.

In the combination based on min-max rule, a series of minimization and maximization processes are used. Referring to Figure 5.5, the membership function $M_B(y)$ of the fuzzy result is defined by the minimum value of the predicted true value, also known as the **minimum correlation** $M_A(x)$. Then, the membership function of the fuzzy result is updated point by point to obtain the fuzzy membership function

$$M(y) = \min\{M_A(x),\ M_B(y)\} \tag{5.9}$$

If there are N rules, do this for each rule (two rules are used as an example in Figure 5.5). Finally, the fuzzy membership function of the solution is obtained by taking the maximum point by point of the minimized fuzzy set:

$$M_S(y) = \max_n\{M_n(y)\} \tag{5.10}$$

Another method, called the **correlation product**, scales rather than truncates the original resulting fuzzy membership function. The minimum correlation calculation is simple, the defuzzification is simple, while the correlation product preserves the shape of the original fuzzy set (see Figure 5.6).

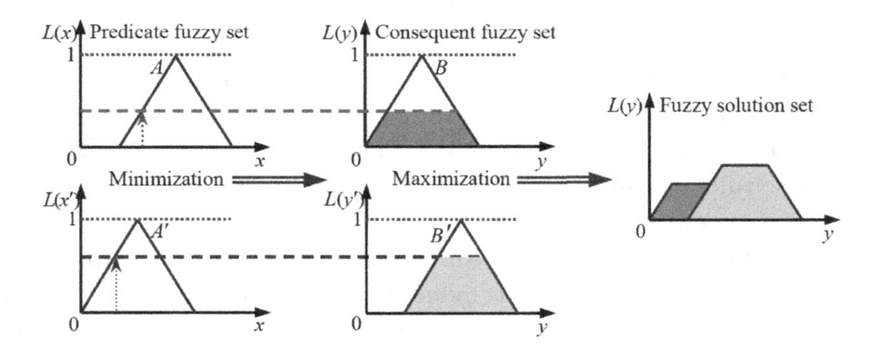

FIGURE 5.5 Using correlation minimum for fuzzy min-max composition.

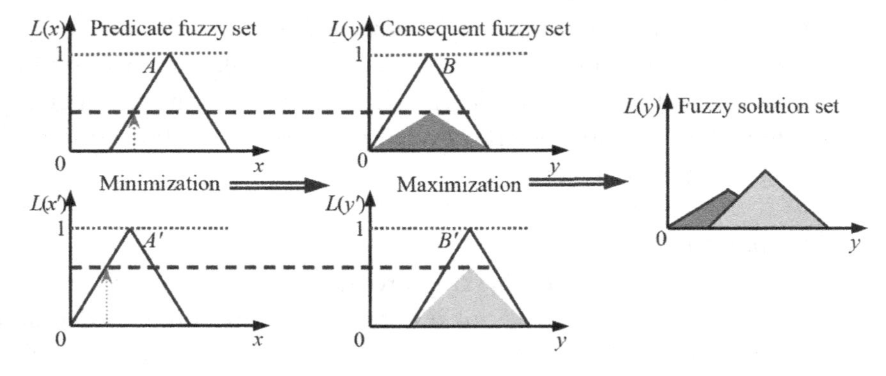

FIGURE 5.6 Using correlation product for fuzzy min-max composition.

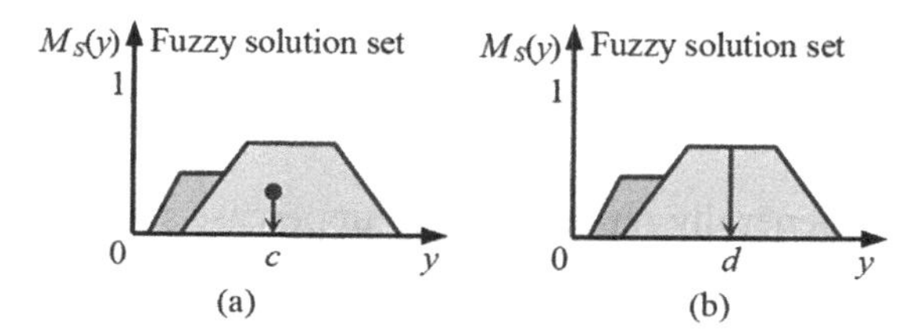

FIGURE 5.7 Two methods for defuzzification.

5.2.2.3 Defuzzification

Fuzzy combination gives a fuzzy membership function of a single solution for each solution variable. In order to determine the exact solution for making a decision, a vector containing multiple scalars (one for each solution variable) that best represents the information in the fuzzy solution set needs to be determined. This process is performed independently for each solution variable and is called **defuzzification**. Two commonly used defuzzification methods are the combining moment method and the maximum combining method.

The **combining moment** method first determine the centroid c of the membership function of the fuzzy solution, and converts the fuzzy solution into a clear solution c (see Figure 5.7(a)). The **maximum combining** method determines the domain point with the largest membership value in the membership function of the fuzzy solution. If the maximum value is on a platform, the center of the platform gives a clear solution d (see Figure 5.7(b)). The result of the method of combining moments is sensitive to all rules (a weighted average is taken), while the result of the maximum combining method depends on the single rule with the highest predicted true value. The combining moment method is often used in control applications, while the maximum combining method is often used in identification applications.

5.3 PREDICATE LOGICAL SYSTEM

Predicate logic, also known as first-order logic, has a history hundreds of years old and is a well-organized and widely used type of knowledge. It is useful for expressing propositions and deriving new facts from a knowledge base of facts, one of the most powerful elements of which is the **predicate calculus**. In most cases, logical systems are based on first-order predicate calculus, which can express almost anything. A first-order predicate calculus is a symbolic formal language (symbolic logic) that can be used to express a wide range of numerical formulas or statements in various natural languages, as well as statements ranging from simple facts to complex expressions. With it, logical demonstration in mathematics can be symbolized, knowledge can be represented by logical rules, and these rules can be used to prove whether logical representations are true or not. This is a natural way of representing knowledge in the form of formulas. Its characteristic is that it can express knowledge accurately and flexibly (meaning that the method of knowledge representation can be independent of the method of inference).

5.3.1 Predicate Calculus Rules

There are four basic elements of predicate calculus:

(1) Predicate symbols: generally represented by uppercase strings (including letters and numbers);

(2) Function symbols: generally represented by lowercase strings;

(3) Variable symbols: generally represented by lowercase characters;

(4) Constant symbols: generally represented by uppercase strings.

A predicate symbol represents a relation within the domain in question. For example, the proposition "1 is less than 2" can be represented as LESSTHAN(1, 2), where LESSTHAN is the predicate symbol, and 1 and 2 are both constant symbols.

Some examples of the basic elements of predicate calculus are shown in Table 5.1. In these examples, a predicate consists of a predicate symbol and one or more of its arguments, which may be constants or functions of other arguments.

Predicates are also known as atoms. Combining atoms with logical conjunctions can obtain clauses. Commonly used logical conjunctions are: "$\wedge$" (AND), "$\vee$" (OR), "~" (NOT), "$\Rightarrow$" (IMPLIES). In addition, there are two kinds of quantifiers that represent quantity: "$\forall$" is called a **universal quantifier**, $\forall x$ stands for all x; "$\exists$" is called an existential quantifier, and $\exists x$ stands for the existence of an x. For logical representations, representations obtained by connecting other representations with $\wedge$ or $\vee$ are called **conjunctive representations** or **disjunctive representations**, respectively. **Legal predicate calculus** representations are called **well fitted formulas** (WFFs) (Gonzalez and Woods 1992).

Some examples of clauses obtained by combining logical conjunctions with atoms are shown in Table 5.2. The first four examples are related to constant symbols, and the last two examples also include variable symbols.

There are two types of logical expressions. A logical expression is said to follow **clause-form syntax** if it is of the form $(\forall x_1\, x_2 \ldots x_k)\ [A_1 \wedge A_2 \wedge \ldots \wedge A_n \Rightarrow B_1 \vee B_2 \vee \ldots \vee B_m]$, where each A and B are atoms. The left and right parts of a clause are called the condition of the clause and conclusion of the clause, respectively. If a logical expression includes atoms, logical conjunctions, existential quantifiers, and universal quantifiers, the expression is said to follow **non-clause form syntax**.

TABLE 5.1 Predicate Examples

Statement	Predicate
Image I is a digital image	DIGITAL(I)
Image J is a scanned image	SCAN(J)
Combining digital image I and scanned image J	COMBINE [DIGITAL(I), SCAN(J)]
There is a pixel p in image I	INSIDE(p, I)
Object x is behind object y	BEHIND(x, y)

TABLE 5.2 Clause Examples

Statement	Clause
Image I is both digital image and a scanned image	DIGITAL(I) ∧ SCAN(I)
Image I is a digital image or an analog image	DIGITAL(I) ∨ ANALOG(I)
Image I is not a digital image	~ DIGITAL(I)
If image I is a scanned image, then image I is a digital image	SCAN(I) ⇒ DIGITAL(I)
An image is either a digital image or an analog image	(∀x) DIGITAL(x) ∨ ANALOG(x)
There is an object in the image	(∃x) INSIDE(x, I)

Now consider the proposition: for each x, if x represents both images and numbers, then x is either black and white or color. In clause-form syntax, this proposition can be written as the following expression:

$$(\forall x)\left[\text{IMAGE}(x) \wedge \text{DIGITAL}(x) \Rightarrow \text{GRAY}(x) \vee \text{COLOR}(x)\right]$$

In non-clause form syntax, this proposition can be written as the following expression:

$$(\forall x)\left[\text{IMAGE}(x) \wedge \text{DIGITAL}(x) \vee \text{GRAY}(x) \vee \text{COLOR}(x)\right]$$

It is easy to verify that the above two expressions are equivalent, or that the above two expressions have the same expressive power (which can be proved with the help of the truth table of logical connectors (Table 5.3). In fact, it is always possible to convert from clause form to non-clause form, or vice versa.

Table 5.3 shows the relationship between the above logical connectors. The first five columns are the basic logical operation contents, and the sixth column is the implicit operation. For an implicit (operation), the left part is called the **premise** and the right part is called the **result**. If the premise is empty, the expression "⇒ P" can be regarded as representing P; conversely, if the result is empty, the expression "P ⇒" represents the "not" of P, i.e., "~ P". Table 5.3 indicates that if the result is T (regardless of the premise at this time) or the premise is F (regardless of the result at this time), the implied value is T; otherwise, the implied value is F. In the above definition, for an implicit (operation), as long as the premise is F, the implicit value is T. This definition often leads to confusion and strange propositions. For example, consider a meaningless sentence: "if the image is round, then all objects are green". Because the premise is F, the predicate calculus expression result of

TABLE 5.3 Truth Table for Logical Connectors

A	B	~ A	A ∧ B	A ∨ B	A ⇒ B
T	T	F	T	T	T
T	F	F	F	T	F
F	T	T	F	T	T
F	F	T	F	F	T

the sentence will be T, but it is obviously not true here. However, in practice, considering that the logical implicit operation is not always meaningful in natural language, the above problems do not always arise.

The following examples of logical representations can help explain the concepts discussed above.

(1) If the image is digital, it has discrete pixels:

$$\text{DIGITAL}(\text{image}) \Rightarrow \text{DISCRETE}(x)$$

(2) All digital images have discrete pixels:

$$(\forall x)\{[\text{IMAGE}(x) \wedge \text{DIGITAL}(x)] \Rightarrow (\exists y)[\text{PIXEL\_IN}(y,x) \wedge \text{DISCRETE}(y)]\}$$

The above representation reads: For all x, where x is an image, and is digital, then there is always y, where y is a pixel in x, and is discrete.

(3) Not all images are digital:

$$(\forall x)[\text{IMAGE}(x)] \Rightarrow (\exists y)[\text{IMAGE}(y) \wedge \sim \text{DIGITAL}(y)]$$

The above representation reads: For all x, if x is an image, then there is y, y is an image, but not digital.

(4) Color digital images carry more information than monochrome digital images:

$$\begin{aligned}&(\forall x)(\forall y)\{[\text{IMAGE}(x) \wedge \text{DIGITAL}(x) \wedge \text{COLOR} \times (x)] \wedge \\ &\quad [\text{IMAGE}(y) \wedge \text{DIGITAL}(y) \wedge \text{MONOCHROME}(y)] \Rightarrow \text{MOREINFO}(x,y)\}\end{aligned}$$

The above representation reads: For all x and all y, if x is a color digital image and y is a monochrome digital image, then x carries more information than y.

Some important equivalence relations (here $\Leftrightarrow$ stands for equivalence) are given in Table 5.4, which help to convert between clause forms and non-clause forms. The rationality of these equivalence relations can be verified with the help of the truth table of logical connectors in Table 5.4.

5.3.2 Inference by Theorem Proving

In predicate logic, inference rules can be applied to certain WFFs and sets of WFFs to generate new WFFs. Table 5.5 gives some examples of inference rules (W stands for WFFs) (Gonzalez and Woods 1992). In the table, c is a constant symbol, and the general statement

TABLE 5.4 Some Important Equivalence Relations

Relation	Definition		
	$\sim(\sim A)$	$\Leftrightarrow$	A
Basic logic	$A \vee B$	$\Leftrightarrow$	$\sim A \Rightarrow B$
	$A \Rightarrow B$	$\Leftrightarrow$	$\sim B \Rightarrow \sim A$
De Morgan's law	$\sim(A \wedge B)$	$\Leftrightarrow$	$\sim A \vee \sim B$
	$\sim(A \vee B)$	$\Leftrightarrow$	$\sim A \wedge \sim B$
Distributive law	$A \wedge (B \vee C)$	$\Leftrightarrow$	$(A \wedge B) \vee (A \wedge C)$
	$A \vee (B \wedge C)$	$\Leftrightarrow$	$(A \vee B) \wedge (A \vee C)$
Commutative law	$A \wedge B$	$\Leftrightarrow$	$B \wedge A$
	$A \vee B$	$\Leftrightarrow$	$B \vee A$
Combination law	$(A \wedge B) \wedge C$	$\Leftrightarrow$	$A \wedge (B \wedge C)$
	$(A \vee B) \vee C$	$\Leftrightarrow$	$A \vee (B \vee C)$
Others	$\sim(\forall x)\ P(x)$	$\Leftrightarrow$	$(\exists x)\ [\sim P(x)]$
	$\sim(\exists x)\ P(x)$	$\Leftrightarrow$	$(\forall x)\ [\sim P(x)]$

TABLE 5.5 Examples of Inference Rules

Inference Rules	Definition		
Modus Ponens	From $W_1 \wedge (W_1 \Rightarrow W_2)$	infer	W_2
Modus Tollens	From $\sim W_2 \wedge (\sim W_1 \Rightarrow W_2)$	infer	W_1
Projection	From $W_1 \wedge W_2$	infer	W_1
Universal specification	From $(\forall x)\ W(x)$	infer	$W(c)$

"deduce G from F" means that $F \Rightarrow G$ is always true, so that G can be used instead of F in logical representations.

Inference rules can generate "derived WFFs" from given WFFs. In predicate calculus, the derived WFFs are called theorems, and the sequential application of inference rules in the derivation constitutes the proof of the theorem. Much image understanding work can be represented in the form of theorem proving via predicate calculus. In this way, a set of known facts and some inference rules can be used to derive new facts or to prove the rationality (correctness) of the hypothesis.

In predicate calculus, in order to prove the correctness of logical representation, two basic methods can be used: the first is to directly operate the non-clause form with a process similar to proving a mathematical expression, and the second is to match the items in the clause form of representation.

How to prove the correctness of logical representations is discussed in the following.

It is assumed that the following facts are known: (i) there is a pixel p in image I, (ii) image I is a digital image. Let the following "physical" law hold; (iii) if the image is digital, its pixels are discrete. The facts (i) and (ii) vary with the application problem, but the condition (iii) is knowledge irrelevant to the application.

The above two facts can be written as INSIDE(p, I) and DIGITAL(I). According to the description of the problem, the above two facts are connected by the logical conjunction $\wedge$, that is, INSIDE(p, I) Λ DIGITAL(I). The "physical" law (i.e., condition (iii)) expressed in the clause is: $(\forall x, y)$[INSIDE(x, y) Λ digital (y) Λ discrete (x)].

Now use the clause representation to prove that the pixel p is indeed discrete. The idea of the proof is to first prove that the negation of the clause is inconsistent with the facts, so that the clause that needs to be proved holds. Based on the preceding definitions, knowledge about the problem can be represented in the following clause form:

(1) ⇒ INSIDE(p, I);

(2) ⇒ DIGITAL(I);

(3) $(\forall x, y)$[DIGITAL(y) ⇒ DISCRETE(x)];

(4) DISCRETE(p) ⇒;

Note that the negation of the predicate DISCRETE(p) here can be represented as DISCRETE(p) ⇒.

After the basic elements of the question are represented in clause form, the resulting contradiction can be used to obtain proofs by matching the left and right sides of each implicit to reach the empty clause. The matching process is performed by variable substitution to make atoms identical. After matching, a clause called a **resolvent** can be obtained, which contains left and right sides that do not match. If y is replaced with I and x with p, the left side of (3) matches the right side of (2), so the resolvent is:

(5) ⇒ DISCRETE(p).

However, since the left side of (4) is congruent with the right side of (5), the solution to (4) and (5) is an empty clause. This result is contradictory; it shows that DISCRETE(p) ⇒ cannot hold, thus proving the correctness of DISCRETE(p).

Now use the non-clause representation to prove that the pixel p is indeed discrete. First, condition (iii) is transformed into a non-clause form according to the relation $\sim A \Rightarrow B \Leftrightarrow A \vee B$ introduced in Table 5.5, i.e., $(\forall x, y)$[~INSIDE(x, y) ∧ ~DIGITAL(y) ∨ DISCRETE(x)].

The following conjunctive form is used to represent knowledge about this problem:

(1) $(\forall x, y)$[INSIDE(x, y) ∧ DIGITAL(y)] ∧ [~INSIDE(x, y) ∧ ~DIGITAL(y) ∨ DISCRETE(x)];

Substituting I for y and p for x yields:

(2) [INSIDE(p, I) ∧ DIGITAL(I)] ∧ [~INSIDE(p, I) ∧ ~DIGITAL(I) ∨ DISCRETE(p)];

Using the projection rules, it can be deduced that:

(3) INSIDE(p, I) ∧ [~INSIDE(p, I) ∨ DISCRETE(p)];

Using the law of distribution again, we get $A \wedge (\sim A \vee B) = (A \wedge B)$. This results in a simplified representation:

(4) INSIDE(p, I) ∧ DISCRETE(p);

Using the projection rule again, we get:

(5) DISCRETE(p).

This proves that the original representation in (1) is completely equivalent to the representation in (5). In other words, this infers the conclusion that pixel p is discrete based on the given information.

A fundamental conclusion of predicate calculus is that all theorems can be proved in finite time. An inference rule called **resolution** has long been proposed to prove this conclusion (Robison 1965). The basic steps of this parsing rule are to first express the basic elements of the problem in the form of clauses, then seek the premise and result of the implicit representation that can be matched, and then perform matching by substituting variables to make the atoms identical. This clause (called **resolvent**) includes the left and right sides of the mismatch. Theorem proofs now translate to solving clauses to produce empty clauses, which give contradictory results. From the point of view that all correct theorems can be proved, this resolution rule is complete; from the point of view that all false theorems are impossible to prove, this resolution rule is correct.

An example is given below to illustrate how to interpret an image based on knowledge base solving.

Suppose the knowledge base in an aerial image interpretation system has the following information: (i) all commercial airport images have runways; (ii) all commercial airport images have planes; (iii) all commercial airport images have buildings; (iv) at least one building in a commercial airport is a terminal building; (v) a building surrounded and directed by aircraft is a terminal building. This information can be put into a "model" of a commercial airport in clause form:

$$(\forall x)[\text{CONTAINS}(x,\text{runways}) \wedge \text{CONTAINS}(x,\text{airplanes}) \wedge \text{CONTAINS}(x,\text{buildings}) \\ \wedge \text{POINT}-\text{TO}(\text{airplanes, buildings})] \Rightarrow \text{COM}-\text{AIRPORT}(x)$$

Note that the information in (iv) is not directly used in the model, but its meaning is implicit in the two conditions of the existence of the building in the model and the aircraft pointing to the building; condition (v) clearly indicates what kind of building the terminal building is.

There is an aerial image and there is a recognition engine that can distinguish different objects in the aerial image. From the perspective of image interpretation, two types of questions can be asked:

(1) What kind of image is this?

(2) Is this an image of a commercial airport?

In general, the first question cannot be answered with current technology. The second question is generally more difficult to answer, but it becomes easier if the discussion is narrowed down. Specifically, the model-driven approach shown earlier has the distinct advantage that it can be used to guide the work of the recognition engine. The recognition engine in this example should be able to recognize three types of objects, namely runways, aircraft and buildings. If, as is common, the height of the acquired image is known, the task of finding the object can be further simplified, since the relative dimensions of the object can be used to guide the recognition process.

A recognition engine working from the above model will have outputs of the form: CONTAINS(image, runway), CONTAINS(image, aircrafts), and CONTAINS(image, buildings). On the basis of object recognition, it is further possible to determine whether the clause POINT-TO (aircraft, buildings) is true or false. If the clause is false, the process stops; if the clause is true, the process continues to determine whether the given image is an image of a commercial airport by deciding on the correctness of the clause COM-AIRPORT(image).

If you want to solve this problem using the resolution method of proving the theorem, you can start working according to the following four pieces of information obtained from the image:

(1) ⇒ CONTAINS(image, runway);

(2) ⇒ CONTAINS(image, aircraft);

(3) ⇒ CONTAINS(image, buildings);

(4) ⇒ POINT-TO(aircraft, buildings).

The negation of the clause to be proved here is:

(5) COM-AIRPORT(image) ⇒.

First notice that if you replace x with image, one of the clauses on the left side of the model will match the right side of (1). The resolvent is

$$[\text{CONTAINS}(\text{image, airplanes}) \wedge \text{CONTAINS}(\text{image, buildings}) \\ \wedge \text{POINT}-\text{TO}(\text{airplanes, buildings})] \Rightarrow \text{COM}-\text{AIRPORT}(\text{image})$$

Similarly, one of the clauses on the left side of the resolvent above will match the right side of (2), and the new resolvent is

$$[\text{CONTAINS}(\text{image, buildings}) \wedge \text{POINT}-\text{TO}(\text{airplanes, buildings})] \\ \Rightarrow \text{COM}-\text{AIRPORT}(\text{image})$$

Next, the resolvent obtained using (3) and (4) is ⇒ COM-AIRPORT(image).

Finally, the resolvent of this result and (5) give empty clauses, thus creating a contradiction. This proves that COM-AIRPORT(image) is correct, that the given image is indeed an image of a commercial airport (it matches the model of a commercial airport).

5.4 SCENE OBJECT LABELING

Scene object labeling is the **semantic labeling** of the object region in the image, that is, giving the object a semantic symbol. It is assumed here that regions corresponding to objects in the scene image have been detected, and objects and their connections have been described with the aid of region adjacency graphs or the semantic web. The nature of the object itself can be described by a univariate relationship, and the relationship between the objects needs to be described by a binary or multivariate relationship. The purpose of scene object labeling is to assign a label (marker with semantic meaning) to the object in each scene image to obtain an appropriate interpretation of the scene image. The interpretations thus obtained should be coherent with scene knowledge. Labels need to be consistent (meaning that any two objects in the image are in a reasonable structure or relationship) and tend to be the most likely explanation when there are multiple possibilities.

5.4.1 Labeling Methods and Key Elements

There are two main approaches to labeling scene objects (Sonka et al. 2014).

(1) **Discrete labeling**: It assigns only one label to each object, and mainly considers the consistency of labels for each object in the image.

(2) **Probabilistic labeling**: It allows multiple labels to be jointly assigned to existing object sets. The labels are weighted with probability, and each label has a degree of trust (confidence).

The difference between the two methods is mainly reflected in the robustness to scene interpretation. There can be two results of discrete labeling: one is that consistent labels are obtained, and the other is that the impossibility of assigning scene-consistent labels is detected. Due to imperfect segmentation results, discrete labels will generally always yield results that do not give consistent interpretations (even when only a few local inconsistencies are detected). On the other hand, probabilistic labeling always gives the labeling result and the corresponding confidence. Although local inconsistencies may also be possible, it often gives better results than the consistent but highly unlikely explanations given by discrete labels. Discrete labels can be viewed in extreme cases as a special case of probabilistic labels where one label has a probability of 1 and the other labels have a probability of 0.

The scene labeling method includes the following key elements:

(1) A set of objects R_i, $i = 1, 2, \ldots, N$;

(2) For each object R_i, there is a finite set of labels Q_i (this same set can also be applied to all objects);

(3) A limited set of relationships between objects;

(4) A compatibility function between related objects (reflecting constraints on the relationship between objects).

If the direct connection between all objects in the image is considered to solve the labeling problem, it will require a very large amount of calculation, so the method of **constraint propagation** is generally used, that is, local constraints are used to obtain local consistency (local optimization), and then the iterative method is used, which adjusts local consistency to global consistency over the entire image (global optimum).

5.4.2 Discrete Relaxation Labeling

Consider the scene shown in Figure 5.8(a) with five objects (including the background) (Sonka et al. 2014). The five labels are *B* (background), *W* (window), *T* (table), *D* (drawer), *P* (phone). The unary properties of the object explanation are: (i) the window is square; (ii) the table is rectangular; (iii) the drawer is rectangular. The binary (relational) constraints are: (i) the window is above the table; (ii) the phone is on the table; (iii) the drawer is inside the table; (iv) the background is connected to the edge of the image. Under these constraints, the labeling results shown in Figure 5.8(b) are inconsistent for some objects (such as the drawer that is square and runs over the phone, the window that is rectangular and is instead inside the phone, the table that is not rectangular and placed on the phone, and the phone that is placed under the table and contained to the window).

An example of a **discrete relaxed labeling** process is shown in Figure 5.9, which first assigns all existing labels to each object, as shown in Figure 5.9(a), and then iteratively performs a consistency check on each object to identify and remove those labels that may not satisfy the constraints. For example, considering that the background is connected to the

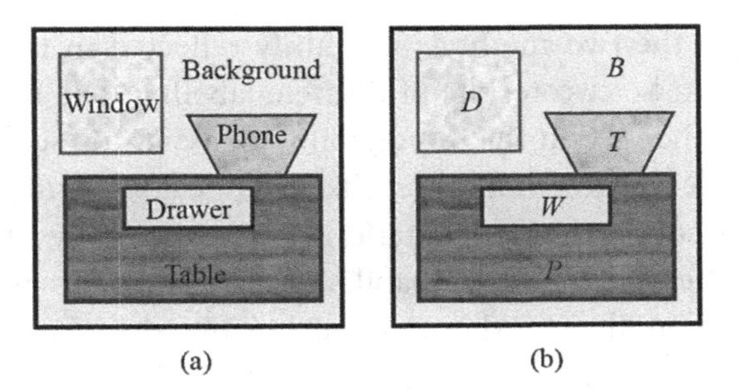

FIGURE 5.8 Labeling of objects in scene.

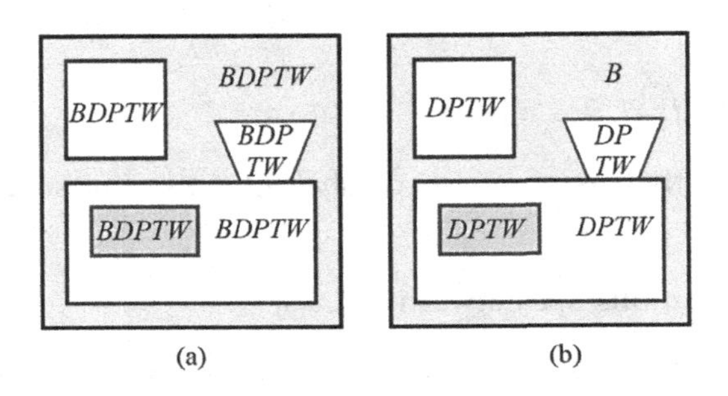

FIGURE 5.9 Discrete relaxed labeling process and results.

edge of the image, the background can be determined, then other objects cannot be labeled as the background, and Figure 5.9(b) is obtained after removing the local inconsistency labels. Then consider that the window, table and drawer are all rectangular, so the phone can be determined (Figure 5.9(c)); then consider that the drawer is inside the table, then the drawer and the table can be determined; and so on. Finally, consistency can be obtained, and the labeling result is shown in Figure 5.9(d).

5.4.3 Probabilistic Relaxation Labels

As a bottom-up interpretation method, discrete relaxed labeling encounters difficulties when object segmentation is incomplete or incorrect. The **probabilistic relaxed labeling** method has the potential to overcome problems caused by missing or extra objects in segmentation, but also has the potential to give ambiguous inconsistent interpretations.

Considering the local structure shown on the left of Figure 5.10 (its region adjacency graph is shown on the right), the object R_j is labeled q_j; $q_j \in Q$, $Q = \{w_1, w_2, \ldots, w_T\}$.

Let the compatibility function of two objects R_i and R_j with labels q_i and q_j be $r(q_i = w_k, q_j = w_l)$. The algorithm iteratively searches for the local most consistent across the entire graph. Assuming that in the l-th step of the iterative process, the marker q_i can be obtained according to the binary relationship between the objects R_i and R_j, then its support (w_k for q_i) can be expressed as

$$s_j^{(l)}(q_i = w_k) = \sum_{t=1}^{T} r(q_i = w_k, q_j = w_t) P^{(l)}(q_j = w_t) \tag{5.11}$$

where $P^{(l)}(q_j = w_t)$ is the probability that the region R_j is marked as w_t at this time.

The support obtained by considering all N objects R_i (marked w_i) associated with object R_j (marked w_j) is

$$S^{(b)}(q_i = w_k) = \sum_{j=1}^{N} c_{ij} s_j^{(l)}(q_i = w_k) = \sum_{j=1}^{N} c_{ij} \sum_{t=1}^{T} r(q_i = w_k, q_j = w_t) P^{(l)}(q_j = w_t) \tag{5.12}$$

where c_{ij} is a positive weight, satisfying $\sum_{j=1}^{N} c_{ij} = 1$, which represents the binary connection strength between the objects R_i and R_j.

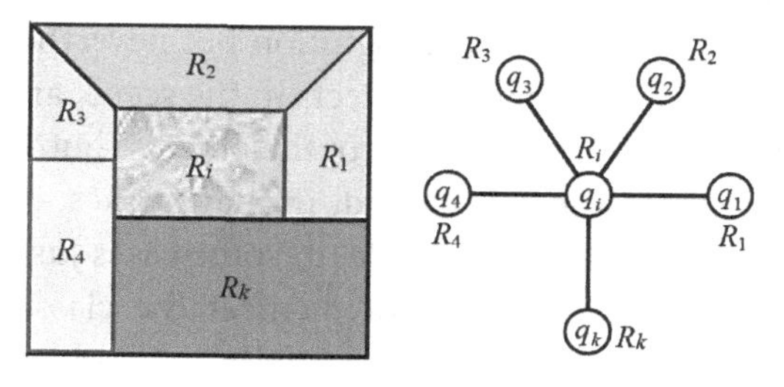

FIGURE 5.10 Local structure and region adjacency graph.

The iterative update rule is

$$P^{(l+1)}\left(q_i = w_k\right) = \frac{1}{K} P^{(l)}\left(q_i = w_k\right) S^{(l)}\left(q_i = w_k\right) \tag{5.13}$$

where K is the normalization constant:

$$K = \sum_{l=1}^{T} P^{(b)}\left(q_i = w_l\right) S^{(b)}\left(q_i = w_l\right) \tag{5.14}$$

This is a nonlinear relaxation problem. Substitute Equation (5.12) into Equation (5.13) to obtain a globally optimized function:

$$F = \sum_{k=1}^{T} \sum_{i=1}^{N} P\left(q_i = w_k\right) \sum_{j=1}^{N} c_{ij} \sum_{t=1}^{T} r\left(q_i = w_k, q_j = w_t\right) P\left(q_j = w_t\right) \tag{5.15}$$

The constraints on its solution are

$$\sum_{k=1}^{T} P\left(q_i = w_k\right) = 1 \quad \forall i \ \ P\left(q_i = w_k\right) \geq 0 \quad \forall i, k \tag{5.16}$$

When actually using the probabilistic relaxation labeling method, first determine a labeled (initial) conditional probability for all objects, and then iteratively repeat the following two steps: (i) calculate the objective function of Equation (5.15) representing the quality of scene labeling; and (ii) update the label probabilities to increase the value of the objective function (improving the quality of the scene labels); the optimal label is obtained until the objective function value is maximized.

5.5 SCENE CLASSIFICATION

Scene classification determines various specific regions (including positions, interrelationships, attributes/properties, etc.) existing in an image according to the principles of visual perception organization, and gives semantic and conceptual explanations of the scene. Its specific means and goal are to automatically classify and label images according to a given set of semantic categories, so as to provide effective contextual information for object recognition and interpretation of scene content.

Scene classification is related to object recognition but nevertheless differs from it. On the one hand, there are often many types of objects in the scene, and to achieve scene classification, it is often necessary to identify some of the objects (but generally it is not necessary to identify all objects). On the other hand, in many cases, objects can be classified based on only a partial perception (for example, in some cases just the low-level information, such as color, texture, etc., is sufficient to meet the classification requirements). Referring to the human visual cognition process, preliminary object recognition can often meet the specific classification requirements of the scene. At this time, it is necessary to establish the connection between the low-level features and high-level cognition, and to determine and explain the semantic category of the scene.

The classified scene has a certain guiding effect on the recognition of the object. In nature, most objects only appear in specific scenes, and the correct judgment of the global scene can provide a reasonable context constraint mechanism for local analysis of images (including object recognition).

5.5.1 Bag of Words/Bag-of-Features Models

The **bag-of-words model** is derived from the processing of natural language, and is often referred to as the **bag-of-features model** after it is introduced into the image field. The bag-of-features model is named after the category of features belonging to the same object set to form a bag (Sivic and Zisserman 2003). The model usually adopts a directed graph structure (the relationship between undirected graph nodes is a probabilistic constraint, the relationship between directed graph nodes is causal, and an undirected graph can be regarded as a special directed graph/symmetric directed graph). The conditional independence between images and visual vocabulary in the bag-of-features model is the theoretical basis of the model, but there is no strict geometric information about the object components in the model.

The original bag-of-words model only considers the symbiotic relationship and **topic** logic relationship between the features corresponding to the words, ignoring the spatial relationship between the features. But in the image domain, not only the image features themselves but also the spatial distribution of the image features can be important. In recent years, many feature descriptors (such as SIFT, see Section 3.4.2) have had relatively high dimensions, which can comprehensively and explicitly express the special properties of key points and their surrounding small regions in the image (different from the corner points that only express location information but implicitly express their own nature), and are also clearly different from other key points and their surrounding small regions. Moreover, these feature descriptors can overlap and cover each other in the image space, which can better preserve the relationship between them. The use of these feature descriptors improves the ability to describe the spatial distribution of image features.

Representing and describing a scene with a bag-of-features model requires local region description features, which can be called **visual vocabulary**, to be extracted from the scene. Scenes have some basic components, so scenes can be decomposed. To apply the concept of a document, a book is made up of many words. Returning to the realm of images, an image of a scene can be thought of as consisting of many visual words. From a cognitive point of view, each visual word corresponds to a feature in the image (more precisely, a feature that describes the local characteristics of the scene), and is the basic unit that reflects the image content or scene meaning. Building a visual vocabulary set (dictionary) can include the following aspects: (i) extracting features; (ii) learning visual vocabulary; (iii) obtaining quantitative features of visual vocabulary; and (iv) using the frequency of visual vocabulary to represent images.

A specific example is shown in Figure 5.11. First, the image regions (neighborhoods of key points) are detected, and different types of regions are divided and extracted (see

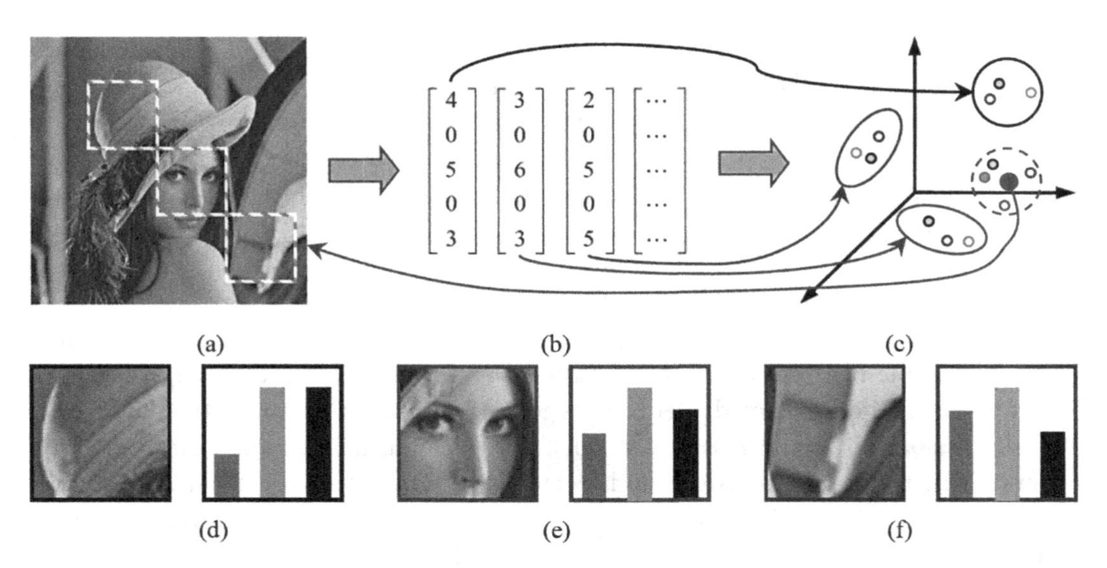

FIGURE 5.11 The acquisition process of the local region description features in the image.

Figure 5.11(a), where small squares are used for simplicity), and then a feature vector is calculated to represent each region (Figure 5.11(b)), then the feature vector is quantified into visual words and a codebook is built (Figure 5.11(c)), and finally the frequency of occurrence of a specific word is counted for each image. A few examples using the histogram are shown in Figures 5.11(d)–5.11(f), which are combined to get a representation of the entire image.

After the image is divided into multiple sub-regions, each sub-region can be assigned a semantic concept, that is, each sub-region can be regarded as a visual unit, so that it has a unique semantic meaning. Since similar scenes should have similar concept sets and distributions, the scenes can be classified into specific semantic categories according to the regional distribution of semantic concepts. If semantic concepts can be associated with visual vocabulary, then the classification of scenes can be carried out with the help of lexical representation and description models.

Visual vocabulary can either be used to represent the object directly, or it can only represent intermediate concepts in the neighborhood of key points. The former needs to detect or segment objects in the scene, and further classify the scene by classifying the objects. For example, if the sky is detected, the image should be outdoor. The latter may not need to segment the object directly, but use trained local descriptors to determine the label of the scene. There are generally three steps:

(1) *Feature point detection.* Commonly used methods include the image grid and Gaussian difference methods. The former divides the image according to the grid, and takes the center position of the grid to determine the feature points. The latter uses the **difference of Gaussian** (DoG) operator (see Section 3.4.2) to detect local feature points of interest, such as corners.

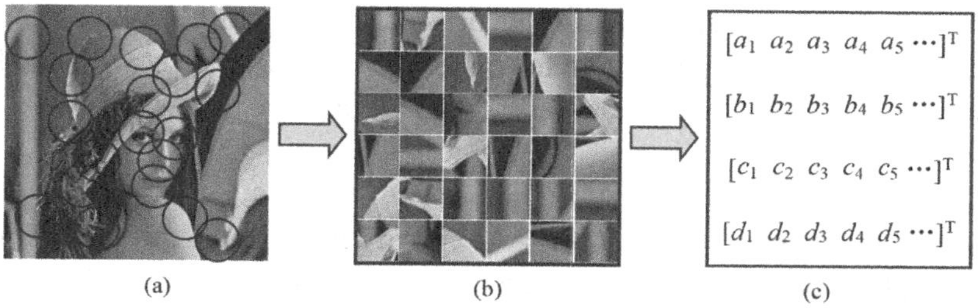

$[a_1\ a_2\ a_3\ a_4\ a_5 \cdots]^T$

$[b_1\ b_2\ b_3\ b_4\ b_5 \cdots]^T$

$[c_1\ c_2\ c_3\ c_4\ c_5 \cdots]^T$

$[d_1\ d_2\ d_3\ d_4\ d_5 \cdots]^T$

(a) (b) (c)

FIGURE 5.12 Visual vocabulary acquisition with SIFT local descriptors.

(2) *Feature representation and description.* This uses the properties of the feature points themselves and the properties of their neighborhoods. The **scale-invariant feature transform** (SIFT) operator has often been used in recent years (see Section 3.4.2), which actually combines feature point detection and feature representation and description.

(3) *Dictionary generation.* Cluster the local description results (such as using the k-means clustering method), and take the cluster center to form a dictionary.

In practice, the selection of local regions can be carried out with the help of SIFT local descriptors. The selected local regions are circular regions centered on key points and have some invariant characteristics, as shown in Figure 5.12(a). The **visual vocabulary** dictionary thus constructed is shown in Figure 5.12(b), where each sub-image represents a basic visual word (a key point feature cluster) and can be represented by a vector, as shown in Figure 5.12(c). Using the visual vocabulary dictionary, the original image can be represented by a combination of visual vocabulary, and the frequency of use of various visual vocabulary reflects the characteristics of the image.

In the actual application process, the image is first represented in visual vocabulary through feature detection operators and feature descriptors, then the parameter estimation and probabilistic reasoning of the visual vocabulary dictionary model are constructed to obtain parameter iteration formulas and probabilistic analysis results, and finally the model obtained from training is analyzed and explained.

Bayesian correlation models are most commonly used in modeling, such as probabilistic latent semantic analysis (pLSA) models and latent Dirichlet assignment (LDA) models. According to the framework of the bag-of-feature model, the image is regarded as text, and the topic found in the image is regarded as the object class (such as classroom, sports field), then a scene containing multiple objects is regarded as a mixture of a set of topics. It is composed of the probability model of the scene, and its semantic categories can be divided by analyzing the distribution of scene topics.

5.5.2 pLSA Model

The **probabilistic latent semantic analysis** (pLSA) model is derived from **probabilistic latent semantic indexing** (pLSI), which is a graph model established to solve object and scene classification problems (Sivic et al. 2005). The pLSA model is derived from the learning of natural language and text, and its original noun definitions use the concepts in the text, but it is also easy to generalize to the image field (especially with the help of the framework of the bag-of-feature model).

5.5.2.1 Model Description

Suppose there is an image set $T = \{t_i\}$, $i = 1, \ldots, N$;N is the total number of images; the visual words contained in T come from the word set, that is, dictionary (visual vocabulary) $S = \{s_j\}$, $j = 1, \ldots, M$, M is the total number of words; the properties of the image set can be described by a statistical co-occurrence matrix $\boldsymbol{P}$ of size $N \times M$; and each element $p_{ij} = p(t_i, s_j)$ in the matrix represents the occurrence frequency of the word s_j in the image t_i. The matrix is actually a sparse matrix.

The pLSA model utilizes a latent variable model to describe the data in the co-occurrence matrix. It associates each observation (the word s_j appears in the image t_i) with a latent variable (called the topic variable) $z \in Z = \{z_k\}$, $k = 1, \ldots, K$. Let $p(t_i)$ represent the probability that the word appears in the image t_i; $p(z_k|t_i)$ represents the probability that the topic z_k appears in the image t_i (that is, the image probability distribution in the topic space); and $p(s_j|z_k)$ represents the probability of word s_j appearing under a specific topic z_k (that is, the topic probability distribution in the dictionary). Then by selecting an image t_i with probability $p(t_i)$ and a topic with probability $p(z_k|t_i)$, the probability $p(s_j|z_k)$ for the word s_j can be generated. In this way, the conditional probability model based on the co-occurrence matrix of topics and words can be defined as

$$p(s_j \mid t_i) = \sum_{k=1}^{K} p(s_j \mid z_k) p(z_k \mid t_i) \tag{5.17}$$

That is, the words in each image can be mixed by K implicit topic variables $p(s_j|z_k)$ according to the coefficient $p(z_k|t_i)$. Thus, the elements of the co-occurrence matrix $\boldsymbol{P}$ are

$$p(t_i, s_j) = p(t_i) p(s_j \mid t_i) \tag{5.18}$$

The pLSA model is illustrated graphically in Figure 5.13, where the box represents the set (the large box represents the image set, and the small box represents the repeated selection of topics and words in the image); the arrows represent the dependencies between nodes; the nodes are random variables; the left observation node t (shaded) corresponds to

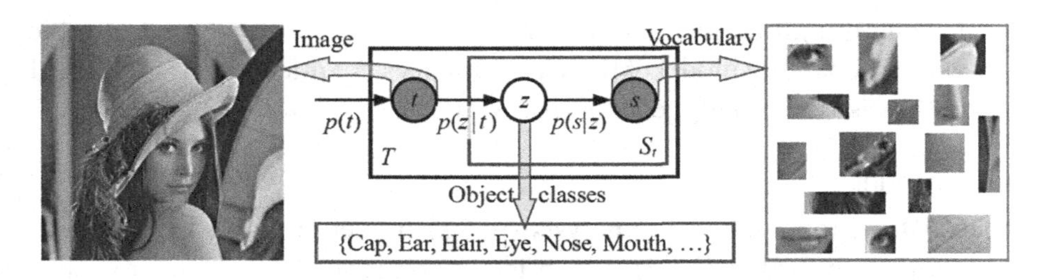

FIGURE 5.13 pLSA schematic model.

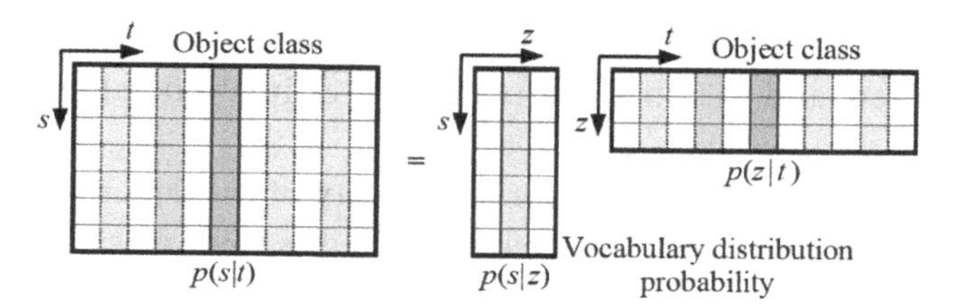

FIGURE 5.14 Decomposition of the co-occurrence matrix.

the image; the right observation node s (shaded) corresponds to the visual vocabulary described by the descriptor; and the middle node z is the (unobserved) latent node, indicating the object category corresponding to image pixels, i.e., the topic. The model establishes a probability mapping relationship between topic z, image t and visual vocabulary s through training, and select the category corresponding to the maximum posterior probability as the result of the final classification decision.

The goal of the pLSA model is to search for the vocabulary distribution probability $p(s_j|z_k)$ under a specific topic z_k and the corresponding mixing ratio $p(z_k|t_i)$ in a specific image, so as to obtain the vocabulary distribution $p(s_j|t_i)$ in a specific image. Equation (5.17) represents each image as a convex combination of K theme vectors, which can be illustrated by matrix operations (see Figure 5.14), in which each column in the left matrix represents a visual vocabulary in a given image, each column in the middle matrix represents a visual vocabulary in a given topic, and each column in the right matrix represents a topic (object category) in a given image.

5.5.2.2 Model Calculation

Here, it is necessary to determine the topic vector common to all images and the special mixing scale coefficient for each image. The purpose is to determine the model that gives high probability to the words appearing in the image, so that the category corresponding to the maximum posterior probability can be selected as the final object category. This can be obtained by optimizing the following objective function to obtain maximum-likelihood estimates of the parameters:

$$L = \prod_{j=1}^{M} \prod_{i=1}^{N} p\left(s_j \mid t_i\right)^{p(s_j|t_i)} \tag{5.19}$$

The maximum-likelihood estimate for the latent variable model can be computed using an **expectation-maximization** (EM) algorithm. The EM algorithm is an algorithm for finding maximum-likelihood estimates or maximum *a posteriori* estimates of parameters in a probability model (depending on unobservable latent variables) in statistical computing. It is an iterative technique for estimating unknown variables given the known partially correlated variables. The algorithm has two alternate iterative computational steps: (i) compute the expectation (E-step), that is, use the existing estimates of the latent variables to compute their maximum-likelihood estimates; (ii) maximize (M-step), that is, the value of the required parameter is estimated on the basis of the maximum-likelihood value obtained in the E-step, and the obtained parameter estimation value is used in the next E-step.

Here, the E-step calculates the posterior probability of the latent variable based on the known parameter estimates, which can be represented as (by Bayesian formula):

$$p(z_k \mid t_i, s_j) = \frac{p(s_j \mid z_k)p(z_k \mid t_i)}{\sum_{l=1}^{K} p(s_j \mid z_l)p(z_l \mid t_i)} \tag{5.20}$$

The M-step is to maximize the likelihood of the fully expected data in the posterior probability obtained from the E-step, and its iterative formula is

$$p(s_j \mid z_k) = \frac{\sum_{i=1}^{N} p(s_j \mid z_k)p(z_k \mid t_i)}{\sum_{l=1}^{K} p(s_j \mid z_l)p(z_l \mid t_i)} \tag{5.21}$$

The E-step and M-step formulas are operated alternately until the termination condition is satisfied. The final decision on the category can be made with the help of the following formula:

$$z^* = \arg\max_{z} \{p(z \mid t)\} \tag{5.22}$$

5.5.2.3 Model Application Example

Consider an image classification problem based on emotional semantics (Li et al. 2010). The image contains not only intuitive scene information but also various emotional semantic information, which can not only express the scene, state and environment of the objective world, but also evoke a strong emotional response. Different emotion categories can generally be expressed by adjectives. There is an emotion classification framework that divides all emotions into ten categories, including five positive ones (amusement, contentment, excitement, awe and undifferentiated positive) and five negative ones (anger, sadness, disgust, fear and undifferentiated negative). An **International Affective Picture System** (IAPS) database (Lang et al. 1997) has been established, in which there are a total of 1182 color pictures, which contain rich object categories. Some pictures belonging to the above ten emotional categories are shown in Figure 5.15. Figures 5.15(a)–5.15(e) correspond to the five positive emotions and Figures 5.15(f)–5.15(j) correspond to the five negative emotions.

In image classification based on emotional semantics, the image is the picture in the database, the words are selected from the emotional category vocabulary, and the topic is the **latent emotional semantic factor** (representing an intermediate semantic layer concept between the underlying image features and high-level emotional categories). First, the k-means algorithm is used to cluster the underlying image features obtained by the SIFT operator into an emotion dictionary. Then, the pLSA model is used to learn the latent

FIGURE 5.15 Examples of pictures of ten emotion categories in the International Affective Picture System database.

emotional semantic factors, so as to obtain the probability distribution $p(s_j|z_k)$ of each latent emotional semantic factor on the emotional word and the probability distribution $p(z_k|t_i)$ of each image on the latent emotional semantic factor. Finally, a **support vector machine** (SVM) method is used to train an emotional image classifier and use it to classify different emotional categories.

Table 5.6 shows some experimental results of classification using the above method, in which 70% of the pictures in each emotion category are used as the training set, and the remaining 30% of the pictures are used as the test set. The training and testing process was repeated ten times, and the average correct classification rate (%) for the ten categories is shown in the table. The value of emotional word s is between 200 and 800 (interval of 100), and the value of the latent emotional semantic factor z is between 10 and 70 (interval of 10).

It can be seen from Table 5.6 that different numbers of latent emotional semantic factors and emotional vocabulary affect the image classification effect. When the value of the latent emotional semantic factor is fixed, with the increase of the number of emotional words, the classification performance gradually improves first and then gradually decreases, and the best value is 500. Similarly, when the number of emotional words is fixed, with the

TABLE 5.6 Examples of Classification

				s			
z	200	300	400	500	600	700	800
10	24.3	29.0	33.3	41.7	35.4	36.1	25.5
20	38.9	45.0	52.1	69.5	62.4	58.4	45.8
30	34.0	36.8	43.8	58.4	55.4	49.1	35.7
40	28.4	30.7	37.5	48.7	41.3	40.9	29.8
50	26.5	30.8	40.7	48.9	39.5	37.1	30.8
60	23.5	27.2	31.5	42.0	37.7	38.3	26.7
70	20.9	22.6	29.8	35.8	32.1	23.1	21.9

increase of the implicit emotional semantic factor, the classification performance gradually improves at first and then gradually decreases, and the best value is achieved when the value of z is 20. Therefore, when $s = 500$ and $z = 20$, the best classification effect can be obtained.

5.5.3 LDA Model

The **latent Dirichlet allocation (LDA)** model is an ensemble probability model. It can be seen as adding a hyper-parameter layer to the pLSA model and establishing the probability distribution of the latent variable z (Blei et al. 2003).

5.5.3.1 Basic LDA Model

The basic LDA model is represented by Figure 5.16(a), where the squares represent repetitions/collections. (The large squares represent a collection of images, M represents the number of images in it; the small squares represent repeated selection of topics and words in the image; and N represents the number of words in an image. It is generally considered that N is independent of q and z.) The leftmost hidden node a corresponds to the Dirichlet prior parameter of the topic distribution in each image; the second hidden node q on the left represents the topic distribution in the image (q_i is the topic distribution in image i), q also is called the mixing probability parameter; the third hidden node z on the left is the topic node, and z_{ij} represents the topic of the word j in the image i; the fourth node s on the left is the only observation node (shaded), and s is the observation variable, s_{ij} represents the j-th word in image i. The upper-right node b is the multinomial distribution parameter of the topic-word layer, that is, the Dirichlet prior parameter of the word distribution in each topic.

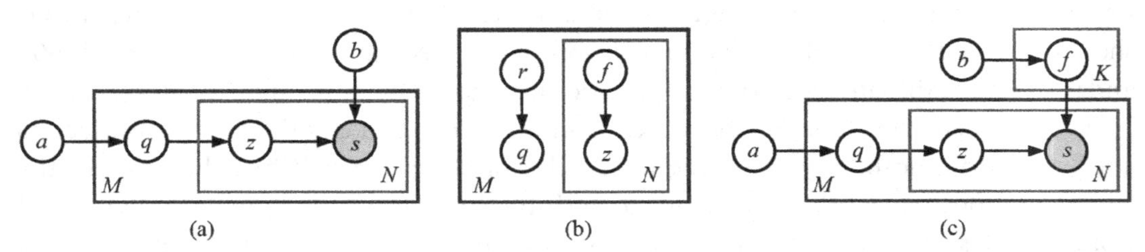

FIGURE 5.16 Schematic diagram of LDA model.

As can be seen from Figure 5.16(a), the basic LDA model is a three-layer Bayesian model, in which a and b are hyper-parameters belonging to the image collection layer, q belongs to the image layer, and z and s belong to the visual vocabulary layer.

The LDA model contains K latent topics $z = \{z_1, z_2, \ldots, z_K\}$, and each word in the image is generated by its corresponding topic. Each image is composed of K subjects mixed with a specific probability q. The model parameter N obeys the Poisson distribution; q obeys the Dirichlet distribution, that is, q ~ Dirichlet(a), a is the Dirichlet distribution prior (the subject distribution of the image conforms to the Dirichlet probability distribution). Each word is an item in the dictionary (visual vocabulary), which can be represented by a K-D vector containing only one 1 and all other 0s. The word s_j is selected from the dictionary with probability $p(s_j|q, b)$.

The solution of the LDA model consists of two processes: approximate **variational inference** (Gibbs sampling can also be used, see Griffiths and Steyvers (2004)) and parameter learning. Variational inference refers to determining the topic mixture probability q of an image and the probability that each word is generated by topic z when given hyper-parameters a and b as well as observation variable s, namely

$$p(q,z \mid s,a,b) = \frac{p(q,z,s \mid a,b)}{p(s \mid a,b)} = \frac{p(q \mid a)\left[\prod_{i=1}^{N} p(z_i \mid q)p(s_i \mid z_i,b)\right]}{\int p(q \mid a)\left[\prod_{i=1}^{N}\sum_{z_i} p(z_i \mid q)p(s_i \mid z_i,b)\right]\mathrm{d}q} \tag{5.23}$$

where the denominator $p(s|a, b)$ is the likelihood function of the word.

Due to the coupling relationship between q and b, $p(s|a, b)$ cannot be calculated directly. According to the LDA model diagram, this coupling relationship is caused by the conditional relationship between q, z and s. Therefore, by deleting the connecting line between q and z and the observation node s in the graph, the simplified model can be obtained, as shown in Figure 5.16(b), and the approximate distribution $p'(q, z|r, f)$ of $p(q, z|s, a, b)$ can be obtained as follows

$$p'(q,z \mid r,f) = p(q \mid r)\prod_{i=1}^{N} p(z_i \mid f_i) \tag{5.24}$$

where r is the Dirichlet distribution parameter of q and f is the polynomial distribution parameter of z.

Further, take the logarithm of $p(s|a, b)$:

$$\log p(s \mid a,b) = L(r,f;a,b) + \mathrm{KL}\left[p'(q,z \mid r,f) \| p(q,z \mid s,a,b)\right] \tag{5.25}$$

The second term on the right of the above equation represents the KL divergence between the approximate distribution model p' and the LDA model p. The smaller the KL divergence, the closer p' is to p. The model parameters R and F can be solved by minimizing the KL divergence by maximizing the lower bound $L(r, f; a, b)$ of the likelihood function. After r and f are determined, q and z can be solved by sampling.

The parameter learning process is the process of determining the super-parameters a and b under the condition of given observation variable set $S = \{s_1, s_2, \ldots, s_M\}$. This can be achieved by variational EM iteration. In E-step, the variational parameters r and f of each image are calculated by using the previous variational reasoning algorithm; in M-step, the variational parameters of all images are collected, the partial derivatives of the super-parameters a and b are obtained, and the lower bound $L(r, f; a, b)$ of the likelihood function is maximized to estimate the super-parameters.

In practice, the basic LDA model is generally extended to a smooth LDA model to obtain better results (to overcome the sparsity problem in large data sets). The diagram of smooth LDA model is shown in Figure 5.16(c), where k represents the number of topics in the model, and f corresponds to a Markov matrix of $K \times V$ (V is the dimension of word vector), in which each line represents the word distribution in the topic.

Here, the image is represented as a random mixture of implicit topics, in which each topic is characterized by the distribution of words. For each image i in the image set, the generation process of LDA is as follows:

(1) Select q_i to satisfy the Dirichlet distribution $q_i \sim$ Dirichlet(a), where $i \in \{1, \ldots, M\}$, Dirichlet(a) is the Dirichlet distribution of parameter a;

(2) Select f_k to satisfy the Dirichlet distribution $f_k \sim$ Dirichlet(b), where $k \in \{1, \ldots, K\}$, Dirichlet(b) is the Dirichlet distribution of parameter b;

(3) For each word s_{ij}, where $j \in \{1, \ldots, N_i\}$, select a topic $z_{ij} \sim$ Multinomial(q_i) and a word $s_{ij} \sim$ Multinomial(f_{zij}), where Multinomial represents polynomial distribution.

5.5.3.2 SLDA Model

In order to further improve the classification performance of the LDA model, we can introduce category information into it, so as to obtain the **supervised LDA model** or **SLDA model** (Wang et al. 2009a). The graph model is shown in Figure 5.17(a). The meaning of each node in the upper part is the same as that in Figure 5.16(a), and a category label node l related to topic z is added in the lower part. The label l corresponding to the topic $z \in Z = \{z_k\}$, $k = 1, \ldots, K$ can be predicted by the parameter h in the Softmax classifier. The reasoning of topic z in SLDA model is affected by category label l, which makes the learned word topic distribution super-parameter d more suitable for classification tasks (also used for annotation).

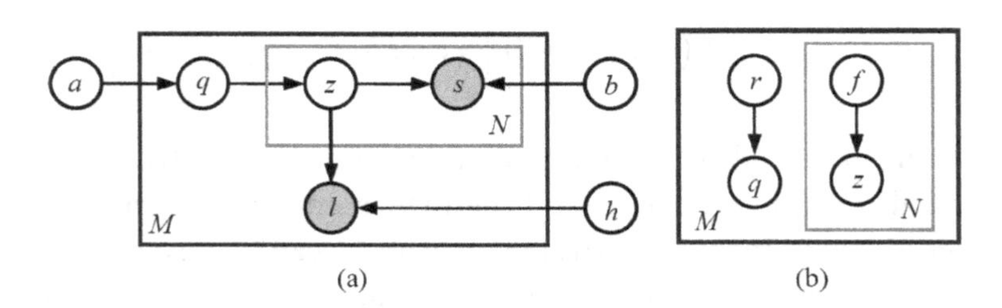

FIGURE 5.17 Schematic diagram of SLDA model.

The likelihood function of the image in the SLDA model is:

$$p(s,l \mid a,b,h) = \int p(q \mid a) \sum_{z_i} \left[\prod_{i=1}^{N} p(z_i \mid q) p(s_i \mid z_i, b) \right] p(l \mid z_{1:N}, h) \mathrm{d}q \tag{5.26}$$

where h is the control parameter of category label l. The parameter learning process is the process of determining the super-parameters a, b and h under the condition of given observation variable set $S = \{s_1, s_2, \ldots, s_M\}$ and category information $\{l_i\}_{i=1:M}$. This can be achieved by a variational EM algorithm. The variational inference of SLDA refers to the process of determining the theme-mixing probability q of the image, the theme probability z of each word and the image category label l under the condition of given super-parameters a, b, h and observation variable s. Compared with LDA, SLDA's variational EM algorithm and variational inference method are more complex (Wang et al. 2009a).

The simplified SLDA model is shown in Figure 5.17(b), which is the same as the simplified LDA model in Figure 5.16(b).

5.6 SOME RECENT DEVELOPMENTS AND FURTHER RESEARCH

In the following sections, some technical developments and promising research directions from the last few years are briefly reviewed.

5.6.1 Interpretation of Remote Sensing Images

Remote sensing imagery is the largest application field of image engineering technology in recent years (Zhang 2018, 2022). At present, remote sensing images are mostly high-resolution optical remote sensing images, which have the characteristics of high spatial resolution, noticeable geometric structure of ground objects, clear texture information, and a large amount of data. Remote sensing image interpretation is the core and key of remote sensing image application.

5.6.1.1 Classification of Remote Sensing Image Interpretation Methods

Over the years, **remote sensing image interpretation** has developed from traditional supervised and unsupervised classifiers to integrated learning and deep learning, and many methods have been proposed. A classification of remote sensing image interpretation methods can be seen in Table 5.7 (Zhang et al. 2021).

In recent years, deep learning technology has shown significant advantages in remote sensing image classification due to its massive data learning ability and high feature abstraction ability. The basic idea of the object recognition method is to use deep convolutional neural networks such as AlexNet and VGGNet to train the labeled data set to obtain a training model, and then use the model to test images with unknown labels. The final-layer output vector of the deep convolutional neural network represents the probability that the input image belongs to each category. The basic idea of the object classification method is to use image segmentation to segment remote sensing images or directly use the surrounding pixels of each pixel to generate objects, take the object as the input of CNN,

TABLE 5.7 Classification of Remote Sensing Image Interpretation Methods

Method Type	Technical Examples	Characteristics
Unsupervised	K-means, fuzzy C-means, mean shift, iterative self-organizing data analysis (ISODATA) methods.	Not suitable for high-resolution images.
Supervised	Maximum likelihood, support vector machines, decision trees.	According to the application purpose and domain, prior knowledge can be fully utilized to selectively determine the classification category; classification accuracy can be improved by repeatedly testing the training samples. However, the determination of the classification system and the selection of training samples are highly subjective, labor-intensive and costly in time.
Integrated learning	Self-service aggregation: such as random forest. Boosting methods: adaptive boosting, gradient boosting iterative decision tree (GBDT), extreme gradient boosting (XGBOOST). Stacking method, model blending (blending) method.	By combining multiple classifiers, it is possible to obtain a better and more comprehensive classification model to compensate for the shortcomings of a single classifier.
Deep learning	Object recognition: using AlexNet, VGGNet. Object-based classification: take objects as input to CNN and identify the class of each object. Semantic segmentation network-based classification: using fully CNN (FCN), deep convolutional networks (DeepLab).	Only the recognition of remote sensing slices can be realized, that is, each slice corresponds to a label, and the pixel-level classification of images cannot be realized. The classification effect depends on the image segmentation accuracy, and the steps are cumbersome. Generalizes the original CNN structure, enabling dense prediction without fully connected layers.

identify the category of each object, and obtain the corresponding label of each object. The basic idea of the network classification method based on semantic segmentation is to use **fully convolutional neural network** (FCN), **deep convolutional network** (DeepLab), etc. to perform semantic segmentation on remote sensing images, and obtain a classification map of the same size as the original image. These methods generalize the structure of the original convolutional neural network and constitute a true end-to-end network capable of dense prediction without fully connected layers.

Semantic segmentation networks are currently developing towards weakly supervised, lightweight, and semantic reasoning. The backbone network has been developed from LeNet, ResNet, DenseNet, VGGNet to MobileNet, NasNet, EfficientNet, etc. The loss functions include **Lovász-Softmax** loss, **Hausdorff distance** (HD) loss, **sensitivity-specificity** (SS) loss, **color-aware** loss, etc. The convolution structure introduces residual block, recurrent convolution, group convolution, ghost convolution, etc. The pooling structure introduces max pooling, average pooling, random pooling, mixed pooling, etc.

5.6.1.2 Knowledge Graph for Remote Sensing Image Interpretation

The knowledge graph was officially named in 2012, when Google proposed the Google Knowledge Graph. Its predecessors include the semantic web proposed in the late 1950s and early 1960s, the "expert system" that appeared in the 1970s, the use of ontology in the philosophical field to create computer models in the mid-to-late 1970s, and the "semantic web" and "linked data" proposed by the World Wide Web.

The knowledge graph describes the concepts, entities and their relationships in the objective world in a structured form, expresses internet information in a form closer to the human cognitive world, and provides a better way to organize, manage and understand the internet's massive information capacity (Wang et al. 2019a).

A **knowledge graph** is essentially a semantic network that expresses various entities, concepts and the semantic relationships between them. It has the following advantages:

(1) High coverage of entity/concept, diverse semantic relationships, friendly structure, and high quality.

(2) It is rich in information such as entities, concepts, attributes, relationships, etc., which can be considered as the cornerstone of cognitive intelligence, enabling machines to fully reproduce the human understanding and interpretation process.

(3) Combined with the learning model under knowledge enhancement, the dependence of the machine learning model on large samples can be reduced, and the learning ability and utilization of prior knowledge can be improved.

Knowledge graphs can be divided into general knowledge graphs, domain knowledge graphs and enterprise knowledge graphs, according to general or special purpose. They can be divided into fully automatic, semi-automatic and manual-based knowledge graphs, according to their construction methods, or into concept graphs, encyclopedia graphs, commonsense graphs, or vocabulary graphs, according to knowledge types.

In the intelligent interpretation of remote sensing images, the knowledge graph can play the role of a knowledge base. The organic integration of relevant knowledge and the realization of in-depth knowledge reasoning through the knowledge graph can greatly improve the ability to accurately and finely analyze and understand remote sensing images.

At present, the construction of the **geographic knowledge graph** has received extensive attention in remote sensing image interpretation. The construction of the geographic knowledge graph is based on the expression of geographic knowledge ontology and the geographic knowledge graph serves geographic knowledge. The construction of the geographic knowledge graph is represented in Figure 5.18. The process includes four main steps – data acquisition, information extraction, knowledge fusion, knowledge processing – that is, starting from the most primitive data, using a series of automatic or semi-automatic technical means to extract geographic knowledge from the original database and third-party databases. The facts are stored in the data layer and model layer of the geographic knowledge base, and finally the geographic knowledge graph is formed.

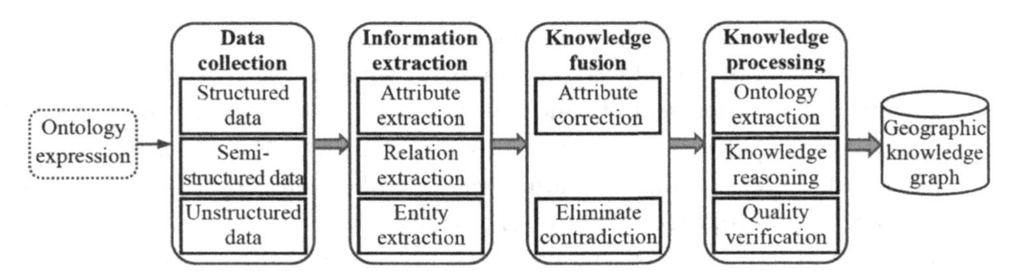

FIGURE 5.18 The process of building a geographic knowledge graph.

After the geographic knowledge graph is constructed, it can further participate in the construction of the "encoding–feature enhancement–decoding" network structure model. Finally, in collaboration with the deep learning model, the semantic classification of remote sensing images is realized, and the interpretation of remote sensing images is completed.

5.6.2 Hybrid Enhanced Visual Cognition

With the deepening of research and the development of technology, some computational intelligence systems have surpassed humans, and some perceptual intelligence systems are comparable to humans. However, there are still many deficiencies in the development of cognitive intelligence, such as poor communication skills, weak understanding ability and high cost of learning. One of the typical challenges is the understanding and reasoning of visual information.

Many researchers have sought to address these issues. In this context, a hybrid enhanced visual cognition architecture that integrates machine perception, human cognition, and computer operations is proposed (Wang and Guan 2021).

Several basic concepts are involved in the architecture:

(1) **Visual perception**. A process of finding out what an object is from an image, understanding and predicting changes in object motion (Shi 2017).

(2) **Visual cognition**. An abstract description of the psychological mechanisms in humans performing visual tasks (such as understanding and recognizing shapes), the processes of which rely on perception, memory, imagination, and logical judgment (Liu et al. 2016).

(3) **Intelligent visual perception**. Refers to the process of image information perception, understanding and prediction with computer operation at the core, referring to optical sensor information, drawing on the biological visual perception mechanism, integrating computer vision processing methods and related intelligent algorithms.

(4) **Computer vision cognition**. Refers to the basic environmental information understood by the perception system, and further expresses internal knowledge through program operations such as reasoning, decision making and learning, and then provides support for changing the environmental state, which is also the key to future state/situational cognition (Ma et al. 2015).

5.6.2.1 From Computer Vision Perception to Computer Vision Cognition

The proposed concept of hybrid enhanced visual cognition is based on the fusion of the perceptual advantages of computer processing large-scale data and the cognitive advantages of human reasoning and decision making, in order to overcome the current technical level of the weak artificial intelligence stage, and realize the completion of complex visual cognitive tasks with computer operation at the core. The purpose of hybrid enhanced visual cognition is, on one side, to enhance people's understanding and cognition of complex data with intelligent visual perception centered on computer operations, and on the other side, to enhance the cognitive learning ability of computers with human reasoning and auxiliary decision support, so as to perform the process of image information detection, understanding, reasoning, prediction, decision making and learning with deep human–computer interaction at the core

This can be seen as a deepening process from computer vision perception to computer vision cognition, as shown in Figure 5.19 (Wang and Guan 2021). The human visual system can complete the natural transition from visual perception to visual cognition. It is still difficult for a computer system to realize the functions of the human visual system and to complete the evolution from computer vision perception to computer vision cognition. The computer vision perception stage with human–computer interaction at its core mainly relies on people to complete advanced operations such as understanding and interpretation of image information, and computers are mainly used as auxiliary tools. In the stage of intelligent visual perception centered on computer operation, algorithms with a certain degree of intelligence can effectively perceive large-scale visual information, but the limited level of cognitive computing makes reasoning and making decisions about the knowledge contained in complex information difficult. Hybrid enhanced visual cognition takes deep human–computer integration as its core, and uses human cognition as an auxiliary element to promote the learning and evolution of computers, so as to finally achieve the level of computer vision cognition with computer cognitive computing as its core.

Based on human–machine collaboration to enhance intelligence, visual perception intelligence will be able to integrate and develop with deeper hybrid enhanced visual cognitive intelligence, and continue to develop visual cognitive reasoning, emotional interaction and auxiliary decision-making cognitive applications on the basis of logical judgment.

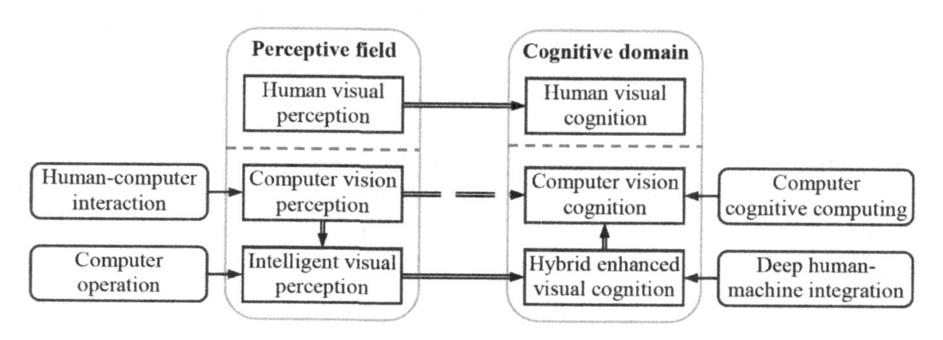

FIGURE 5.19 From computer vision perception to computer vision cognition.

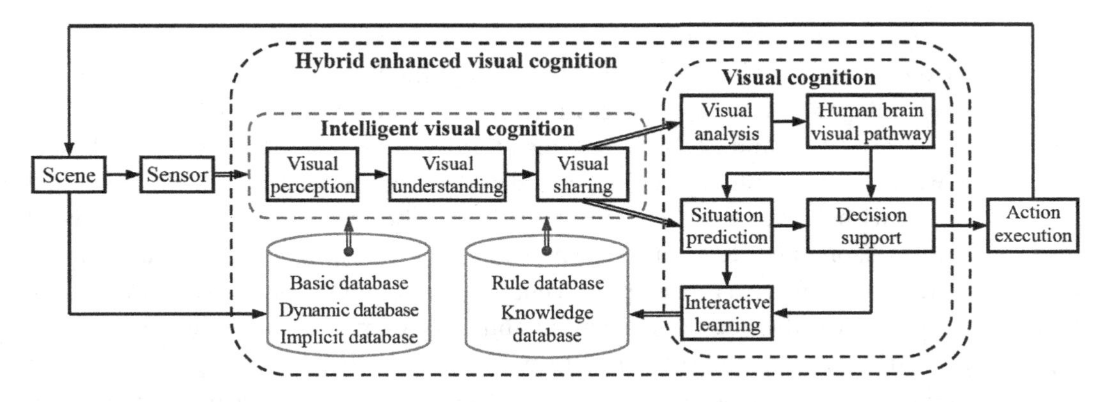

FIGURE 5.20 Basic hybrid enhanced visual cognition framework.

A framework for enhancing visual cognition with human-in-loop mixing is shown in Figure 5.20 (Wang and Guan 2021). First, the scene target and environment data are collected by the vision sensor. Then, computer-based image processing algorithms are used to complete low-level visual element perception, and middle-level visual understanding is completed based on intelligent algorithms, and a joint situation is formed through a visual sharing mechanism to complete intelligent visual perception of large-scale complex data. Next, based on the visualization technology, the feature information is connected with the information processing pathway of the human brain, the human brain and the computer are associated to perform situational prediction and assist decision making, and complete the high-level mixing of complex data to enhance visual cognition. Finally, the human–machine collaboration process is stored in the database, and interaction weights are assigned to complete the learning of visual cognitive strategies.

5.6.2.2 Hybrid Enhanced Visual Cognition Related Technologies

The technologies related to the hybrid enhanced visual cognition architecture mainly include visual analysis, visual enhancement, visual attention, visual understanding, visual reasoning, interactive learning, and cognitive assessment, etc. The relationship between these related technologies and hybrid enhanced visual cognition is shown in Figure 5.21 (Wang and Guan 2021).

The following is a brief description and explanation of the seven types of technologies in Figure 5.21.

(1) Visual analysis

Human beings cannot directly extract useful information from high-dimensional complex data with the naked eye. In order to directly carry out the dialog between people and data as well as help people analyze the uncertainty of data from different angles, it is necessary to carry out visual analysis of complex data (Kovalerchuk 2017). Current research work mainly focuses on lossless visualization and adaptive visualization of high-dimensional data.

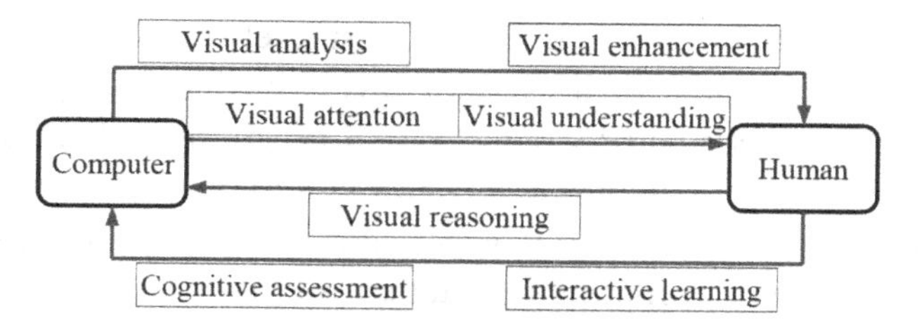

FIGURE 5.21 Technologies related to hybrid enhanced visual cognition.

(2) Visual enhancement

First, **virtual reality** (VR) technology enhances human perception of virtual space (Wu et al. 2018). Second, **augmented reality** (AR) technology transforms synthetic sensor information into human perception of the real environment. Finally, **mixed reality** (MR) uses technologies such as real-time video, 3-D modeling, and multi-sensor information fusion to organically integrate the real world with virtual information, allowing people to experience the virtual world and real objects at the same time, and carrying out surreal hybrid enhanced visual cognition. The research on the existing visual enhancement system also needs to overcome the problem of relatively single functionality, to improve processing ability and human–computer interaction ability.

(3) Visual attention

Attention or attention mechanism is an important psychological adjustment mechanism of human beings, and it is also the basis of human beings' exceptionally prominent data-screening ability. Visual attention can help humans quickly select the most important, useful and relevant visual information. The visual attention mechanism is based on a rough rapid analysis of the image, according to the visual attention model (Wang et al. 2019c, Zibafar et al. 2019), to extract representative informative labels as the basis for subsequent work (Shi et al. 2019).

(4) Visual understanding

Understanding falls more into the realm of neuroscience and cognitive theory, where thinking is based on general human experience. If we want to understand and interpret complex activities in vision, we need to filter out image data of general activities to obtain more essential and in-depth information. In hybrid enhanced visual cognition, the manner, order, and effects of human interaction with scene elements are necessary for visual understanding. In recent years, the attention of visual understanding has gradually shifted to behavioral analysis of high-level visual tasks (Zhang 2013, 2022).

(5) Visual reasoning

Visual reasoning is the process of analyzing visual information and resolving its main contradiction (Daw 2020). Humans have powerful reasoning abilities, based on vision, common sense, and background knowledge, to correctly understand ambiguous information, and to infer the most likely outcomes based on their own experience. The expert knowledge system thus formed is the key to data understanding in the process of visual reasoning. Visual reasoning in hybrid enhanced visual cognition requires a situation where cognitive systems help humans process multiple interpretations and are able to generate essentially the same interpretations as humans under human–computer interaction conditions. For example, a visual reasoning engine (Feest 2021) based on the principles of Gestalt psychology can simulate experienced human decision making, and when ambiguity arises, the most reasonable explanation can be selected in the interaction, making visual cognitive systems have an ability similar to human reasoning.

(6) Interactive learning

Existing interactive learning models integrate machine learning, knowledge bases, and human decision making to learn and predict new data from training data or a small number of samples. When the prediction confidence is low, human-assisted judgment can also be added. This helps to overcome the problems that pure machine learning relies too much on rules, the system has poor portability and scalability so can only work in environments with strict constraints and limited goals, and cannot deal with dynamic, incomplete and unstructured information. Adding active learning and interactive learning capabilities to existing visual systems is not only a challenge to system cognition, but also a practical requirement for autonomous adaptation to environmental changes and continuous intelligence mining. Hybrid enhanced visual cognition enables computers to acquire various empirical data using interactive learning based on imitation (Hou 2019).

(7) Cognitive assessment

Image data with uncertainty, complexity and time constraints will affect the visual cognitive system, resulting in cognitive bias and inability to make the optimal decision. In this case, the hybrid enhanced visual cognition system will generally use heuristic methods to evaluate the information value with the help of empirical rules. At present, visual cognition is mainly evaluated by two modes: automatic calculation prediction and subjective human judgment (Fan et al. 2018).

REFERENCES

Blei, D., A. Ng and M. Jordan. 2003. Latent Dirichlet allocation. *Journal of Machine Learning Research*, 3: 993–1022.

Daw, E. 2020. What is visual reasoning? https://www.wisegeek.com/what-is-visual-reasoning.htm

Fan, S.J., T.T. Ng, B. L. Koenig. 2018. Image visual realism: From human perception to machine computation. *IEEE Transactions on Pattern Analysis and Machine Intelligence*, 40(9): 2180–2193.

Feest, U. 2021. Gestalt psychology, frontloading phenomenology, and psychophysics. *Synthese*, 198: 2153–2173.

Gonzalez, R.C. and R.E. Woods. 1992. *Digital Image Processing*. 3rd Ed. Reading, MA: Addison-Wesley.

Griffiths, T.L. and M. Steyvers. 2004. Finding scientific topics. *Proceedings of the National Academy of Sciences*, 101 (s1): 5228–5235.

Hou, R.L. 2019. Marching ahead in exchanges and mutual learning. *China Today*, 6: 2.

Kovalerchuk, B. 2017. Visual cognitive algorithms for high-dimensional data and super-intelligence challenges. *Cognitive Systems Research*, 45: 95–108.

Lang, P.J., M.M. Bradley and B. N. Cuthbert. 1997. *International Affective Picture System (IAPS): Technical Manual and Affective Ratings*, NIMH Center for the Study of Emotion and Attention. Gainsville, FL: University of Florida.

Li, S., Y.-J. Zhang and H.C. Tan. 2010. Discovering latent semantic factors for emotional picture categorization. *Proceedings of the 17th ICIP*, 1065–1068.

Liu, Y.J., M.J. Yu, Q.F. Fu, et al. 2016. Cognitive mechanism related to line drawings and its applications in intelligent process of visual media: A survey. *Frontiers of Computer Science*, 10(2): 216–232.

Ma, G., X. Yang, B. Zhang, et al. 2015. An environment visual awareness approach in cognitive model ABGP. *Proceedings of the 27th IEEE International Conference on Tools with Artificial Intelligence*, 744–751.

Robison, J.A. 1965. A machine-oriented logic based on the resolution principle. *Journal of ACM*, 12(1): 23–41.

Shi, J. W., Q.G. Zhu and Y.J. Chen. 2019. Human visual perception based image quality assessment for video prediction. *Proceedings of 2019 Chinese Automation Congress (CAC)*, 3205–3210.

Shi, Z. 2017. *Visual Perception, in Mind Computation. New Jersey: World Scientific, Chapter* 5, 183–217.

Sivic, J., B.C. Russell, A.A. Efros, et al. 2005. Discovering objects and their location in images. *Proceedings of the ICCV*, 370–377.

Sivic, J. and A. Zisserman. 2003. Video Google: A text retrieval approach to object matching in videos. *Proceedings of the ICCV*, II: 1470–1477.

Sonka, M., V. Hlavac and R. Boyle. 2014. *Image Processing, Analysis, and Machine Vision*. 4th Ed., Singapore: Cengage Learning.

Wang, C., D. Blei and F.F. Li. 2019a. Simultaneous image classification and annotation. *Proceedings of the CVPR*, 1903–1910.

Wang, H.F., G.L. Qi and H.J. Chen. 2019b. *Knowledge Graph: Method, Practice and Application*. Beijing: Electronic Industry Press.

Wang, P.Y. and X. Guan. 2021. Hybrid enhanced visual cognition framework and its key technologies. *Journal of Image and Graphics*, 26(11):2619–2629.

Wang, W.G., J.B. Shen and Y.D. Jia. 2019c. Review of visual attention detection. *Journal of Software*, 30(2): 416–439.

Wu, W.C., Y.X. Zheng, K.Y. Chen, et al. 2018. A visual analytics approach for equipment condition monitoring in smart factories of process industry. *Proceedings of 2018 IEEE Pacific Visualization Symposium*, 140–149.

Zhang, J.X., H.Y. Gu, Y. Yang, et al. 2021. Research progress and trend of high-resolution remote sensing imagery intelligent interpretation. *National Remote Sensing Bulletin*, 25(11): 2198–2210.

Zhang, Y.-J. 2013. Understanding spatial-temporal behaviors. *Journal of Image and Graphics*, 18(2): 141–151.

Zhang, Y.-J. 2018. An overview of image engineering in recent years. *Proceedings of the 21st IEEE International Conference on Computational Science and Engineering*, 119–122.

Zhang, Y.-J. 2022. Image engineering in China: 2021. *Journal of Image and Graphics*, 27(4): 1009–1020.

Zibafar, A., E. Saffari, M. Alemi, et al. 2019. State-of-the-art visual merchandising using a fashionable social robot: RoMa. *International Journal of Social Robotics*, 13: 509–523.

CHAPTER 6

Multi-Sensor Image Information Fusion

IMAGE INFORMATION FUSION IS a special kind of multi-sensor fusion, which takes the image as the processing operation object. Multi-sensor information fusion is an information technology and process that processes information data from multiple sensors to obtain more comprehensive, accurate and reliable results than can be achieved using a single sensor. **Fusion** can be defined as the technology and process of comprehensively processing and analyzing the data obtained by different sensors, and coordinating, optimizing, and integrating, extracting more information or obtaining new and effective information, thereby improving decision-making capabilities. Fusion can expand the coverage of spatial and temporal information detection, enhance and improve detection capabilities, reduce ambiguity, and increase the credibility of decision making and system reliability.

The chapter is organized as follows. Section 6.1 introduces the categories of information, classification and level of fusion, and the concepts of active vision and active fusion. Section 6.2 explains the main steps of image fusion (preprocessing, registration, and fusion), characteristics and typical technical methods of three fusion levels (pixel-level, feature-level, and decision-level), and subjective and objective evaluation methods for fusion effects. Section 6.3 discusses pixel-level fusion and the techniques for combining various basic methods, briefly introduces the process of image fusion based on compressed sensing, and provides example applications of pixel-level fusion. Section 6.4 discusses Bayesian and evidence-based inference, both of which can be used in both feature-level fusion and decision-level fusion. Section 6.5 introduces the rough set theory method, which is only used for decision-level fusion, as well as the rough set-based fusion process. Section 6.6 reviews technique developments and promising research directions from the last year.

6.1 OVERVIEW OF INFORMATION FUSION

Human perception of the objective world is the result of the combined action of the brain and multiple sense organs, including not only visual information, but also a variety of

DOI: 10.1201/9781003362388-6

non-visual information. For example, intelligent robots currently being studied can have sensors such as vision, hearing, smell, taste, touch (pain), heat (temperature), force, sliding, and proximity, etc. (Luo and Jiang 2002). They perceive information on different sides of the scene in the same environment, and the pieces of information must be related to each other. Therefore, it is necessary to use equivalent information-processing technology to coordinate work among sensors, and this involves the theory and method of multi-sensor fusion.

6.1.1 Multi-Sensor Information Fusion

Multi-sensor information fusion is a basic human ability. The information provided by a single sensor is often incomplete, inaccurate, vague or even contradictory, that is, there is considerable uncertainty. Humans have the instinctive ability to synthesize the information detected by various functional organs of the body with prior knowledge, and to estimate, interpret and judge the environment and the events in it. Using computers and electronic devices to achieve multi-sensor information fusion can be regarded as a functional simulation of how the human brain treats complex problems.

In multi-sensor information fusion, the information obtained by the sensors is put together for comprehensive processing, which has the following advantages in solving problems such as object detection, tracking and recognition, as well as scene classification and interpretation:

(1) The spatial coverage of the system can be enhanced by using multiple sensors to observe different regions.

(2) The spatial resolution of observations can be improved by using multiple sensors to observe the same region.

(3) The reliability and credibility of the system can be enhanced by using multiple sensors to detect the same region.

(4) The amount of information can be increased and the ambiguity can be reduced by using multiple types of sensors to observe the same object or scene.

Information collected from the outside can be divided into the following three categories:

(1) *Redundant information.* This refers to information about the same feature provided by multiple independent sensors (often homogeneous), or to information obtained by a particular sensor multiple times over a period of time. Redundant information can be used to improve fault tolerance and reliability. The fusion of redundant information can reduce the uncertainty caused by measurement noise and improve the accuracy of the system.

(2) *Complementary information.* This refers to the information provided by multiple independent sensors (often heterogeneous) about different features in environmental

information. This information can be combined to form a more complete description of the environment. The fusion of complementary information reduces the ambiguity of environmental understanding caused by the lack of certain environmental features, improves the completeness and accuracy of the environment description, and enhances the ability to make correct decisions.

(3) *Collaborative information.* For example, in a multi-sensor system, when the acquisition of a certain item of sensor information must depend on information from other sensors, the information is called collaborative information. The fusion of collaborative information is often related to the time sequence of the sensors used.

6.1.2 Information Fusion Level

There are many different methods of dividing information fusion into different levels. For example, according to the level of information abstraction (mainly combined with national defense applications, taking the strategic early warning of the battlefield as the background, and considering the military C^3I system – command, control, communication and information), information fusion can be divided into five levels (He et al. 2000):

(1) *Detection-level fusion.* Fusion directly performed at the multi-sensor signal detection level, that is, the signal detected by a single sensor is pre-processed and then transmitted to the central processor.

(2) *Position-level fusion.* Fusion on the output signals of individual sensors. This can include both temporal and spatial fusion. The state of the observed object is obtained through temporal fusion, and the motion trajectory of the object is obtained through spatial fusion.

(3) *Object recognition- level fusion.* Fusion of detected object attributes according to the purpose of object recognition and classification. The methods adopted can be divided into three types. (i) Decision fusion: the estimation classification of object attributes given by each single sensor is fused to obtain consistent estimation. (ii) Feature fusion: the object feature description vector given by each single sensor is fused and estimated by using the joint vector. (iii) Data fusion: the original data given by each single sensor is directly fused, and then the feature extraction and object estimation are carried out based on the data fusion results.

(4) *Situation assessment-level fusion.* The analysis and evaluation of the whole scene based on object recognition, which needs to combine the attributes and behaviors of various objects and events to describe the activities in the scene.

(5) *Threat estimation-level fusion.* The situation emphasizes the state, while the threat emphasizes the trend. Threat estimation-level fusion should not only consider the state information, but also combine prior knowledge, so as to obtain the state change trend and the possible consequences of events.

6.1.3 Active Vision and Active Fusion

In order to solve the ill posed problem of using a single still image to understand a complex scene, a framework called **active perception** or **active vision** was proposed, and was extended to "active, qualitative and purpose vision" (Andreu et al. 2001). The basic idea is to let an active observer collect multiple images of the scene and make comprehensive use of them to transform the ill-posed problem into a clearly defined and easy-to-solve problem.

If an active vision system acquires more than one image for a scene, or more generally, when there are moving observers and objects in the scene and the system is equipped with multiple sensors, the basic problem to be solved is how to integrate the information obtained from multiple sensors or at multiple times. Specifically, we should: (i) select information from different information sources; (ii) register information in space and time; (iii) integrate different information into a new state; and (iv) integrate information at different levels of abstraction (pixels, features, symbols).

Building on the meaning of "active" in active vision, the concept of active fusion has been proposed. Figure 6.1 presents a representation and control framework for image understanding with active fusion at the core (Andreu et al. 2001). The upper part of the figure corresponds to the real-world situation, and the lower part reflects mapping it to the computer. The rectangular solid boxes in the figure indicate processing boxes and the dashed boxes indicate presentation boxes; solid arrows indicate control flow and dashed arrows indicate data flow; and boldface letters and bold arrows emphasize the role of fusion in this framework. The fusion process actively selects the information sources to be analyzed and controls the processing of the data when combining information, so it is called **active fusion**. Fusion can be performed at isolated levels (e.g., fusing multiple input images to produce an output image), or it can combine information from different representation levels (e.g., generating a heat distribution map from surveying maps, digital elevation models, and image information). Processing at all levels can be performed and controlled as desired (e.g., selection of input images, selection of classification algorithms, refinement of results for selected regions).

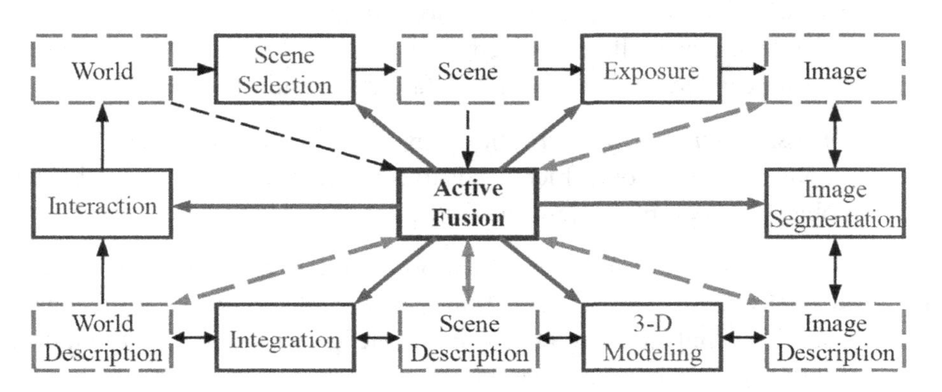

FIGURE 6.1 The framework of general image understanding based on active fusion.

6.2 IMAGE FUSION

The following sections mainly consider the fusion of image and video information, or the fusion of visual information. Some early works can be found in Stathaki (2008).

At present, there are many imaging modes, among which many sensors that can collect different types of images/videos are used, such as visible light sensor (CCD, CMOS), infrared sensor, depth sensor, tomography (CT), magnetic resonance imaging (MRI), synthetic aperture radar (SAR) and millimeter-wave radar sensor (MMWR), etc.

6.2.1 Main Steps of Image Fusion

In order to realize image fusion, we need to use a variety of image technologies to complete three steps.

6.2.1.1 Image Preprocessing

Image preprocessing often includes image normalization (grayscale equalization, resampling, grayscale interpolation), image filtering, enhancing image color and image edge, etc. Image fusion is often performed between images with different sizes, resolutions and gray/color dynamic ranges. The purpose of image normalization is to normalize these parameters. In addition to geometric correction, each image may also be resampled to make them have the same resolution. Image filtering is performed to obtain high-frequency texture information from high-resolution images, so as to retain the information when it is fused with a low-resolution image to produce a high-resolution image. The purpose of image color enhancement is to enhance the color of the low-resolution image, increase its color contrast, and make the image color brighter without changing the original spectral information from the low-resolution image, so as to fully reflect the spectral information of the low-resolution image on the fused image. Image edge enhancement is for high-resolution images; it not only reduces the noise as much as possible, but also makes the image boundary clear and structured, so as to effectively fuse the spatial texture information of high-resolution images into low-resolution images.

6.2.1.2 Image Registration

Image registration is performed to spatially register each image to be fused. Image fusion has high registration accuracy requirements. If the spatial error exceeds one pixel, ghosting will appear in the fusion result, which will seriously affect the quality of the fusion image.

Image registration can be divided into relative registration and absolute registration. Relative registration means selecting a certain (band of) image from multiple images of the same class as a reference image, and then registering another (band of) images with the reference image. Absolute registration means using the same spatial coordinate system as the reference system, and registering multiple images to be fused with this reference system.

Image registration can also be technically divided into region-based and feature-based registration (see discussion in Sections 3.3 and 3.4, respectively). Control points are the typical features commonly used in feature-based registration. There are many image registration algorithms based on control points, which need to first determine the corresponding matching point pairs, and then use the least-squares method to calculate the registration parameters to register the images. To obtain registration point pairs and registration parameters simultaneously, a generalized Hough transform can be used. The generalized Hough transform can be regarded as a voting mechanism. When the correct corresponding point pairs exceed half of the total control points, a better registration result can be obtained. However, the computational complexity of the generalized Hough transform can be considerable, due to the need to perform a global search for scale, rotation parameters and possible control point pairs. In order to reduce the computational complexity of the generalized Hough transform, an iterative method can be used to separate the parameters of the Hough transform and gradually obtain the optimal transformation parameters.

However, iterative Hough transform is easily affected by the initial transformation parameters and parameter value ranges, and often converges to a local maximum. To solve this problem, the robustness of the generalized Hough transform and the efficiency of the iterative Hough transform can be combined, and a multi-scale Hough transform decomposition method can be used (see framework in Figure 6.2) (Li and Zhang 2005).

The image registration algorithm based on multi-scale Hough transform combines the advantages of generalized Hough transform and iterative Hough transform. It only uses a small number of control points in the low-resolution layer, and uses the generalized Hough transform with a larger parameter precision interval to obtain the initial values of the transformation parameters. Since only a few control points are required, the computational complexity of the generalized Hough transform can be reduced. It adopts the iterative Hough transform with faster calculation speed in the high-resolution layer, and uses more control points and combines the initial registration parameters obtained by using the generalized Hough transform to obtain more precise transformation parameters.

6.2.1.3 Image Information Fusion

After preprocessing and registration, the obtained images can be fused in terms of information. Quantitative information fusion fuses a group of similar data to give consistent data, representing the transformation from data to data, while qualitative information fusion can fuse multiple single-sensor decisions into collectively consistent decisions,

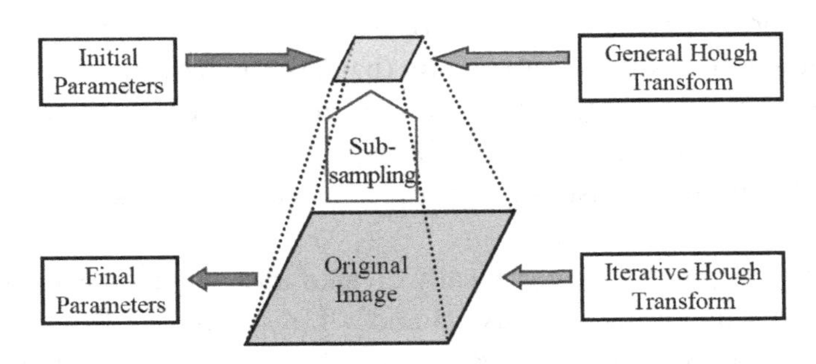

FIGURE 6.2 Framework for image registration using the control point-based multi-scale Hough transform.

representing the conversion between the representation of various uncertainties and the representation of relative consistency. Quantitative information fusion mainly deals with numerical information, while qualitative information fusion mainly deals with non-numerical information. Some specific techniques can be found in the following sections.

6.2.2 Three Levels of Image Fusion

Generally, multi-sensor image fusion methods are divided into three levels from low to high: **pixel-level fusion**, **feature-level fusion** and **decision-level fusion** (Polhl and Genderen 1998). This division is related to the overall work steps to be done with the help of image fusion. In the three-level flow chart of multi-sensor image fusion shown in Figure 6.3, there are three steps from collecting scene images to making judgment decisions: feature extraction, object recognition, and judgment and decision. The three levels of image fusion correspond to these three steps respectively. Pixel-level fusion is performed before feature extraction, feature-level fusion is performed before object recognition (attribute description), and decision-level fusion is performed before judgment and decision (according to the independent attributes of each sensor data). They will be discussed separately below.

(1) Pixel-level fusion

Pixel-level fusion, performed at the underlying data layer, is the processing and analysis of the physical signal data (two or more images) originally collected by the image sensor and the generation of object features to obtain a single fusion image. The advantage of pixel-level fusion is that it retains as much original information as possible, so accuracy is higher than with the other two fusion levels. The main disadvantages of pixel-level fusion are the large amount of processing information, poor real-time performance, high computational cost (high requirements for data transmission bandwidth and registration accuracy), and the requirement for fusion data to be acquired by sensors of the same type or little difference.

(2) Feature-level fusion

Feature-level fusion is fusion at the intermediate level. Features need to be extracted from the original image, object information obtained (such as the edge, contour, shape, surface orientation and mutual distance of the object, etc.) and then integrated to obtain judgment results with higher confidence. Feature-level fusion

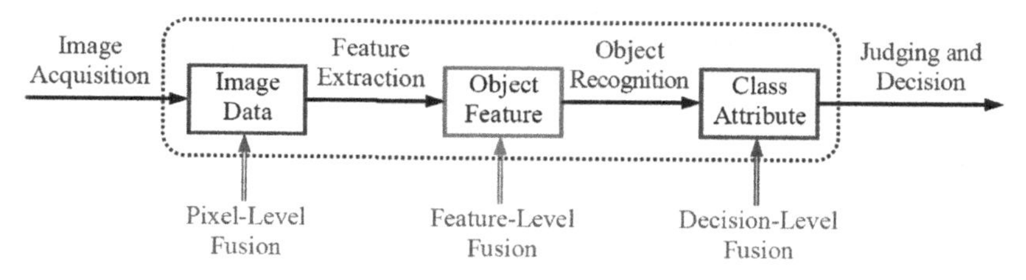

FIGURE 6.3 Flowchart of multi-sensor image fusion.

not only retains important information, but also compresses the amount of data, which is more suitable for heterogeneous or very different sensors. The advantage of feature-level fusion is that the amount of data involved is less than in pixel-level fusion, which is beneficial for real-time processing, and the features provided are directly related to decision analysis. The disadvantage of feature-level fusion is that it is less accurate than pixel-level fusion.

(3) Decision-level fusion

Decision-level fusion is fusion at the highest level, which can directly make optimal decisions according to certain criteria and the credibility of each decision. Before decision-level fusion, the corresponding processing components of each sensor have completed the work of object classification or recognition. Decision-level fusion is often carried out by means of symbolic operations, which has the advantages of strong fault tolerance, good openness and high real-time performance. The disadvantage of decision-level fusion is that there may be a relatively large loss of original information before fusion, so accuracy in space and time is often relatively low.

Some of the main properties of the above three fusion approaches are summarized in Table 6.1 (Jia et al. 2000).

The techniques used for different fusion methods often differ, but some techniques can also be used for fusion of different fusion levels. Some of the fusion techniques commonly used in the above three fusion levels are listed in Table 6.2 (Jia et al. 2000, Luo and Jiang 2002). Different fusion techniques are at different levels from an abstract point of view, and

TABLE 6.1 Principal Properties of Three Fusion Modes

Fusion Level	**Abstraction Level**	**Information Loss**	**tolerance**	**Anti-disturb**	**Precision**	**Real Time**	**Computation Complexity**
Pixel Level	Low	Small	Bad	Bad	High	Poor	Big
Feature Level	Middle	Moderate	Middle	Middle	Middle	Moderate	Middle
Decision Level	High	Big	Good	Good	Low	Good	Small

TABLE 6.2 Commonly Used Fusion Techniques in the Three Fusion Modes

Pixel-Layer Fusion	**Feature-Layer Fusion**	**Decision-Layer fusion**
Weighted Average	Weighted Average	Knowledge-Based
Pyramid Fusion	Bayesian Method	Bayesian Method
HSI Transformation	Evidence Reasoning (D-S Theory)	Evidence Reasoning (D-S Theory)
PCA Operation	Neuron Network	Neuron Network
Wavelet Transformation	Cluster Analysis	Fuzzy Set Theory
High-Pass Filtering	Entropy	Reliability Theory
Kalman Filtering	Vote-based	Logical Module
Regression Model		Production Rule
Parameter Estimation		Rough Set Theory

the fusion development trend is from pixels to regions (objects) (Piella 2003). An introduction to some of these typical techniques (shaded) is given in the next three sections.

6.2.3 Evaluation of Image Fusion Effect

Evaluation of the image fusion effect is an important part of research in fusion. Different methods and indicators are often used to evaluate the effect of fusion at different levels. The fusion of lower levels often analyzes and compares more from the visual effect, while the fusion of higher levels emphasizes the help of fusion in completing tasks. Although there are many kinds of image fusion techniques, the enrichment of fusion image information and the improvement of image quality are always the fundamental purpose of image fusion, and also the basic criteria for measuring the effect of various fusion techniques. The ideal fusion process should not only excavate and introduce new information, but also retain the original useful information. Evaluation of the fusion effect should include innovation and inheritance.

Evaluation of the image fusion effect is different from the distortion evaluation of encoded and decoded images. The difference between the input and output images is considered in the image coding evaluation, and the smaller the difference (less distortion), the better. The input in image fusion has at least two images, and the fusion effect is not completely determined by the difference between them and the output image. Methods for systematic and comprehensive evaluation of the fusion effect are still being researched.

Evaluation of the image fusion effect includes subjective evaluation and objective evaluation. The former relies on the subjective feeling of the observer, but it should be noted that this varies not only with the observers themselves, but also with the interests of the observers and the requirements of application fields and occasions. Objective evaluation is often carried out according to some computable indicators, which has a certain correlation with subjective visual effects but is not equivalent. In the basic evaluation metrics introduced below, $f(x, y)$ represents the original image, $g(x, y)$ represents the fused image, and their dimensions are $N \times N$. In the discussion, both $f(x, y)$ and $g(x, y)$ take grayscale images as an example, but the evaluation indicators obtained are also easy to generalize to other types of images.

6.2.3.1 Subjective Evaluation

Subjective evaluation, usually carried out by visual inspection, often includes the following:

(1) Judging the accuracy of image registration – if the registration is not good, then the fusion image will appear ghosting.

(2) Judging the overall color distribution of the fused image – if it is consistent with the natural color, the color of the fused image is true.

(3) Judging the overall brightness and color contrast of the fused image – if the overall brightness and color contrast are not appropriate, fog or patches will appear.

(4) Judging the richness of texture and color information of the fused image – if the spectral and spatial information is lost during the fusion process, the fused image will appear dull.

(5) Judging the sharpness of the fused image – if the sharpness of the image decreases, the edges of the object image will become blurred.

The subjective evaluation method is more intuitive, fast and convenient when evaluating obvious image information, but it is generally only a qualitative analysis and description, with relatively strong subjectivity.

6.2.3.2 Objective Evaluation Based on Statistical Characteristics

(1) Grayscale mean value

The average gray level of an image is reflected as the average brightness perceived by the human eye, which has a great impact on the visual effect of the image. The grayscale mean of a fused image is

$$\mu = \frac{1}{N \times N} \sum_{x=0}^{N-1} \sum_{y=0}^{N-1} g(x,y) \tag{6.1}$$

If the average gray level of an image is moderate, the subjective visual effect of the image will be better.

(2) Grayscale standard deviation

The grayscale standard deviation of an image reflects the dispersion of each gray level relative to the average gray level, and can be used to judge the amplitude of the image contrast. The formula for calculating the grayscale standard deviation of a fused image is:

$$\sigma = \frac{1}{N \times N} \sqrt{\sum_{x=0}^{N-1} \sum_{y=0}^{N-1} \left[g(x,y) - \mu \right]^2} \tag{6.2}$$

If the grayscale standard deviation of an image is small, it means that the contrast of the image is small (the contrast between adjacent pixels is small), the overall tone of the image is relatively uniform and single, and less information can be observed. The opposite is true if the grayscale standard deviation is larger.

(3) Average grayscale gradient

The average grayscale gradient of an image is similar to the grayscale standard deviation and also reflects the contrast in the image. Since the calculation of gradients is often carried out around the local region, the average gradient more especially reflects the small detail changes and texture features of the local image. A formula for calculating the average grayscale gradient of a fused image is

$$A = \frac{1}{N \times N} \sum_{x=0}^{N-1} \sum_{y=0}^{N-1} \sqrt{G_X^2(x,y) + G_Y^2(x,y)} \tag{6.3}$$

where $G_X(x, y)y)$ and $G_Y(x, y)$ are the differences (gradients) of $g(x, y)$ along the X and Y directions, respectively. If the average grayscale gradient of an image is small, it means there are fewer image levels; otherwise, the average grayscale gradient is large, and the general image will be clearer.

(4) Grayscale deviation

The grayscale deviation between the fused image and the original image reflects their difference in spectral information (also known as spectral distortion), which is calculated as

$$D = \frac{1}{N \times N} \sum_{x=0}^{N-1} \sum_{y=0}^{N-1} \frac{|g(x,y) - f(x,y)|}{f(x,y)} \tag{6.4}$$

If the grayscale deviation is small, it indicates that the fused image better retains the grayscale information of the original image.

(5) Mean square error

When the ideal image (that is, the expected result of fusion, or the true value) is known, the mean square error between the ideal image and the fused image can be used to evaluate the fusion result. If the ideal image is represented by $i(x, y)$, then the mean square error between the ideal image and the fused image is

$$E_{rms} = \left\{ \frac{1}{N \times N} \sum_{x=0}^{N-1} \sum_{y=0}^{N-1} \left[g(x,y) - i(x,y) \right]^2 \right\}^{1/2} \tag{6.5}$$

6.2.3.3 Objective Evaluation Based on the Amount of Information

(1) Entropy

Consider the entropy of an image, which is a measure of the richness of information in that image. The entropy of an image can be calculated from the histogram of the image. Let the histogram of the image be $h(l)$, $l = 1, 2, \ldots, L$, then the entropy is

$$H = -\sum_{l=0}^{L} h(l) \log\left[h(l)\right] \tag{6.6}$$

If the entropy of the fused image is greater than that of the original image, it means that the information content of the fused image has increased relative to that of the original image.

(2) Cross-entropy

The cross-entropy (also called directional divergence) between the fused image and the original image directly reflects the relative difference in the amount of information contained in the two images. The symmetric form of cross-entropy is called **symmetric cross-entropy**. If the histograms of the fused image and the original image are $h_g(l)$ and $h_f(l)$, respectively, and $l = 1, 2, \ldots, L$, the formula for calculating the symmetric cross-entropy between them is:

$$K(f:g) = -\sum_{l=0}^{L} h_g(l)\log\left[\frac{h_g(l)}{h_f(l)}\right] - \sum_{l=0}^{L} h_f(l)\log\left[\frac{h_f(l)}{h_g(l)}\right] \tag{6.7}$$

The smaller the cross-entropy, the more information the fused image gets from the original image.

(3) Correlation entropy/joint entropy

The correlation entropy between the fused image and the original image is also a measure that reflects the correlation between the two images. The correlation entropy between two images can be calculated as:

$$C(f:g) = -\sum_{l_2=0}^{L}\sum_{l_1=0}^{L} P_{fg}(l_1,l_2)\log P_{fg}(l_1,l_2) \tag{6.8}$$

where $P_{fg}(l_1, l_2)$ represents the joint probability that the pixel at the same position in the two images has a gray value of l_1 in the original image and a gray value of l_2 in the fused image. Generally speaking, the larger the correlation entropy between the fused image and the original image, the better the fusion effect.

(4) Mutual information

The mutual information between two images reflects the information connection between them, which can be calculated using the probabilistic meaning of the image histogram. With the help of $h_f(l)$, $h_g(l)$ and $P_{fg}(l_1, l_2)$ defined for correlation entropy, the mutual information between the two images is

$$H(f,g) = \sum_{l_2=0}^{L}\sum_{l_1=0}^{L} P_{fg}(l_1,l_2)\log\frac{P_{fg}(l_1,l_2)}{h_f(l_1)h_g(l_2)} \tag{6.9}$$

If the fusion image $g(x, y)$ is obtained from two original images $f_1(x, y)$ and $f_2(x, y)$, then the mutual information of $g(x, y)$ with $f_1(x, y)$ and $f_2(x, y)$ is (if the mutual information of f_1 and g is related to the mutual information of f_2 and g, the relevant part $H(f_1, f_2)$ needs to be subtracted)

$$H(f_1,f_2,g) = H(f_1,g) + H(f_2,g) - H(f_1,f_2) \tag{6.10}$$

It reflects the amount of original image information contained in the final fused image. Equation (6.10) can be generalized and used for the fusion of a greater number of images.

6.2.3.4 Evaluation According to the Purpose of Fusion

The selection of fusion evaluation indicators is often carried out according to the purpose of fusion. Here are a few examples:

(1) If the purpose of fusion is to remove the noise in the image, the evaluation index based on the signal-to-noise ratio can be used.

(2) If the purpose of fusion is to improve the image resolution, the evaluation index based on statistical characteristics and spectral information can be used.

(3) If the purpose of fusion is to increase the amount of image information, the evaluation index based on the amount of information can be used.

(4) If the purpose of fusion is to improve the clarity of the image, the evaluation index based on the average grayscale gradient can be used.

Generally, improved image clarity greatly helps to enrich the detailed information and improve the visual effect. Therefore, to reflect the fusion effect from the local and subjective point of view, fusion effect evaluation can often be carried out with the help of indicators based on statistical characteristics. In general, the improvement of image information will help in image feature extraction and object recognition. Therefore, to reflect the fusion effect from the overall and objective points of view, fusion effect evaluation can often be carried out with the help of indicators based on information content.

6.3 PIXEL-LEVEL FUSION METHODS

In **pixel-level fusion,** the original images to be fused often have some different but complementary characteristics, so the method chosen to fuse them should also take these different characteristics into account.

6.3.1 Basic Fusion Methods

The fusion of the **thematic mapper** (TM) multispectral image obtained by the LANDSAT earth resources satellite and the **earth observation system** (SPOT, *système probatoire d'observation de la terre*) panchromatic image obtained by the SPOT remote sensing satellite provides examples of several basic and typical pixel-level fusion methods (see the following images for some experimental results in Subsection 6.3.5) (Bian et al. 2005), (Li and Zhang 2005). Current TM multispectral images cover seven bands from blue to thermal infrared (wavelength 0.45 ~ 12.5 μm), while SPOT panchromatic images cover five bands from visible and near infrared spectrum (wavelength 0.5 ~ 1.75 μm). The spatial resolution

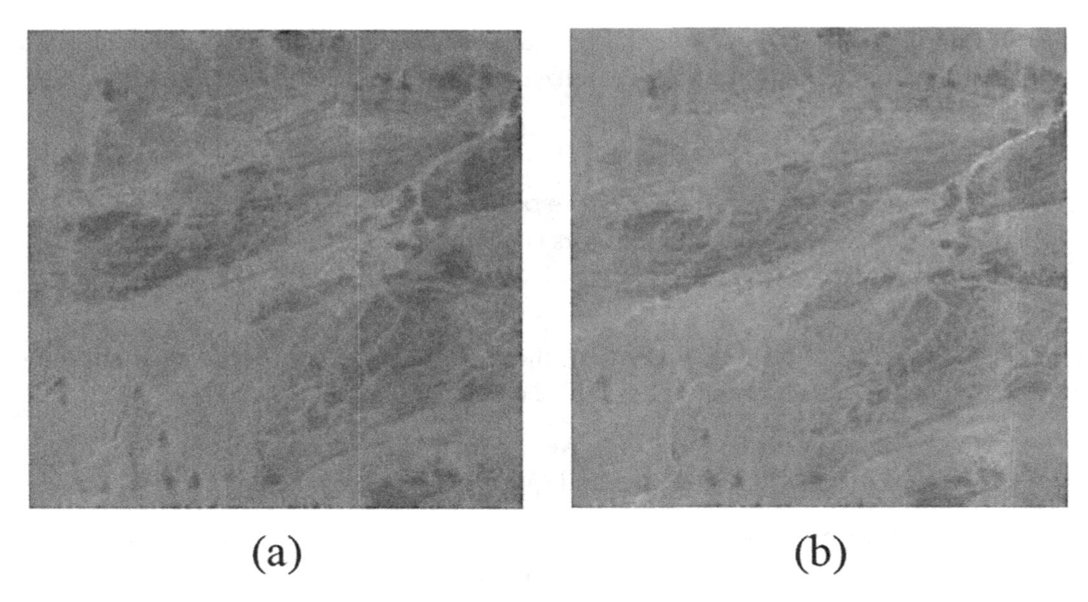

FIGURE 6.4 Example of TM original image and SPOT original image.

of SPOT images is higher than that of TM images, but the spectral coverage of TM images is larger than that of SPOT images. Figures 6.4(a) and 6.4(b) respectively show the TM images $f_t(x, y)$ with wavelength 0.5 ~ 0.73 μm (TM5 band) and SPOT panchromatic image $f_s(x, y)$ with wavelength 1.55 ~ 1.75 μm, both for the same scene.

6.3.1.1 Weighted Average Fusion Method

The **weighted average fusion method** is a relatively simple and intuitive method, and its specific steps are as follows:

(1) Select the region of interest in $f_t(x, y)$.

(2) The image of each band in the region is expanded into a high-resolution image by resampling.

(3) Select $f_s(x, y)$ corresponding to the same region and register it with $f_t(x, y)$.

(4) Perform algebraic operations as follows to obtain a weighted average fused image:

$$g(x,y) = w_s f_s(x,y) + w_t f_t(x,y) \tag{6.11}$$

where w_s and w_t represent the weighted values of $f_s(x, y)$ and $f_t(x, y)$, respectively.

6.3.1.2 Pyramid Fusion Method

A pyramid is a data structure that spatially represents an image. The **pyramid fusion method** uses the pyramid structure obtained by the multi-scale decomposition of the image to fuse the image. The specific steps are as follows:

(1) Select the region of interest in $f_t(x, y)$.

(2) The image of each band in the region is expanded into a high-resolution image by resampling.

(3) Pyramid decomposition of each image $f_t(x, y)$ and $f_s(x, y)$ participating in the fusion.

(4) Fusion of the corresponding $f_t(x, y)$ decomposition results and $f_s(x, y)$ decomposition results at each layer of the pyramid to obtain a fused pyramid.

(5) The fused image is reconstructed from the fused pyramid using the inverse process of pyramid generation.

6.3.1.3 *Wavelet Transform Fusion Method*

Wavelet transform can decompose an image into low-frequency sub-images and high-frequency sub-images corresponding to different structures in the image. The **wavelet transform fusion method** uses various sub-images obtained by wavelet decomposition for fusion. The specific steps are as follows:

(1) Perform wavelet transform on $f_t(x, y)$ and $f_s(x, y)$, respectively, to obtain their individual low-frequency sub-images and high-frequency sub-images.

(2) Replace the low-frequency sub-image of $f_s(x, y)$ with the low-frequency sub-image of $f_t(x, y)$.

(3) Perform wavelet inverse transformation on the replaced $f_t(x, y)$ low-frequency sub-image and $f_s(x, y)$ high-frequency sub-image to obtain the fusion image.

The above wavelet transform fusion result effectively retains the low-frequency part of $f_t(x, y)$ representing spectral information, and adds the high-frequency part containing detailed information in $f_s(x, y)$, so the fusion result image will have improvements in both the visual effect and the statistical results.

6.3.1.4 *HSI Transform Fusion Method*

HSI transform is the transform from RGB color space to HSI color space. The **HSI transform fusion method** is a method of performing fusion operations with the help of HSI transform. The specific steps are as follows:

(1) Select the three band images in $f_t(x, y)$ as RGB images and transform them into HSI space.

(2) Replace the I component obtained after HSI transform with $f_s(x, y)$; this component mainly determines the details of the image.

(3) Perform HSI inverse transform to obtain a new RGB image and use it as a fusion image.

6.3.1.5 PCA Transform Fusion Method

The basis of the **PCA transform fusion method** is **principal component analysis** (PCA). The specific steps are as follows:

(1) Select images of three or more bands in $f_t(x, y)$ for PCA transform.

(2) Histogram matching is performed between $f_s(x, y)$ and the first principal component image obtained after the above PCA transform, so that their gray mean and variance are consistent.

(3) Use the above-matched $f_s(x, y)$ to replace the first principal component image obtained by performing PCA transform on $f_t(x, y)$, and then perform inverse PCA transform to obtain the fusion image.

6.3.2 Combining Various Fusion Methods

Each of the various (single-type) fusion methods described above has its own advantages and disadvantages. In order to overcome the problems caused by the use of each fusion method alone, they are often used in combination to benefit from each other's strengths.

6.3.2.1 Problems with a Single-Type Fusion Method

The weighted average fusion method is the simplest and requires the least amount of computation, but the anti-interference ability is poor, and the quality of the fusion image is often not ideal. A typical problem is that the fused image is not sharp enough due to the average smoothing effect.

The pyramid fusion method is simple and easy to implement, and the fusion image has high definition. However, the data between different levels of the pyramid has correlation, or there is redundancy between images of different scales, resulting in a large amount of processed data. In addition, the reconstruction of the pyramid is unstable, especially when there is a significant difference between the two images.

The wavelet transform decomposes the image into high-frequency detail parts and low-frequency approximate parts, which correspond to different structures in the image respectively, so it is easy to extract the structural information and detail information of the image. Wavelet transform has perfect reconstruction ability, which can ensure that there is no information loss or redundant information produced in the decomposition process. The wavelet transform fusion method effectively retains the low-frequency part of the multispectral image representing spectral information, and adds the high-frequency part of the detailed information contained in the panchromatic image, so the fusion result image will be better both in the visual effect and the statistical result. However, there are two problems for the standard wavelet transform: (i) since the standard wavelet transform is equivalent to performing high-pass and low-pass filtering, using the low-frequency part of the TM image to replace the low-frequency part of the SPOT image for inverse wavelet transform will inevitably cause the loss of the original information in the TM image; and (ii) since the gray values between the TM and SPOT images are generally significantly different, the spectral information in the fused TM image will change and even lead to the appearance of noise.

When the HSI transform fusion method fuses the TM multispectral image and the SPOT panchromatic image, the fused image can have the spatial resolution of the original high-resolution panchromatic image, so the spatial detail information of the image is enhanced. However, if the SPOT panchromatic image is used to completely replace the I component of the TM multispectral image, a great deal of spectral information will also be lost and the spectral distortion will be considerable.

The PCA transform fusion method can make the fused image contain both high spatial resolution and high spectral resolution features in the original image, and the detailed features of the object in the fused image will be clearer. However, as before, simply replacing the first principal component of the TM image with the SPOT image during fusion will cause some useful information reflecting the spectral characteristics in the first principal component of the TM image to be lost, affecting the spectral resolution of the fusion result image.

6.3.2.2 Fusion by Combining HSI Transform and Wavelet Transform

The characteristic feature of HSI transform fusion images is that the high-frequency information is rich, but there is a big loss of spectral information. The wavelet transform fusion image preserves the spectral information of the multispectral image well, but because the low-frequency part of the high spatial resolution image is discarded, the block effect will appear, which affects the visual effect. The specific steps of a fusion method combining HSI transform and wavelet transform are as follows:

(1) Select three band (such as TM3, TM4, TM5) images in $f_t(x, y)$ (here $f_t(x, y)$ can be regarded as a vector image) and transform them from RGB space to HSI space.

(2) Perform wavelet transform decomposition on the obtained I component and $f_s(x, y)$, respectively.

(3) Replace the high-frequency coefficients obtained by the wavelet decomposition of the I component with the high-frequency coefficients obtained from the wavelet decomposition of $f_s(x, y)$.

(4) All wavelet decomposition coefficients obtained after replacement are obtained by inverse wavelet transform to obtain a new luminance component I'.

(5) Use H, S and I' components to transform from HSI space to RGB space to obtain the fused image.

Figures 6.5(a) and 6.5(b) show the results obtained by performing HSI transform fusion and wavelet transform fusion on the original images of Figures 6.4(a) and 6.4(b), respectively (Bian et al. 2005). Figure 6.5(c) shows the result obtained by the fusion method combining HSI transform and wavelet transform. As can be seen from these figures, because the fusion result retains both the high-frequency information of the SPOT panchromatic

(a) (b) (c)

FIGURE 6.5 Fusion image obtained by combining HSI transform and wavelet transform.

image and the texture information of the TM multispectral image, the details are clearer, and the visual effect is better than that of the HSI transform fusion or wavelet transform fusion alone. There has been improvement.

6.3.2.3 Fusion by Combining PCA Transform and Wavelet Transform

The PCA transformation method introduced in Subsection 6.3.1 directly uses $f_s(x, y)$ to replace the first principal component of the TM image. The disadvantage is that some spectral information reflecting the spectral characteristics in the first principal component of the TM image is lost (the spectral characteristics in the first principal component of the TM image are not completely consistent with the spectral characteristics of the SPOT image), thus making the spectral characteristics of the fusion result image. Resolution is greatly affected and some spectral distortion is caused. Combining wavelet transform and PCA transform can improve the effect (both the detail information of the SPOT image is integrated and the spectral information in the first principal component of the TM image is preserved). The specific steps are as follows:

(1) Perform PCA transformation on all bands of $f_t(x, y)$.

(2) Histogram matching of the first principal component image obtained by transformation and $f_s(x, y)$ (to make the grayscale mean and variance of the two images consistent).

(3) Perform wavelet transform decomposition on both matched images.

(4) Replace the high-frequency coefficients obtained from the first principal component wavelet decomposition of $f_t(x, y)$ with the high-frequency coefficients obtained from the wavelet decomposition of $f_s(x, y)$.

(5) Obtain the new first principal component of $f_t(x, y)$ by inverse wavelet transform of all the wavelet decomposition coefficients obtained after substitution.

(6) Use the new $f_t(x, y)$ first principal component and the original components to jointly perform the inverse PCA transformation.

(a)

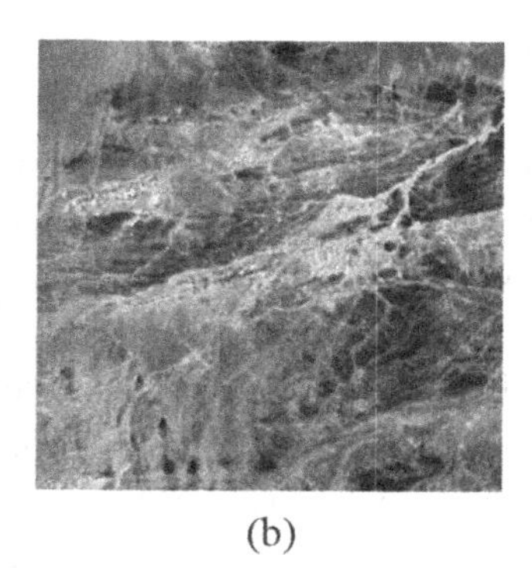

(b)

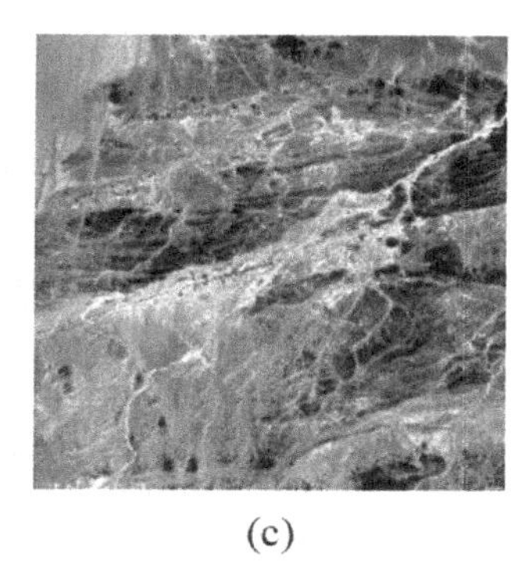

(c)

FIGURE 6.6 Fusion image obtained by combining PCA transform and wavelet transform.

Figures 6.6(a) and 6.6(b) show the results obtained by performing PCA transform fusion and wavelet transform fusion on the original images of Figures 6.4(a) and 6.4(b), respectively (Bian et al. 2005). Figure 6.6(c) shows the result obtained by the fusion method combining PCA transform and wavelet transform. Because the fusion result enhances the texture features in the image, the spectral information is more abundant, the object details are more refined, and the object outline is clearly distinguishable, so the visual effect of the fusion image is improved and better than that obtained by using PCA transform fusion or wavelet transform fusion alone.

6.3.2.4 *Performance of Combined Fusions*

The effect of the above two types of combined fusion and the effect of single-type fusion can be compared using the various evaluation indicators discussed in Subsection 6.2.3. Figures 6.7(a), 6.7(b), and 6.7(c) show the (normalized) comparison results of the three single-type fusion methods and the two combined methods using the grayscale standard deviation, average grayscale gradient and entropy, respectively. For each fusion method, the average obtained from the fusion results of the three bands was considered.

As can be seen from Figure 6.7, the fusion result of the combination of HSI transform and wavelet transform, compared with the fusion result of the single HSI transform, the average grayscale gradient decreases slightly, but the entropy has some increases, which indicates that the image contains more information. In addition, compared with the fusion result of single wavelet transform, the fusion results of HSI transform and wavelet transform have increased values of grayscale standard deviation, average grayscale gradient and

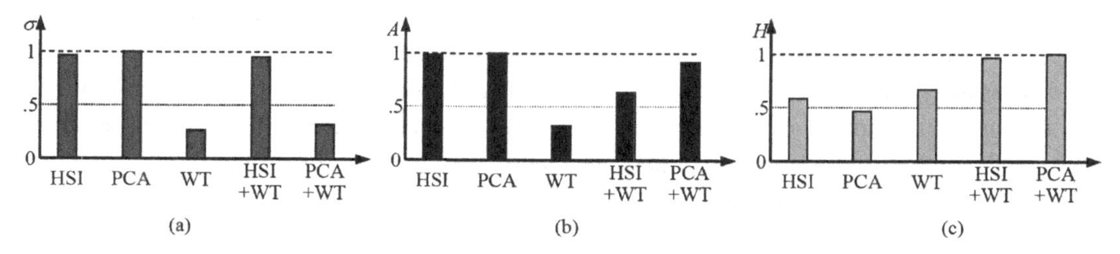

FIGURE 6.7 Comparison of the results obtained by the five fusion methods.

entropy, indicating that the effect of this fusion method is better than that of the wavelet transform fusion method.

It can also be seen from Figure 6.7 that the fusion result of the combination of PCA transform and wavelet transform, compared with the fusion result of single PCA transform, the grayscale standard deviation is relatively reduced, the average grayscale gradient is basically the same (but relatively consistent among various bands), and the entropy has a larger increase, which also indicates that the image contains more information. In addition, compared with the fusion result of single wavelet transform, the fusion result of combining PCA transform and wavelet transform has a slight increase in the grayscale standard deviation, but the average grayscale gradient increases significantly, indicating that the clarity of details has been greatly improved, and the entropy also has a certain increase, indicating that the effect of such fusion is better than that of the wavelet transform fusion method.

6.3.3 The Optimal Number of Decomposition Layers for Wavelet Fusion

When using wavelet transform for fusion, if fewer layers of wavelet decomposition is selected, performance for the spatial details of the fused image will be poor, but the spectral characteristics will be better retained; if more wavelet decomposition layers are selected, the performance for the spatial detail representation of the image is better, but fewer spectral properties will be retained. This is because when the number of layers of wavelet decomposition is small, after the panchromatic image is decomposed, it still contains a lot of detail information in its low-frequency part, but this detail is ignored in the fusion process, so the performance for the spatial detail of the fusion image will be reduced. However, after the multispectral image is decomposed, it does not contain much spectral information in its high-frequency part, so the spectral characteristics of the fusion image remain better. When there are more layers of wavelet decomposition, the result is opposite. If we analyze it from the perspective of computational cost, when the number of decomposition layers is too large, the rapid increase of computational complexity will far exceed the improvement of fusion effect, and increasing the number of decomposition layers will not be worth the loss. Therefore, in practical applications, an appropriate value for the wavelet decomposition layer number should be selected, so that the fused image can achieve a compromise between the enhancement of spatial detail information, the preservation of multispectral information, and the computational complexity.

An analysis of choosing the optimal number of decomposition layers for multi-scale fusion using wavelet transform can be found in Li and Zhang (2005), in which the situation with an ideal reference image is derived and verified experimentally. First, images with different focus are combined to form experimental images, and then they are gradually decomposed and fused layer by layer. By comparing the fusion results at two adjacent layers, the optimal number of decomposition layers is directly estimated according to their closeness. The following only introduces its specific experimental results.

A total of ten test images were used in the experiment, as shown in Table 6.3. For each image, Gaussian filter with 7×7 windows was used to blur the center part and surrounding part of the image separately (corresponding to the effect of different focus), and get two

TABLE 6.3 A Comparison of the Actual Best Level L_{best} and Estimated Best Level L_{est}

Decomposition Methods	**LPT**			**DWT**			**DWF**			**DT-CWT**		
Variances of Blurring Filter	**0.5**	**1**	**5**	**0.5**	**1**	**5**	**0.5**	**1**	**5**	**0.5**	**1**	**5**
Airplane	1↔1	1↔1	2↔2	1↔1	3↔3	4↔4	1↔1	2↔2	4↔4	1↔1	2 ≠ 1	3↔3
Baboon	1↔1	1↔1	2↔2	2 ≠ 1	3↔3	4↔4	1↔1	2↔2	3↔3	1↔1	2↔2	3↔3
Boat	1↔1	1↔1	2↔2	1↔1	3 ≠ 2	4↔4	1↔1	2↔2	3↔3	1↔1	2↔2	3↔3
Bridge	1↔1	1↔1	2↔2	1↔1	3↔3	4↔4	1↔1	2↔2	4↔4	1↔1	2↔2	3 ≠ 4
Cameraman	1↔1	2 ≠ 1	2↔2	2 ≠ 1	3↔3	4 ≠ 5	1↔1	3 ≠ 2	4↔4	1↔1	2 ≠ 3	3 ≠ 4
Facial	1↔1	1↔1	2↔2	1↔1	2↔2	3 ≠ 4	1↔1	2 ≠ 1	3↔3	1↔1	1↔1	3 ≠ 4
Lena	1↔1	1↔1	2 ≠ 1	1↔1	2↔2	3 ≠ 4	1↔1	2 ≠ 1	3↔3	1↔1	2 ≠ 1	3↔3
Light	1↔1	2 ≠ 1	2↔2	2 ≠ 1	3↔3	4↔4	1↔1	3 ≠ 2	3 ≠ 4	1↔1	2↔2	3↔3
Peppers	1↔1	1↔1	2 ≠ 1	1↔1	2↔2	3 ≠ 4	1↔1	2 ≠ 1	3↔3	1↔1	1↔1	3↔3
Rice	1↔1	1↔1	2↔2	1↔1	1 ≠ 2	4↔4	1↔1	2 ≠ 1	4↔4	1↔1	1↔1	4↔4

original images for image fusion. The standard deviations of the Gaussian filters used are 0.5, 1, and 5, respectively, so three pairs of original images for image fusion can be obtained from each test image.

The experiments compared four commonly used multi-scale decomposition methods: Laplace pyramid technique (LPT), discrete wavelet transform (DWT), discrete wavelet frame (DWF) and dual tree complex wavelet (DT-CWT). Table 6.3 gives both the optimal decomposition level S obtained by the exhaustive method and the optimal decomposition level T estimated by comparing the results of two adjacent levels. If the two levels are the same, they are marked as S $\leftrightarrow$ T, and if the two levels are different, they are marked as $S \neq T$ and shaded. It can be seen from the table that in most cases, the two are consistent (close to 80%), and the time difference is only 1, which indicates that the optimal number of decomposition levels can be determined by automatic estimation instead of manual exhaustion. In addition, it can be seen from Table 6.3 that for the same method, the number of optimal decomposition layers increases with the increase of blur, because at this time a finer decomposition is required to utilize more detail components.

6.3.4 Image Fusion Based on Compressed Sensing

Effective image fusion can also be achieved with the help of compressed sensing theory. The main steps are as follows:

(1) Perform sparse transformation on the input images $f_s(x, y)$ and $f_t(x, y)$:

$$\begin{cases} \boldsymbol{X}_A = T\left[f_s(x,y)\right] \\ \boldsymbol{X}_B = T\left[f_t(x,y)\right] \end{cases} \tag{6.12}$$

where sparse transform $T(\cdot)$ can be various transforms, such as Fourier transform, cosine transform and wavelet transform, etc.; $X\cdot$ is the coefficient matrix after sparse transformation;

(2) Under-sampling the transformed coefficients:

$$\begin{cases} \boldsymbol{Y}_A = \boldsymbol{\Phi} \cdot \boldsymbol{X}_A \\ \boldsymbol{Y}_B = \boldsymbol{\Phi} \cdot \boldsymbol{X}_B \end{cases} \tag{6.13}$$

where $\boldsymbol{\Phi}$ is the under-sampling matrix, $\cdot$ represents the dot product of the matrices.

According to the characteristics of coefficient distribution in $X\cdot$ (for example, in the coefficients of Fourier transform, the low-frequency coefficients are in the center of the spectrum, the high-frequency coefficients are around, and the higher the distance from the center; among the coefficients of cosine transform and wavelet transform, the low-frequency coefficients are in the upper left corner of the spectrum, the rest are high-frequency coefficients, the higher the lower right), the commonly used under-sampling patterns are star, double star, star circle, and single radial, as shown in Figures 6.8(a)–6.8(d).

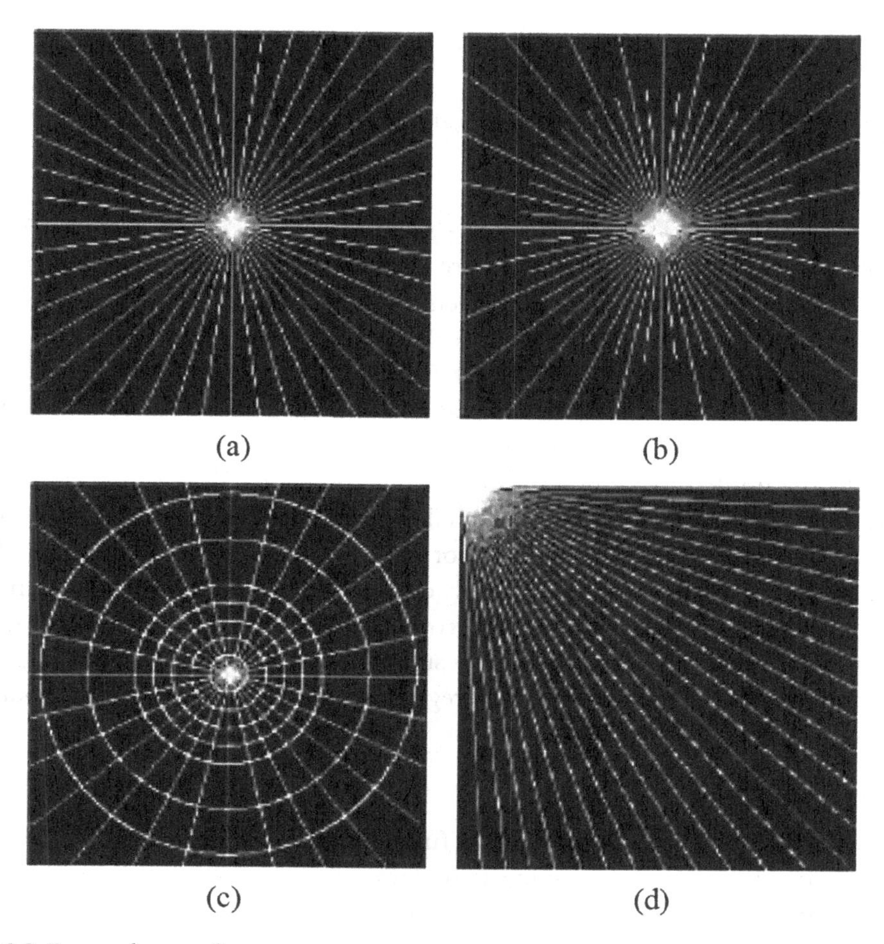

FIGURE 6.8 Four under-sampling patterns.

(3) Use the fusion rule to fuse the sampled coefficients, namely:

$$Y_F = \mathcal{F}(Y_A, Y_B) \tag{6.14}$$

where the fusion rule $\mathcal{F}(\cdot)$ can use the weighted average method, or other methods (such as the absolute value maximum method, etc.).

(4) Perform compressed sensing reconstruction on the fused image $\boldsymbol{X}_F$ according to the fused sampling value $\boldsymbol{Y}_F$ and the under-sampling matrix $\boldsymbol{\Phi}$. The reconstruction problem is a constrained optimization problem, and the common ones are as follows:

$$\min \left\| T(\boldsymbol{X}_F) \right\|_{L_0} \quad \text{s.t.} \quad \Phi \cdot T(\boldsymbol{X}_F) = Y_F \tag{6.15}$$

$$\min \left\| T(\boldsymbol{X}_F) \right\|_{L_1} \quad \text{s.t.} \quad \Phi \cdot T(\boldsymbol{X}_F) = Y_F \tag{6.16}$$

$$\min \mathrm{TV}(\boldsymbol{X}_F) \quad \text{s.t.} \quad \Phi \cdot T(\boldsymbol{X}_F) = \boldsymbol{Y}_F \tag{6.17}$$

where $||\cdot||_{L0}$, $||\cdot||_{L1}$, and TV(·) represent the L_0 norm, L_1 norm and total variation norm of ·, respectively.

6.3.5 Examples of Pixel-Level Fusion

The results and effects of pixel-level fusion are relatively easy to reflect from the image, and several application examples and fusion effects are given below.

6.3.5.1 Fusion of Different Exposure Images

Fusion of images of the same type already has a wide range of uses. For example, for high-contrast scenes, it is often difficult to obtain images that are normally exposed to all sceneries due to the limited dynamic range of the photosensitive device. Figures 6.9(a) and 6.9(b) respectively show two images at normal exposure of different sceneries in the same scene, of which Figure 6.9(a) is an image at normal exposure of the foreground building. At this time, the background building is severely overexposed; Figure 6.9(b) shows a normally exposed image for the building in the background, while the building in the foreground is severely underexposed. Figure 6.9(c) is the result obtained by fusing the two images. Here, the great differences in brightness in the foreground and background scenes produce better results.

6.3.5.2 Fusion of Different Focus Images

Another example of the same type of image fusion is shown in Figure 6.10. Figures 6.10(a) and 6.10(b) are the results of imaging two clocks placed at the front and back of the scene, respectively. Figure 6.10(a) focuses on the small clock placed in the front left, while Figure 6.10(b) focuses on the large clock placed at the rear right. In Figure 6.10(a), the big clock in the rear right is blurred; while in Figure 6.10(b), the small clock in the front left is blurred. In Figure 6.10(c), obtained by fusing the two images, both clocks are quite clear.

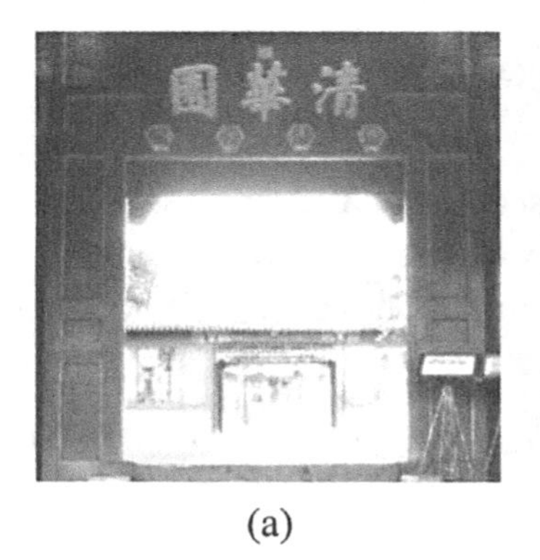

(a)

(b)

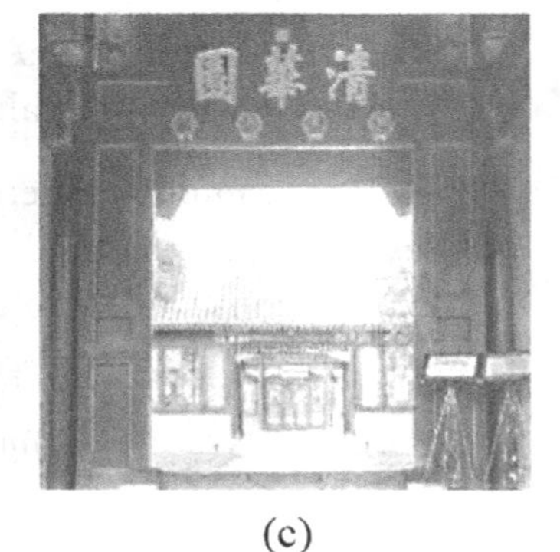

(c)

FIGURE 6.9 Fusion example of images with different exposures.

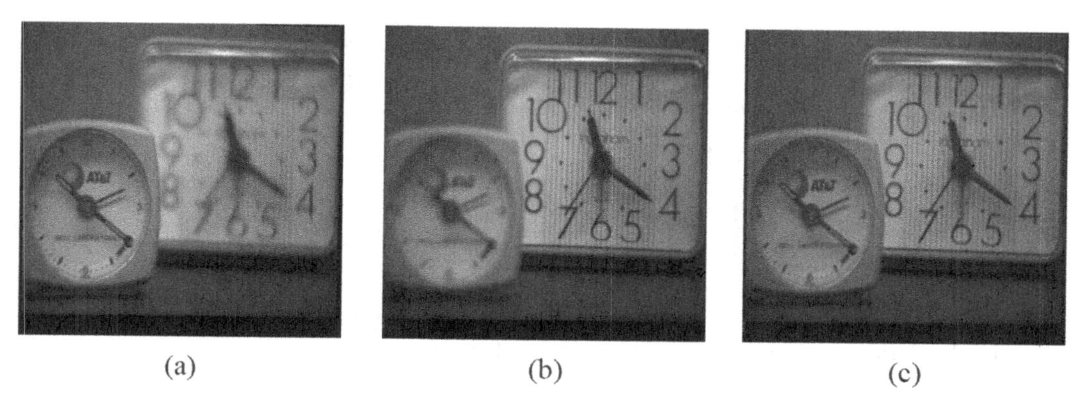

FIGURE 6.10 Fusion example of image with different focus.

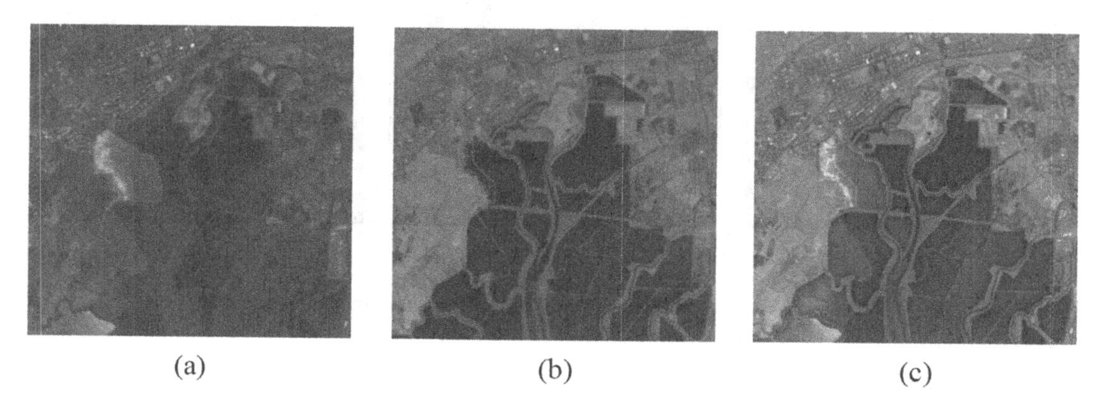

FIGURE 6.11 Example of remote sensing image fusion.

6.3.5.3 Fusion of Remote Sensing Images

The fusion of different bands or different types of remote sensing images is a typical application of image fusion. Figure 6.16 shows a set of examples. The original image is obtained with the Airborne Visible Infra-Red Imaging Spectrometer (AVIRIS, see http://aviris.jpl.nasa.gov). AVIRIS can simultaneously collect 224 bands of spectral information from 400nm (visible light) to 2500nm (infrared). Figures 6.11(a) and 6.11(b) give two images from band 30 and band 100, respectively. Their fusion results are shown in Figure 6.11(c). It can be seen from the figure that the fused image contains the information of both original images at the same time, providing a more comprehensive reflection of the surface situation of the photographed area.

6.3.5.4 Fusion of Visible Light Image and Infrared Image

Figure 6.12 shows an example of the fusion of a set of visible and infrared images. Figures 6.12(a) and 6.12(b) are the visible light image (given a scene of a big fire burning) and infrared image (reflecting the scene with different temperatures in the scene) obtained from the same scene at the same time, and Figure 6.12(c) is their fused image, from which the position of the person in the thick smoke can be clearly seen.

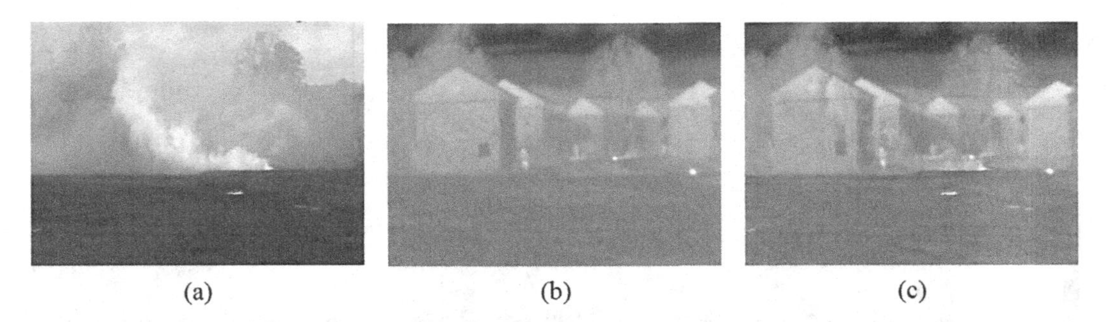

(a) (b) (c)

FIGURE 6.12 Fusion example of visible light image and infrared image.

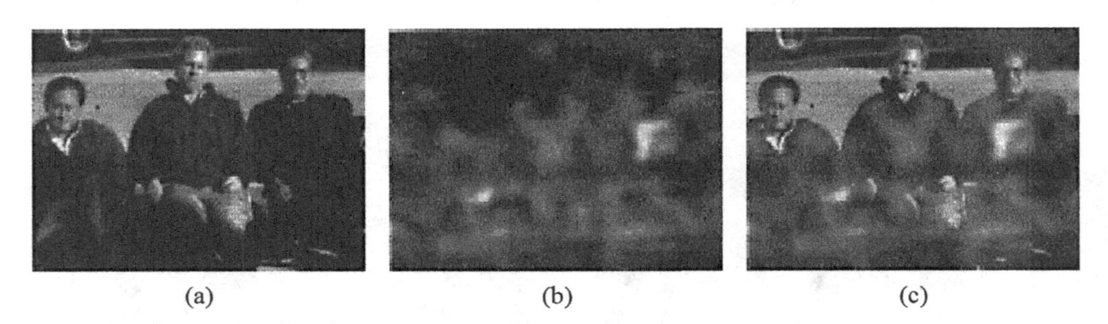

(a) (b) (c)

FIGURE 6.13 Fusion example of visible light image and millimeter-wave radar image.

6.3.5.5 Fusion of Visible Light Image and Millimeter-Wave Radar Image

Figure 6.13 shows an example of the fusion of a set of visible light images and millimeter-wave radar images. Here, Figures 6.13(a) and 6.13(b) are the visible light image and millimeter-wave radar image taken at the same time for the same group of people, respectively. Visible light images of people are clearer and more natural, and due to the sensitivity of millimeter-wave radar to metals, millimeter-wave radar images can detect weapons (highlighted in the figure). Figure 6.13(c) is the image resulting from their fusion, showing both the weapon and the carrier, to achieve the effect of detecting hidden weapons.

6.3.5.6 Fusion of CT Image and PET Image

The fusion of different types of images also has widespread applications in medicine. For example, CT images and PET images can respectively reflect different characteristics of the imaging object. In the case of the human body, CT images mainly give anatomical and structural information about the internal organs and tissues, while PET images mainly provide physiological and composition information about them. Figure 6.14 shows an example of the fusion of a set of CT and PET images. Figures 6.14(a) and 6.14(b) are an abdominal tomographic CT image and PET image taken at the same time, and Figure 6.14(c) is the image resulting from their fusion, showing both the tumor and the location of the tumor, thus aiding effective diagnosis.

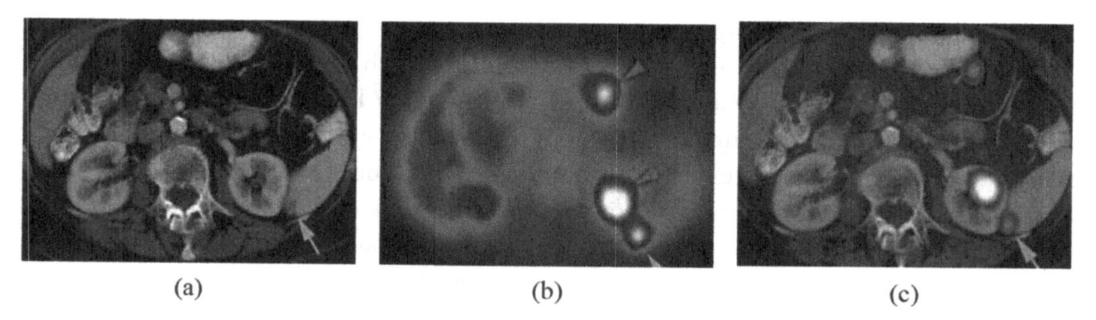

FIGURE 6.14 Fusion example of CT image and PET image.

6.3.5.7 Fusion of Dual-Energy Transmission Image and Compton Backscatter Image

Figure 6.15 gives an example of a set of X-ray dual-energy transmission images fused with Compton backscatter (CBS) images (see Wang et al. (2012) for imaging principle and system) for security screening (Wang et al. 2011). The first is the fusion of low-energy X-ray images (see Figure 6.15(a)) and high-energy X-ray images (see Figure 6.15(b)), and the result is in Figure 6.15(c). Since substances with the same atomic number absorb differently for high-energy and low-energy X-rays, the effective atomic number of the inspected object can be calculated by separately calculating the data of their attenuation projections. However,

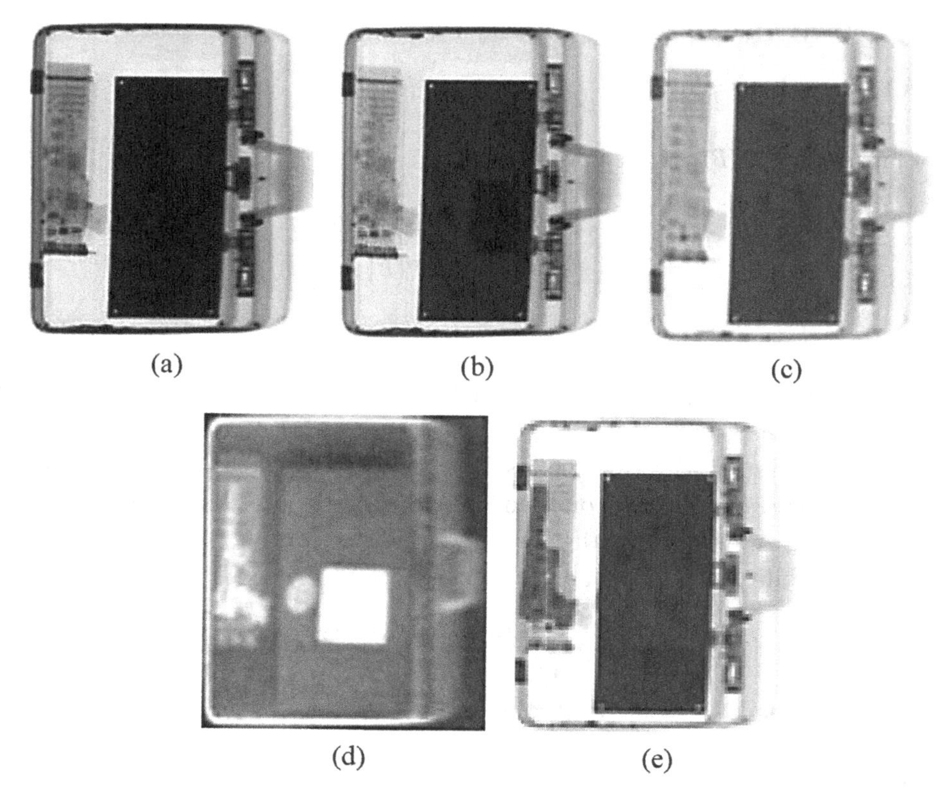

FIGURE 6.15 Fusion example of dual-energy transmission X-ray image and CBS image.

X-ray dual-energy transmission images are more sensitive to substances with high atomic numbers such as metals, but have poor ability to distinguish organic substances with low atomic numbers. Although Compton backscatter imaging has weak penetration ability and poor imaging quality (small grayscale dynamic range and low resolution), it is very sensitive to organic substances with low atomic number but high material density (such as drugs and explosives). More types of contraband can be detected by fusing the X-ray dual-energy transmission image (see Figure 6.15(c)) with the Compton backscatter image (see Figure 6.15(d)). The fusion result in Figure 6.15(e) clearly shows all prohibited items.

6.4 FEATURE-LEVEL AND DECISION-LEVEL FUSION METHODS

It can be seen from Table 6.2 that some of the techniques used in feature-level fusion and decision-level fusion are the same. In other words, these fusion methods can be used for both feature-level fusion and decision-level fusion. The principles of two of these methods are described below.

6.4.1 Bayesian Method

Bayesian methods are based on the **Bayesian conditional probability** formula. Suppose the sample space S is divided into $A_1, A_2, \ldots, A_N$, satisfying

(1) $A_i \cap A_j = \varnothing$,

(2) $A_1 \cup A_2 \cup \cdots \cup A_N = S$,

(3) $P(A_i) > 0$, $i = 1, 2, \cdots, N$.

Then for any event B, $P(B) > 0$, we have

$$P(A_i \mid B) = \frac{P(A_i B)}{P(B)} = \frac{P(B \mid A_i)P(A_i)}{\sum_{j=1}^{N} P(B \mid A_j)P(A_j)} \tag{6.18}$$

Considering the decision of multi-sensor fusion as the division of a sample space, the Bayesian conditional probability formula can be used to solve the decision-making problem of a multi-sensor system.

Consider first a system with two sensors. Let the observation result of the first sensor be B_1, the observation result of the second sensor be B_2, and the possible decisions of the system are $A_1, A_2, \ldots, A_n$. Assuming that each A_i is independent of B_1 and B_2, the Bayesian conditional probability formula can be written as

$$P(A_i \mid B_1 \wedge B_2) = \frac{P(B_1 \mid A_i)P(B_2 \mid A_i)P(A_i)}{\sum_{j=1}^{N} P(B_1 \mid A_j)P(B_2 \mid A_j)P(A_j)} \tag{6.19}$$

The above results can be generalized to the case of multiple sensors. Suppose there are M sensors, and the observation results are $B_1, B_2, \ldots, B_M$. If the sensors are independent of each other and the conditions of the observed object, the total posterior probability of each decision when the system has M sensors is

$$P(A_i \mid B_1 \wedge B_2 \wedge \cdots \wedge B_M) = \frac{\prod_{k=1}^{M} P(B_k \mid A_i) P(A_i)}{\sum_{j=1}^{N} \prod_{k=1}^{M} P(B_k \mid A_j) P(A_j)} \quad i = 1,\ 2,\ \cdots,\ N \tag{6.20}$$

In this case, the decision that gives the system the largest posterior probability can be selected as the final decision.

An example of object classification using fusion is given below.

Suppose there are four types of objects (i = 1, 2, 3, 4), and their probability of occurrence is $P(A_1)$, $P(A_2)$, $P(A_3)$, $P(A_4)$. The measurement value B_j of the two sensors (j = 1, 2) satisfies the Gaussian distribution $N(\mu_{ji}, \sigma_{ji})$, that is, the prior probability density of each measurement value is

$$P(B_j \mid A_i) = \frac{1}{\sqrt{2\pi}\sigma_{ji}} \exp\left(\frac{-(b_j - \mu_{ji})^2}{2\sigma_{ji}^2} \right) \quad i = 1,\ 2,\ 3,\ 4 \quad j = 1,\ 2 \tag{6.21}$$

The above conditional probabilities can be obtained by the following methods: assuming that the probability of the observed value occurring on the mean value of the prior probability distribution is 1, then the probability of the actual observed value is the sum of the probabilities on both sides of the mean center, namely

$$P(B_1 \mid A_1) = 2\int_{B_1}^{\infty} p(b_1 \mid A_1) \mathrm{d}b_1 \qquad P(B_2 \mid A_1) = 2\int_{B_2}^{\infty} p(b_2 \mid A_1) \mathrm{d}b_2 \tag{6.22}$$

$$P(B_1 \mid A_2) = 2\int_{-\infty}^{B_1} p(b_1 \mid A_2) \mathrm{d}b_1 \qquad P(B_2 \mid A_2) = 2\int_{B_2}^{\infty} p(b_2 \mid A_2) \mathrm{d}b_2 \tag{6.23}$$

$$P(B_1 \mid A_3) = 2\int_{-\infty}^{B_1} p(b_1 \mid A_3) \mathrm{d}b_1 \qquad P(B_2 \mid A_3) = 2\int_{-\infty}^{B_2} p(b_2 \mid A_3) \mathrm{d}b_2 \tag{6.24}$$

$$P(B_1 \mid A_4) = 2\int_{-\infty}^{B_1} p(b_1 \mid A_4) \mathrm{d}b_1 \qquad P(B_2 \mid A_4) = 2\int_{-\infty}^{B_2} p(b_2 \mid A_4) \mathrm{d}b_2 \tag{6.25}$$

The final fusion result can be represented as

$$P(A_i \mid B_1 \wedge B_2) = \frac{P(B_1 \mid A_i) P(B_2 \mid A_i) P(A_i)}{\sum_{j=1}^{4} P(B_1 \mid A_j) P(B_2 \mid A_j) P(A_j)} \quad i = 1,\ 2, 3, 4 \tag{6.26}$$

6.4.2 Evidential Reasoning Method

The Bayesian multi-information fusion method is based on probability theory, in which the additivity of probability is a principle that should be followed. But sometimes when the credibilities of two opposing propositions are both very small (cannot be judged according to the current evidence), the Bayesian method is not suitable. The **evidential reasoning method** (proposed by Dempster, perfected by Shafer, also known as **D–S theory**) abandons the principle of additivity and replaces it with a principle called semi-additivity.

The D–S theory uses the recognition frame F to represent the set of propositions of interest, which defines a set function C: $2^F \to [0, 1]$ that satisfies

(1) $C(\varnothing) = 0$, that is, no confidence is generated for the empty set;

(2) $\sum_{A \subset F} C(A) = 1$, that is, although a proposition A can be assigned any credibility value, the sum of the credibility values assigned to all propositions is 1.

Here C is called the basic confidence assignment on the recognition frame F. $\forall A \subset F$, $C(A)$ is called the **basic credibility number** of A, which reflects the credibility of A itself.

Corresponding to any set of propositions, D–S theory also defines a credibility function

$$B(A) = \sum_{E \subset A} C(E) \qquad \forall A \subset F \tag{6.27}$$

That is, the credibility function value of A is the sum of the credibility numbers of each subset in A. According to the credibility function, we can get

$$B(\varnothing) = 0 \qquad B(F) = 1 \tag{6.28}$$

It is not comprehensive to describe the trust of a proposition A only by the credibility function, because $B(A)$ cannot reflect the degree of suspicion about A, that is, the degree of belief that A is not true. To this end, the suspicion of A can also be defined as

$$D(A) = B(\bar{A}) \qquad P(A) = 1 - B(\bar{A}) \tag{6.29}$$

where D is the suspicion function of B, P is the plausibility function of B, $D(a)$ is the suspicion of a, and $P(a)$ is the plausibility of A.

P can be re-expressed by C corresponding to B

$$P(A) = 1 - B(\bar{A}) = \sum_{E \subset F} C(E) - \sum_{E \subset A} C(E) = \sum_{E \cap A \neq \varnothing} C(E) \tag{6.30}$$

A is said to be compatible with E if $A \cap E \neq \varnothing$. Equation (6.30) shows that $P(A)$ contains the basic credibility numbers of all the sets of propositions that are compatible with A.

Since $A \cap \bar{A} = \emptyset$, $A \cup \bar{A} \subset F$, we have

$$B(A) + B(\bar{A}) \leq \sum_{E \subset F} C(E) \tag{6.31}$$

$$B(A) \leq 1 - B(\bar{A}) = P(A) \tag{6.32}$$

$[B(A), P(A)]$ represents the uncertainty interval for A, also known as the upper and lower bounds of the probability, respectively. $[0, B(A)]$ is the fully credible interval, indicating the degree of support for the proposition "A is true". $[0, P(A)]$ is the interval of non-suspicion about the proposition "A is true", indicating the degree to which the evidence cannot refuse that "A is true". The above intervals can be seen in Figure 6.16. Obviously, the larger the interval $[B(A), P(A)]$, the higher the degree of uncertainty.

If the proposition is regarded as an element on the recognition frame F, for $\forall C(A) > 0$, A is called the focal element of the credibility function B. If there are two credibility functions B_1 and B_2 on the same recognition frame F, let C_1 and C_2 be assigned to their corresponding basic credibility, respectively, then the synthesis of credibility value can be seen in Figure 6.17. In Figure 6.17, the vertical bars represent the reliability of C_1 assigned to its focal elements $A_1, A_2, \ldots, A_K$, and the horizontal bars represent the reliability of C_2 assigned to its focal elements $E_1, E_2, \ldots, E_L$. The shaded region represents the intersection of the horizontal and vertical bars, and its measure is $C_1(A_i)C_2(E_j)$. It can be seen that the combined effect of B_1 and B_2 is to exactly assign $C_1(A_i)C_2(E_j)$ to $A_i \cap E_j$.

Given $A \subset F$, if $A_i \cap E_j = A$, then $C_1(A_i)C_2(E_j)$ is the partial credibility assigned to A, and the total credibility assigned to A is $\sum_{A_i \cap E_j = A} C_1(A_i)C_2(E_j)$. But when $A = \emptyset$, according to this understanding, it is obviously unreasonable to assign some credibility $\sum_{A_i \cap E_j = \emptyset} C_1(A_i) C_2(E_j)$ to the empty set. To avoid this problem, each credibility level can be multiplied by a

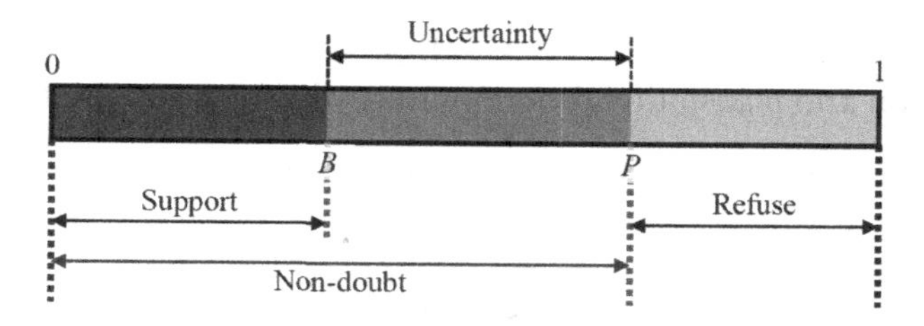

FIGURE 6.16 A partition of information regions.

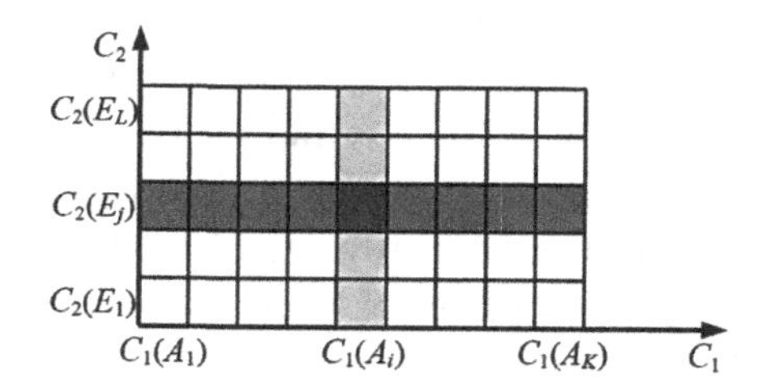

FIGURE 6.17 Synthesis of credibility values.

credibility coefficient $[1-\sum_{A_i \cap E_j = \varnothing} C_1(A_i)C_2(E_j)]^{-1}$, so that the total credibility level satisfies the requirement of equal to 1. In this way, the rule for combining two credibility values is as follows (use ⊕ to represent the operation of synthesis):

$$C(A) = C_1(A) \oplus C_2(A) = \frac{\sum_{A_i \cap E_j = A} C_1(A_i)C_2(E_j)}{1 - \sum_{A_i \cap E_j = \varnothing} C_1(A_i)C_2(E_j)} \tag{6.33}$$

Generalizing the above process, for the synthesis of multiple credibility values (i.e., the fusion of multiple information), if C_1, C_2, …, C_n, respectively, represent the credibility distribution of n pieces of information, when they are derived from independent information, the fused credibility value $C = C_1 \oplus C_2 \oplus \ldots \oplus C_n$ can be represented as

$$C(A) = \frac{\sum_{A_i \cap E_j = A} \prod_{i=1} C_i(A_i)}{1 - \sum_{A_i \cap E_j = \varnothing} \prod_{i=1} C_i(A_i)} \tag{6.34}$$

In actual use, the information collected by each sensor is used as evidence, each sensor provides a set of propositions, and a corresponding credibility function is established. In this way, multi-sensor information fusion becomes a process of combining different evidences into new evidence under the same recognition framework. The main steps of this process are as follows:

(1) Calculate the basic credibility number, credibility function and plausibility function of each sensor respectively.

(2) Using the merging rule of Equation (6.34), obtain the basic credibility number, credibility function and plausibility function under the joint action of all sensors.

(3) Under certain decision rules, select the object with the greatest support.

6.5 ROUGH SET THEORY IN DECISION-LEVEL FUSION

When using the evidential reasoning method for multi-sensor information fusion, the problem of combinatorial explosion may occur. Therefore, it is necessary to study how to analyze complementary sensor information, to compress redundant information, and to obtain a fusion algorithm according to the internal relationship between the data. Rough set theory provides a means to solve this problem. Unlike fuzzy set theory, which can express fuzzy concepts but cannot specifically calculate the number of fuzzy elements, according to rough set theory, the number of fuzzy elements can be calculated with an exact mathematical formula.

The method based on rough set theory is mainly used in decision-level fusion.

6.5.1 Rough Set Definition

Let $L \neq \varnothing$ be a finite set of objects of interest, called the definition domain. For any subset X in L, call it a concept in L. The set of concepts in L is called knowledge about L (often expressed in the form of attributes). Let R be an equivalence relation defined on L (which can represent the attributes of things), then a knowledge base is a relational system $K = \{L, \mathbf{R}\}$, where $\mathbf{R}$ is the set of equivalence relations on L (Zhang et al. 2001). If $\mathbf{R}$ is an equivalence relation set on X, then $\mathbf{R}$ is reflexive, symmetric, and transitive. The $\mathbf{R}$ equivalence class of an element a $[a]\mathbf{R} = \{x|x \in \mathbf{R}, \mathrm{a} \in \mathrm{X}\}$, that is, the equivalence class of a is the set of all elements x that can form an order couple with a.

For any subset X in L, if it can be defined by R, it is called ***R* exact set**; if it cannot be defined by R, it is called ***R* rough set**. A rough set can be described (approximately) by two exact sets, an **upper approximation set** and a **lower approximation set**:

$$R^*(X) = \{X \in L : R(X) \cap X \neq \varnothing\} \tag{6.35}$$

$$R_*(X) = \{X \in L : R(X) \subseteq X\} \tag{6.36}$$

where $R(X)$ is an equivalence class containing X. The R boundary of X is defined as the difference between the upper approximation set of X and the lower approximation set of R, i.e.

$$B_R(X) = R^*(X) - R^*(X) \tag{6.37}$$

For example, suppose a knowledge base $K = (L, \mathbf{R})$ is given, where $L = \{x_1, x_2, \ldots, x_8\}$, and $\mathbf{R}$ is an equivalence set including the equivalence class $E_1 = \{x_1, x_4, x_8\}$, E2 = {x2, x5, x7}, $E_3 = \{x_3\}$, $E_4 = \{x_6\}$. If the set $X = \{x_3, x_5\}$ is considered, then $R_*(X) = \{X \in L: R(X) \subseteq X\} = E_3 = \{x_3\}$, $R^*(X) = \{X \in L: R(X) \cap X \neq \varnothing\} = E_2 \cup E_3 = \{x_2, x_3, x_5, x_7\}$, $B_R(X) = R^*(X) - R_*(X) = \{x_2, x_5, x_7\}$.

For knowledge R, $R^*(X)$ is the set of elements in L that may be classified into X; for knowledge R, $R_*(X)$ is the set of all elements in L that must be classified into X; for knowledge R, $B_R(X)$ is the set of elements in L that can neither be assigned exactly to X nor be assigned exactly to $\bar{X}$ (the complement of X). Further, $R_*(X)$ can be called the R positive domain of X, $L - R^*(X)$ can be called the R negative domain of X, and $B_R(X)$ can be called the boundary domain of X. According to this definition, the positive field is the set of elements that can be assigned to the set X with complete certainty, according to the knowledge R, and the negative field is the set of elements that, according to the knowledge R, do not belong to the set X but undoubtedly belong to the complement of X. A boundary domain is a domain of uncertainty in a sense. For knowledge R, the elements belonging to the bounding domain cannot be divided exactly as belonging to X or $\bar{X}$. It can also be seen from the above that $R^*(X)$ consists of those elements whose possibility of belonging to X cannot be ruled out according to the knowledge of R. In other words, the upper approximation set is the union of the positive and boundary domains.

A 2-D illustration of the above discussion is shown in Figure 6.18. The space shown in the figure consists of rectangles divided into basic regions, each of which represents an equivalence class of R. The region between $R^*(X)$ and $R_*(X)$ represents the R boundary of X, which is the uncertainty region of X.

It can be seen from the above that X is a definable set of R if and only if $R^*(X) = R_*(X)$; X is a rough set of R if and only if $R^*(X) \neq R_*(X)$. In other words, $R_*(X)$ can be described as the largest definable set in X, and $R^*(X)$ can be described as the smallest definable set containing X.

6.5.2 Rough Set Description

The boundary domain is an undetermined region that exists due to incomplete knowledge, that is, the elements on $B_R(X)$ are uncertain. Therefore, the uncertainty relation of subset X on L with respect to L is rough, $B_R(X) \neq \varnothing$. The larger the boundary field of a set X, the more the rough elements this set contains, and the less accurate it is. To express this precisely, the concept of precision is introduced as follows:

$$d_R(X) = \frac{\mathrm{card}\left[R_*(X)\right]}{\mathrm{card}\left[R^*(X)\right]} \tag{6.38}$$

where card(·) represents the cardinality of the set, and $X \neq \varnothing$.

In the example given earlier, $d_R(X) = \mathrm{card}[R_*(X)] / \mathrm{card}[R^*(X)] = 1/4$. The precision $d_R(X)$ reflects how complete the set X is known. For any R and $X \subseteq L$, it has $0 \leq d_R(X) \leq 1$. When $d_R(X) = 1$, the R boundary domain of X is empty, and the set X is R definable. When $d_R(X) < 1$, the set X has a non-empty boundary domain, and the set X is indefinable for R.

The concept corresponding to precision is roughness:

$$h_R(X) = 1 - d_R(X) \tag{6.39}$$

It reflects the incompleteness of knowledge about set X.

It can be seen from the above that, unlike probability theory and fuzzy set theory, the numerical value of imprecision here is not presupposed, but is computationally approximated by a concept expressing the imprecision of knowledge. Such imprecise values represent the result of limited knowledge (object classification ability), so it is not necessary to express imprecise knowledge with precise values, but to use quantitative concepts (classification) to deal with.

The topological properties of rough sets can also be represented by means of upper approximation set and lower approximation set. Four important rough sets are defined below (refer to Figure 6.18 to analyze their geometric meaning):

(1) If $R_*(X) \neq \varnothing$, and $R^*(X) \neq L$, then X is called R rough definable set. At this point, it can be determined that any element in L belongs to X or $\bar{X}$. This is exactly the case in Figure 6.18, both inside of $R_*(X)$ and outside of $R^*(X)$ belong to $\bar{X}$.

(2) If $R_*(X) = \varnothing$, and $R^*(X) \neq L$, then X is called an undefinable set in R. At this time, it can be determined whether some elements in L belong to $\bar{X}$, but it cannot be determined whether any element in L belongs to X. This is equivalent to shrinking $R_*(X)$

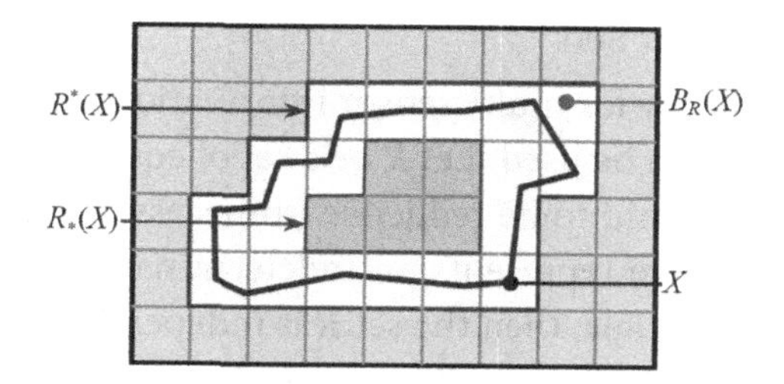

FIGURE 6.18 Schematic diagram of rough set.

in Figure 6.18 into an empty set. At this time, all elements outside of $R^*(X)$ belong to $\bar{X}$, but the elements inside of $R^*(X)$ cannot be determined whether they belong to X.

(3) If $R_*(X) \neq \varnothing$, and $R^*(X) = L$, then X is called an undefinable set outside R. At this time, whether some elements in L belong to X can be determined, whether any element in L belongs to $\bar{X}$ cannot. This is equivalent to expanding $R^*(X)$ in Figure 6.18 to the entire L. At this time, the elements inside $R_*(X)$ do not belong to X, but it cannot be determined whether elements outside $R_*(X)$ belong to $\bar{X}$.

(4) If $R_*(X) = \varnothing$, and $R^*(X) = L$, then X is said to be a completely undefinable set of R. At this point it cannot be determined whether any element in L belongs to X or $\bar{X}$. This is equivalent to the above two cases happening at the same time, the undefinable set is the union of the indefinable sets in the above two cases, or it is not definable from $R_*(X) = \varnothing$ to $R^*(X) = L$.

For example, given a knowledge base $K = (L, \mathbf{R})$, where $L = \{x_0, x_1, \cdots, x_{10}\}$, $\mathbf{R}$ is an equivalence set, and there are some equivalence classes $E_1 = \{x_0, x_1\}$, $E_2 = \{x_2, x_6, x_9\}$, $E_3 = \{x_3, x_5\}$, $E_4 = \{x_4, x_8\}$, $E_5 = \{x_7, x_{10}\}$. In this case, the rough sets are:

The set $X_1 = \{x_0, x_1, x_4, x_8\}$is R definable because $R^*(X_1) = R_*(X_1) = E_1 \cup E_4$.

The set $X_2 = \{x_0, x_3, x_4, x_5, x_8, x_{10}\}$ is R rough definable set, at this time $R_*(X_2) = E_3 \cup E_4 = \{x_3, x_4, x_5, x_8\}$, $R^*(X_2) = E_1 \cup E_3 \cup E_4 \cup E_5 = \{x_0, x_1, x_3, x_4, x_5, x_7, x_8, x_{10}\}$, $B_R(X_2) = E_1 \cup E_5 = \{x_0, x_1, x_7, x_{10}\}$, $d_R(X_2) = 1/2$.

The set $X_3 = \{x_0, x_2, x_3\}$is not definable in R because $R_*(X_3) = \varnothing$, $R^*(X_3) = E_1 \cup E_2 \cup E_3 = \{x_0, x_1, x_2, x_3, x_5, x_6, x_9\} \neq L$.

The set $X_4 = \{x_0, x_1, x_2, x_3, x_4, x_7\}$ is an indefinable set outside R, at this time $R_*(X_4) = E_1 = \{x_0, x_1\}$, $R^*(X_4) = L$, $B_R(X_4) = E_2 \cup E_3 \cup E_4 \cup E_5 = \{x_2, x_3, x_4, x_5, x_6, x_7, x_8, x_9, x_{10}\}$, $d_R(X_4) = 2/11$.

The set $X_5 = \{x_0, x_2, x_3, x_4, x_7\}$ is a completely indefinable set of R, because $R_*(X_5) = \varnothing$, $R^*(X_5) = L$.

6.5.3 Fusion Based on Rough Sets

In order to use rough set theory for multi-sensor information fusion, the concept of kernel and **reduction** of rough set can be used. Let $\boldsymbol{R}$ be a set of equivalence relations, and $R \in \boldsymbol{R}$, if $I(\boldsymbol{R}) = I(\boldsymbol{R} - \{R\})$, then R is said to be reducible (unnecessary) in $\boldsymbol{R}$, otherwise it is non-reducible (necessary). $I(\cdot)$ above represents an indeterminate relationship. When for any $R \in \boldsymbol{R}$, all of $\boldsymbol{R}$ cannot be reducible, then the set $\boldsymbol{R}$ is independent.

If $\boldsymbol{R}$ is independent, and $\boldsymbol{P} \subseteq \boldsymbol{R}$, then $\boldsymbol{P}$ is also independent. If $I(\boldsymbol{P}) = I(\boldsymbol{R})$, then $\boldsymbol{P}$ is a reduction of $\boldsymbol{R}$. The set of all irreducible relations in $\boldsymbol{R}$ is called the kernel of $\boldsymbol{P}$, denoted as $C(\boldsymbol{P})$. The relationship between kernel and reduction is

$$C(\boldsymbol{P}) = \cap J(\boldsymbol{R}) \tag{6.40}$$

where $J(\boldsymbol{R})$ represents the set of all reductions of $\boldsymbol{R}$.

It can be seen from the above that the kernel is included in all the reductions and can be calculated directly from the intersection of the reductions. A kernel is a collection of knowledge features that cannot be eliminated when knowledge is reduced.

Let S and T be the equivalence relation in L, and the positive domain of S in T (the set of equivalence classes in L that can be accurately divided into T) is

$$P_S(T) = \bigcup_{X \in T} S_*(X) \tag{6.41}$$

The dependency between S and T is

$$Q_S(T) = \frac{\text{card}\left[P_S(T)\right]}{\text{card}(L)} \tag{6.42}$$

It can be seen from the above that $0 \leq Q_S(T) \leq 1$. The compatibility of the two equivalence classes of S and T can be determined by using the dependency $Q_S(T)$ of S and T. When $Q_S(T) = 1$, it means that S and T are compatible; and when $Q_S(T) \neq 1$, it means that S and T are incompatible. When the rough set theory is used for multi-sensor information fusion, it uses the dependence relationship $Q_S(T)$ of S and T to analyze a large amount of data to find out the inherent essential relationship and eliminate the compatible information, so as to determine the smallest invariant kernel in the data from a large number of data, and to obtain the fastest fusion method based on the most useful decision information.

6.6 SOME RECENT DEVELOPMENTS AND FURTHER RESEARCH

In the following sections, some technical developments and promising research directions from the last few years are briefly reviewed.

6.6.1 Spatial-Spectral Feature Extraction of Hyperspectral Images

Hyperspectral images (HSI) contain rich spectral, spatial and radiometric information. The spectrum is nearly continuous, and the image and spectrum are integrated, reflecting the scene information better. In addition, it has a large amount of data, multiple bands and strong correlation between bands, bringing many challenges to its application in classification and identification.

For HSI, there may be different matters with the same spectrum, that is, two different ground targets may show the same spectral line characteristics in a certain band; there may also be the phenomenon of different spectra for the same object, that is, the same ground target presents different spectral line characteristics due to the influence of the surrounding environment. In order to classify them accurately, we can consider integrating the spatial information of adjacent locations into the spectral feature extraction method (that is, fully mining the spatial-spectral features of HSI data).

There are many extraction strategies for the spatial-spectral feature extraction of hyperspectral images. Originally, traditional extraction methods were mainly used, but more recently extraction methods based on deep learning have been adopted (Ye et al. 2021).

6.6.1.1 Traditional Spatial-Spectral Feature Extraction Methods

The traditional spatial-spectral feature extraction methods for hyperspectral images (summarized in Table 6.4) can be divided into three categories: spatial texture and

TABLE 6.4 Overview of Traditional Spatial-spectrum Feature Extraction Methods

Category	Methods	Typical Technical Ideas
Spatial texture and morphological feature extraction	Gabor features	Incorporating Gabor features into the principal component extraction subspace
	Local binary pattern (LBP) features	Applying LBP using a rotation-invariant texture structure oriented to HSI local spatial information
	Morphological profile (MP)	Segmenting the light and dark spatial structures of the image through morphological opening and closing operations, and using morphological transformation to construct MP
Spatial neighborhood information acquisition	Sparse represent classification model	In SRC, the class label of a test pixel is determined to provide the class label with the smallest approximation error for its labeled samples.
	Collaborative representation classification models	In collaborative expression, each pixel has an equal opportunity to participate, and all pixels can also cooperate with each other to further improve classification accuracy.
	Nearest neighbor (NN)-based classifiers	Nonparametric classifiers do not require any prior knowledge of data density distribution.
	Support vector machine-based classifiers	Use composite kernel (CK) to combine spectral and contextual features of SVM
Post-processing of spatial information	Markov random field	Combine SVM with MRF, and use MRF to correct the results of SVM rough classification
	Bilateral filter	Consider spectral and spatial information simultaneously using spectral distances and multivariate Gaussian functions to preserve spatial detail and remove noise at the same time
	Binary partition trees	Based on BPT construction and pruning strategy, use local unmixing of regions to find partitions that achieve the global minimum reconstruction error

morphological feature extraction, spatial neighborhood information acquisition, and spatial information post-processing.

(1) Spatial texture and morphological feature extraction

This type of method preprocesses pixel spatial information by extracting spatial texture and morphological features, that is, the spatial features of the pixels to be classified are first extracted through certain structures and rules, then the features obtained are sent to the classifier for classification. These spatial features mainly include texture features and shape features, such as **Gabor features** (Clausi and Jernigan 2000), **local binary pattern** (LBP) features (Ojala et al. 2002), and morphological contour (**morphological profile**, MP) features (Pesaresi and Benediktsson 2001), etc.

(2) Spatial neighborhood information acquisition

This type of method directly combines the relationship between the pixels to be classified and the pixels in the spatial neighborhood in the classifier. By constructing a classification model or improving the classifier, the spatial-spectral information of the pixels to be classified and its neighborhood pixels is directly applied, so that feature extraction can be achieved, and the classification is done synchronously. This type of spatial-spectral feature extraction method directly constructs and reflects spatial information in the classification model through mathematical expressions. Typical representatives include **sparse represent classification** (SRC) models (Wright et al. 2009), **collaborative representation classification** (CRC) models (Zhang et al. 2011), **nearest neighbor** (NN)-based classifiers, and **support vector machine** (SVM)-based classifiers (Camps-Valls et al. 2006), etc.

(3) Post-processing of spatial information

This kind of method uses spatial information to correct the classification results obtained in the post-processing stage to further improve classification accuracy. The main methods include techniques based on **Markov random field** (MRF) (Tarabalka et al. 2010), techniques based on **bilateral filter** (BF) (Peng and Rao 2009), techniques based on **binary partition trees** (BPT) (Veganzones et al. 2014), etc.

6.6.1.2 Deep Learning-Based Methods of Extracting Spatial-Spectral Features

Deep learning-based algorithms directly learn two or more layers of representative and discriminative deep features (deep spatial-spectral features) from HSI data by building a deep network framework. In chronological order, there are three main categories of methods that have emerged sequentially: convolutional neural networks, multi-source data cross-scene models, and graph convolutional networks.

(1) Convolutional neural networks

The first deep network framework to be applied is the **convolutional neural network** (CNN), whose input can be spectral information or spatial-spectral information.

Specifically, it includes **3-D convolutional neural networks** (3D-CNN), **deep convolutional neural networks** (deep CNN, DCNN) (Hu et al. 2015), **recurrent neural networks** (RNN), and **convolutional recurrent neural networks** (CRNN) (Wu et al. 2017), etc.

(2) Multi-source data cross-scene model

Since **light detection and ranging** (LiDAR) data can provide elevation information on the surveyed region and can be acquired at any time of day and under severe weather conditions, it is often fused with hyperspectral images for feature extraction to better achieve portrayal of the scene. Typical methods include **joint feature extraction and classification from multi-source data, unsupervised collaborative classification, semi-supervised graph fusion** (SSGF), **ensemble classifier systems, hyperspectral image segmentation**, etc.

(3) Graph convolutional networks

Graph convolutional networks (GCNs) efficiently process data with a graph structure by modeling the relationships between samples. At this time, the potential relationship between the data is represented by a graph. Each pixel in the image is treated as a node in the graph, and its neighborhood is determined by the filter size (variable neighborhood size), adapting to local regions with various object distributions and geometric appearances. In this way, long-range spatial relationships in HSI can be naturally modeled using GCN. However, GCN is a transductive learning method, which requires all nodes to participate in the training process in order to obtain node embedding, resulting in high memory usage problems; at the same time, GCN needs to construct an adjacency matrix, and the computational cost will increase exponentially as the layer number increases.

In recent years, many different graph convolutional networks have been proposed, such as **edge-conditioned graph convolutional networks** (edge-conditioned GCN) (Sha et al. 2019), **spatial-spectral feature-based graph convolutional networks** (spatial-spectral GCN) (Qin et al. 2019), **nonlocal graph convolutional networks** (nonlocal GCN) (Mou et al. 2020), etc. An overview of spatial-spectrum feature extraction methods based on deep learning is given in Table 6.5.

6.6.2 Multi-Source Remote Sensing Image Fusion

Image fusion is a type of image technology that has been widely studied and applied, and is currently widely used in the field of remote sensing, which has been the largest image technology application field in recent years (Zhang 2021).

6.6.2.1 Nine Multi-Source Remote Sensing Data Sources

There are many remote sensing data sources. The imaging principles for remote sensing images obtained from different sources are different, and the spatial resolution, observation

TABLE 6.5 Overview of Spatial-spectrum Feature Extraction Methods Based On Deep Learning

Category	Methods	Typical Technical Ideas
Convolutional Neural Networks	Deep convolutional neural networks	Include input layer, convolution layer, max pooling layer, fully connected layer and output layer, using spectral information to classify hyperspectral images.
	Recurrent neural networks	Process hyperspectral data as spectral sequences, use recurrent neural networks to model correlations between different spectral bands, and extract contextual information from hyperspectral data.
	Convolutional recurrent neural networks	Extract spectral background information from features generated by convolutional layers for hyperspectral data classification.
Multi-source data cross-scene model	Joint feature extraction and classification from multi-source data	One CNN is used to learn spectral-spatial features from hyperspectral data, another CNN is used to obtain elevation information from LiDAR data, and the last two convolutional layers are coupled together through a parameter sharing strategy.
	Unsupervised collaborative classification	Design patch-to-patch convolutional neural network (PToPCNN) hidden layers to integrate multi-scale features of HSI and LiDAR data.
	Semi-supervised graph fusion	The spectral features, elevation features and spatial features are projected onto a lower subspace to obtain new features, which maximizes the classification ability and preserves the local neighborhood structure.
	Ensemble classifier systems	Split spectral bands, hyperspectral morphological features, and LiDAR features into several disjoint subsets, and concatenate the features extracted in each subset into a random forest classifier for classification.
	Hyperspectral image segmentation	Using the framework of combining CNN and conditional random field, comprehensively considering spectral and spatial information, using spectral cube to learn deep features, and using CNN-based potential function to build deep conditional random fields.
Graph Convolutional Networks	Edge-conditioned GCN	Build a spatial-spectral graph model to combine spatial-spectral information, utilize GCN to extract features from input data, and learn their topological relationships with edge labels.
	Spatial-spectral feature-based GCN	A multispectral spatial graph semi-supervised learning framework, alleviating the inadequacy of labeled samples and improving HSI classification accuracy.
	Nonlocal GCN	A semi-supervised network that can take in the entire image (including labeled and unlabeled data) to balance high-dimensional properties with limited samples.

scale and target characteristics of the images will also be different. Nine remote sensing data sources are summarized by Li et al. (2021):

(1) **Hyperspectral image**

Hyperspectral images can have dozens or even thousands of spectral bands with wide spectral range and rich band information, enabling fine spectral information to be captured from ground targets. This method is often used for fine classification and recognition of ground targets. However, higher spectral resolution means lower spatial resolution, which will limit image quality. In practical applications, hyperspectral images are often fused with multispectral, panchromatic and synthetic aperture radar images to obtain hyperspectral images with higher spatial resolution.

(2) **Panchromatic image**

The panchromatic image has only one band, and the band range is basically 0.50 ~ 0.75 μm. The panchromatic image is displayed as a gray image, which has high spatial resolution and contains many ground target details. Due to the rich information, it can obtain the fine geometric and texture features of ground targets. However, panchromatic images lack spectral information.

(3) **Multispectral image**

A multispectral image has spectral information from multiple bands, and its spatial resolution and spectral resolution are between the panchromatic image and hyperspectral image obtained by optical imaging. A multispectral image in the fusion process can therefore provide spectral information for a panchromatic image and spatial information for a hyperspectral image.

(4) **Thermal infrared remote sensing images**

Thermal infrared remote sensing images reflect the temperature distribution of ground targets. **Infrared** (IR) is between visible light and radio waves in the electromagnetic spectrum. After the infrared imaging system receives the infrared radiation of the scene target, it can be converted into an infrared thermal imaging image through processing. Because all objects in nature whose temperature is higher than absolute zero will radiate infrared rays, it has a wide range of applications.

(5) **Synthetic aperture radar image**

Synthetic aperture radar (SAR) imaging utilizes the Doppler frequency shift and radar coherence principles, and is an active imaging method. SAR has an antenna array, the elements of which interfere with each other to form a narrow beam. SAR emits microwaves as the space-borne or airborne radar travels along its orbit. Due to the relative motion between the ground target and the radar, the radar superimposes the received echo signals and converts them into electrical signals to record them as digital pixels to form an SAR image. The echo signal recorded by the synthetic aperture radar is the backscattered energy of the ground target, which can reflect the surface characteristics and dielectric properties of the ground target.

(6) **LiDAR image**

Light detection and ranging (LiDAR) is also an active imaging method, and its imaging principle is similar to that of SAR. LiDAR works in the optical frequency band from infrared to ultraviolet. LiDAR has good mono-chromaticity, directionality and coherence, concentrated laser energy, high detection sensitivity and resolution, and can accurately track and identify the motion state and position of the target. Compared with SAR, the laser beam of LiDAR is narrower, so the probability of being intercepted is lower, and the concealment is better.

(7) **Night time light remote sensing image**

A night time light (NTL) remote sensing image can capture the weak light radiation of the ground at night, and has considerable advantages for large-scale earth observation.

(8) **Stereo remote sensing image**

Stereo remote sensing images usually obtain 3-D depth information by spatial stereoscopic effect. Human eyes are a typical stereoscopic instrument. Stereo remote sensing imaging draws on the human binocular visual perception function to obtain 3-D depth information from the scene by acquiring remote sensing images from multiple angles. At present, space stereo imaging schemes are mainly used to install multiple optical cameras with different observation angles on the satellite carrier, or to change the attitude of a satellite on an agile platform to obtain ground target images at different angles, enabling further estimation of the 3-D information of the scene.

(9) **Video remote sensing data**

Video remote sensing data is a new type of earth observation data in the field of remote sensing. Compared with traditional satellite data, the biggest advantage of video remote sensing data is that it can "stare" at the same region. Compared with images, video can show the dynamic change information of targets or scenes, and is especially suitable for continuous observation and tracking of targets.

At present, multispectral and panchromatic imaging are the most mature methods of remote sensing imaging. Synthetic aperture radar imaging has all-time and all-weather advantages, and is widely used by the military. There are relatively few ways of obtaining laser, stereo and infrared images.

6.6.2.2 Multi-Source Remote Sensing Image Fusion Literature

Multi-source remote sensing image fusion can make full use of the richness and reliability of multi-source data to obtain more comprehensive information. In recent years, various remote sensing data fusion technologies have developed rapidly, but the development of remote sensing image fusion from different sources is not very balanced. Li et al. (2021) conducted a statistical analysis of publications on paired fusion technology in respect of the multi-source remote sensing data from the above-mentioned nine sources.

Two literature databases are considered in this work: the Web of Science (WOS) database and the China National Knowledge Infrastructure (CNKI) full-text database. In the WOS database, the literature title contains two remote sensing data sources and the word "fusion", and the document subjects containing the word "remote sensing" are considered. In the CNKI database, the document title contains two remote sensing data sources and the document subjects containing the words "fusion" and "remote sensing" are considered. A total of 385 relevant literatures were retrieved from the WOS database and 194 from the

CNKI database. The total number of related literatures is 579.The two databases are similar in the distribution of various remote sensing data source pairs ($C_9^2 = 36$ pairs in total). If the corresponding data are combined for statistical analysis, the top three results can be obtained:

(1) A total of 218 multispectral images and panchromatic images were fused, accounting for 37.65% of the total.

(2) A total of 110 multispectral and hyperspectral images were fused, accounting for 19.00% of the total.

(3) A total of 108 hyperspectral images and LiDAR images were fused, accounting for 18.65% of the total.

Together, these three pairs account for more than three-quarters of the total literature. On the other hand, there is little literature on the fusion of night light remote sensing images, stereo remote sensing images and video remote sensing data, which are new imaging methods for which few relevant tasks have yet been carried out.

6.6.2.3 Spatial-Spectral Fusion of Remote Sensing Images

The spatial resolution and spectral resolution of remote sensing images are mutually restricted. Remote sensing images with high spatial and hyperspectral resolution cannot be obtained using a single imaging method. Fusion of two (or more) images with different spatial resolution and spectral resolution is an effective means to obtain high spatial and high spectral resolution images. This type of image fusion technology is also known as **spatial-spectral fusion of remote sensing image** (Li et al. 2021).

Spatial-spectral fusion mainly includes panchromatic image and multispectral image fusion (also called panchromatic sharpening), panchromatic image and hyperspectral image fusion, and multispectral image and hyperspectral image fusion. A general introduction to the panchromatic and multispectral image fusion methods can be found in Table 6.6. The fusion of panchromatic and hyperspectral images basically adopts the technology of panchromatic image and multispectral image fusion, but because the spatial resolution of hyperspectral images is very low, there will be confusion between pixels, so the model optimization method is often used. A general introduction to multispectral images and multispectral image fusion methods can be found in Table 6.7.

6.6.2.4 Fusion with Deep Recurrent Residual Networks

The spatial-spectrum fusion of remote sensing images can be realized with the help of deep learning methods. Methods for fusing a **multispectral image** (MS) with rich spectral information but low spatial resolution and a **panchromatic image** (PAN) with rich spatial detail but only grayscale information by means of a deep recurrent residual network (Wang et al. 2021) are presented below.

TABLE 6.6 Fusion Methods of Panchromatic Image and Multispectral Image

Category	Principle	Advantage and/or Disadvantage
Spatial information injection	Spatial information of high spatial resolution panchromatic images is extracted by means of spatial transformation and multi-scale analysis, and the extracted spatial information is injected into low spatial resolution multispectral images as lossless as possible.	The advantage is that it can better retain the spectral information. The disadvantage is that the extracted spatial information only contains the spatial structure within a specific spectral range, which does not completely match the spatial structure of the low spatial resolution image, so that spatial structure distortion can easily arise.
Spectral information injection	Through linear or nonlinear image transformation, the hyperspectral resolution image is transformed into a new projection space, decomposed into spectral components and spatial components, the spatial components are replaced by a panchromatic image, and then the fused image is obtained through inverse transformation.	The advantage is that it can better retain spatial information, while the disadvantage is that spectral information may produce certain distortion.
Spatial-spectral sampling modeling	By establishing the relationship model between the source image and the fusion result, the fusion problem is regarded as an inverse reconstruction problem, and the fusion result is obtained by optimization.	The advantage is that the model is more rigorous, which can better maintain the spatial and spectral information of the image. Probability statistics and a priori constraints can also be introduced in the fusion process to further optimize the fusion effect. The disadvantage is that the solution of the model will be more complex.

TABLE 6.7 Fusion Method of Multispectral Image and Hyperspectral Image

Category	Description	Considerations
Based on panchromatic sharpening	The multispectral and panchromatic fusion method is extended to the fusion of multispectral and hyperspectral images.	When extending the application of the panchromatic sharpening fusion method, we need to consider the correspondence of bands.
Based on imaging model	Based on mixed pixel decomposition: using the principle of spectral decomposition, under the constraint or a priori of sensor characteristics, the end element information and high-resolution abundance matrix are obtained from hyperspectral image and multispectral image respectively to reconstruct the fused image.	Estimation of dimensions and matrix coefficients of spectral basis.
	Based on tensor decomposition: by representing hyperspectral images as three-dimensional tensors, spectral dictionary and tensor kernel are used to approximate the fusion results of high spatial and spectral resolution.	Construction of spectral dictionary and estimation of tensor kernel.
Based on deep network	Take the low-resolution image as the input of the depth network, and output the high-resolution image by learning the end-to-end mapping between the low- and high-resolution images.	Construct a more reasonable loss function, deal with image residuals and use a deeper framework to improve the fusion performance.

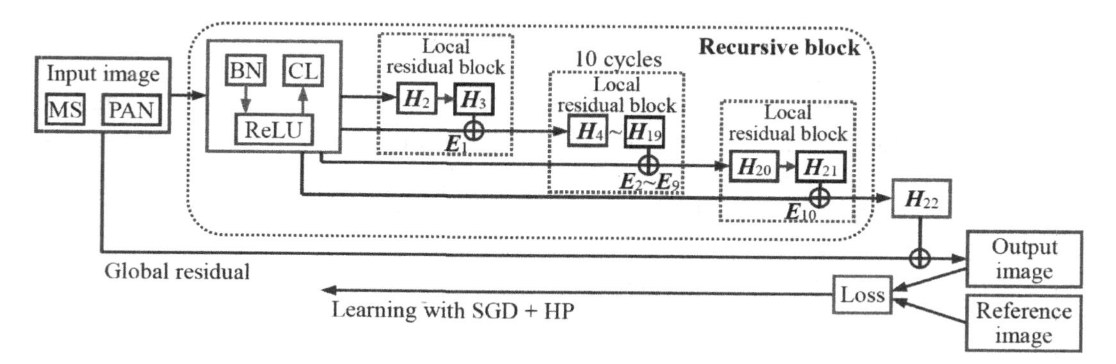

FIGURE 6.19 Deep recursive residual network model.

The end-to-end **deep recurrent residual network** (DRRN) model that has been designed is shown in Figure 6.19. There are three points to consider:

(1) In order to solve the problem of gradient disappearance and gradient explosion, global residual is used in the identity branch from input to output of the network, and recursive learning is introduced into residual learning by constructing a recursive block structure in the residual branch. The final output of the residual network model is the sum of the residual branch and the identity branch.

(2) In order to solve the problem that the deep network requires a lot of network parameters and a large memory space, which makes the model run slowly and with difficulty, a recurrent neural network with parameter sharing is designed, which can reduce network parameters and reduce overfitting. At the same time, the number of network layers is deepened, the deep-level features of the image are fully learned, and the effect of image fusion is improved without a lot of memory consumption.

(3) In order to overcome the problem that image details may be lost after many layers in very deep networks, resulting in performance degradation, a local residual unit using multi-path mode local residual learning is designed. In the recursive block, stacked local residual units are utilized to learn deeper, richer image features.

In Figure 6.19, the input of the network is MS image and PAN image, that is, H_0 = {MS, PAN}. Batch normalization (BN) and activation function ReLU (rectified linear unit) are performed before each **convolution layer** (CL). In this way:

$$\boldsymbol{B}_0 = \mathrm{BN}(\boldsymbol{H}_0) \tag{6.43}$$

$$\boldsymbol{R}_0 = \mathrm{ReLU}(\boldsymbol{B}_0) \tag{6.44}$$

$$\boldsymbol{H}_1 = \boldsymbol{W}_1 \otimes \boldsymbol{R}_0 + b_1 \tag{6.45}$$

W and b are the weight term and bias term of the network, respectively, which need to be obtained by network training.

There are ten local residual units stacked in the recursive block. Input H_1 into the first local residual network unit, and after two convolutions, add the result to H_1 as the input of the next local residual unit. Regardless of BN and ReLU, the learning process of local residual units in the entire recursive block can be expressed as (when $i = 1$, $E_0 = H_1$):

$$H_{2i} = W_{2i} \otimes E_{i-1} + b_{2i} \qquad i = 1,2,\ldots,10 \tag{6.46}$$

$$H_{2i+1} = W_{2i+1} \otimes H_{2i} + b_{2i+1} \qquad i = 1,2,\ldots,10 \tag{6.47}$$

$$E_i = H_{2i+1} + H_1 \qquad i = 1,2,\ldots,10 \tag{6.48}$$

In these equations, i represents the number of cycles of the local residual block.

After ten cycles, E_{10} can be obtained, and the output H_{22} can be obtained after one convolution operation. The final network output image H_{out} is obtained by adding H_{22} to the network input data and then convolution. H_{out} can be compared with the ground truth value G_t as a reference image and optimized by constraining the loss function. The optimization process can use **stochastic gradient descent** (SGD) and **back propagation** (BP) to learn $\{W, b\}$. Here the loss function can be expressed as:

$$\text{Loss} = \frac{1}{2}\|H_{\text{out}} - G_t\|^2 \tag{6.49}$$

REFERENCES

Andreu, J.P., H. Borotsching, H. Ganster, et al. 2001. *Information fusion in image understanding. Digital Image Analysis – Selected Techniques and Applications*. Heidelberg: Springer.

Bian, H., Y.-J. Zhang and W.D. Yan. 2005. The study on wavelet transform for remote sensing image fusion. *Proceedings of the Fourth Joint Conference on Signal and Information Processing*, 109–1139.

Camps-Valls, G.L. Gomez-Chova, J. Munoz-Mari, et al. 2006. Composite kernels for hyperspectral image classification. *IEEE Geoscience and Remote Sensing Letters*, 3(1): 93–97.

Clausi, D.A. and M.E. Jernigan. 2000. Designing Gabor filters for optimal texture separability. *Pattern Recognition*, 33(11): 1835–1849.

He, Y., G.H. Wang, D.J. Lu, et al. 2000. *Multi-Sensor Information Fusion and Applications*. Beijing: Publishing House of Electronics Industry.

Hu, W., Y.Y. Huang, L. Wei, et al. 2015. Deep convolutional neural networks for hyperspectral image classification. *Journal of Sensors*, 258619: 1–12.

Jia, B., Y.-J. Zhang and X.G. Lin. 2000. Genera l and fast algorithm for disparity error detection and correct ion. *Journal of Tsinghua University (Sci & Tech)*, 40(1): 28–31.

Li, R. and Y.-J. Zhang. 2005. Automated image registration using multi-resolution based Hough transform. *SPIE*, 5960: 1363–1370.

Li, S.T., C.Y. Li and X.D. Kang. 2021. Development status and future prospects of multi-source remote sensing image fusion. *National Remote Sensing Bulletin*, 25(1): 148–166.

Luo, Z.Z. and J.P. Jiang. 2002. *Robot Sense and Multi-Information Fusion*. Beijing: China Machine Press.

Mou, L.C., X.Q. Lu, X.L. Li, et al. 2020. Nonlocal graph convolutional networks for hyperspectral image classification. *IEEE Transactions on Geoscience and Remote Sensing*, 58(12): 8246–8257.

Ojala, T., M. Pietikainen and T. Maenpaa. 2002. Multiresolution gray-scale and rotation invariant texture classification with local binary pattern. *IEEE Transactions on Pattern Analysis and Machine Intelligence*, 24(7): 971–987.

Peng, H.H. and R. Rao. 2009. Hyperspectral image enhancement with vector bilateral filtering. *Proceedings of the 16th IEEE International Conference on Image Processing (ICIP)*, 3713–3716.

Pesaresi, M. and J.A. Benediktsson. 2001. A new approach for the morphological segmentation of high-resolution satellite imagery. *IEEE Transactions on Geoscience and Remote Sensing*, 39(2): 309–320.

Piella, G. 2003. A general framework for multiresolution image fusion: from pixels to regions. *Information Fusion*, 4: 259–280.

Polhl, C. and J.L. Genderen. 1998. Multisensor image fusion in remote sensing: Concepts, methods and applications. *International Journal of Remote Sensing*, 19(5): 823–854.

Qin, A.Y., Z.W. Shang, J.Y. Tian, et al. 2019. Spectral-spatial graph convolutional networks for semi-supervised hyperspectral image classification. *IEEE Geoscience and Remote Sensing Letters*, 16(2): 241–245.

Sha, A.S., B. Wang, X.F. Wu, et al. 2019. Semi-supervised classification for hyperspectral images using edge conditioned graph convolutional networks. *Proceedings of the IEEE International Geoscience and Remote Sensing Symposium (IGARSS)*, 2690–2693.

Stathaki, T. 2008. *Image Fusion: Algorithms and Applications*. Amsterdam: Elsevier Ltd.

Tarabalka, Y., M. Fauvel, J. Chanussot, et al. 2010. SVM- and MRF-based method for accurate classification of hyperspectral images. *IEEE Geoscience and Remote Sensing Letters*, 7(4): 736–740.

Wang, F., Q. Guo and X.Q. Ge. 2021. Pan-sharpening by deep recursive residual network. *National Remote Sensing Bulletin*, 25(6): 1244–1256.

Wang, H.Y., L.R. Yang, Y.-J. Zhang, et al. 2011. Study on X-ray security inspection technology based on multi-sensory detection. *Acta Metrologica Sinica*, 32(6): 555–558.

Wang, H.Y., Y.-J. Zhang, L.R. Yang, et al. 2012. Hybrid filter designed for X-ray Compton back-scattering security inspection images. *Acta Metrologica Sinica*, 33(1): 87–90.

Wright, J., A.Y. Yang, A. Ganesh, et al. 2009. Robust face recognition via sparse representation. *IEEE Transactions on Pattern Analysis and Machine Intelligence*, 31(2): 210–227.

Wu, J.F., Z.G. Jiang, H.P. Zhang, et al. 2017. Hyperspectral remote sensing image classification based on semi-supervised conditional random field. *Journal of Remote Sensing*, 21(4): 588–603.

Veganzones, M.A., G. Tochon and M. Dalla-Mura. 2014. Hyperspectral image segmentation using a new spectral unmixing-based binary partition tree representation. *IEEE Transactions on Image Processing*, 23(8): 3574–3589.

Ye, Z., L. Bai and M.Y. He. 2021. Review of spatial-spectral feature extraction for hyperspectral image. *Journal of Image and Graphics*, 26(8): 1737–1763.

Zhang, L., M. Yang and X.C. Feng. 2011. Sparse representation or collaborative representation: which helps face recognition? *Proceedings of 2011 IEEE International Conference on Computer Vision*, 471–478.

Zhang, W.X., W.Z. Wu, J.Y. Liang, et al. 2001. *Rough Set Theory and Methods*. Beijing: Science Publisher.

Zhang, Y.-J. 2021. Twenty-five years of image engineering in China. *Journal of Image and Graphics*, 26(10): 2326–2336.

CHAPTER 7

Content-Based Visual Information Retrieval

CONTENT-BASED IMAGE AND VIDEO retrieval for rapid extraction of useful visual information is becoming increasingly popular with the rapid expansion of data brought about by the development and popularization of science and technology. Much research has been carried out and progress made in the last 30 years (Zhang 2003, 2007a, 2007b, 2008, 2009a, 2009b, 2009c, 2015a, 2015b), and many image techniques have been developed (Zhang 2005, 2006a, 2009d).

This chapter is organized as follows:

Section 7.1 introduces the principle and basic process of content-based image and video retrieval, with its main functional modules, and also discusses the semantic and abstraction levels of related research. Section 7.2 presents matching techniques and criteria in image retrieval based on color, texture, and shape features, with a recent retrieval example utilizing synthetic features, including CNN features. Section 7.3 analyzes video retrieval techniques based on motion features (both global and local). Section 7.4 describes methods for analyzing, retrieving and highlighting news videos and sports videos. Section 7.5 introduces two works in high-level semantic retrieval: image classification with object semantic description and image classification with abstract atmosphere semantics. Section 7.6 reviews technique developments and promising research directions from the last year.

7.1 PRINCIPLES OF IMAGE AND VIDEO RETRIEVAL

Both image and video retrieval represent the retrieval of visual information. The goal of **visual information retrieval** (VIR) is to rapidly extract image collections or image sequences relevant to a query from a visual database (collection). Visual information is characterized by a large amount of information, while the amount of visual information data is also large, but the degree of abstraction is low. With the rapid development of imaging methods and equipment, the rapid expansion of visual information presents many challenges to the real acquisition of useful information.

DOI: 10.1201/9781003362388-7

Not only are technologies for mass storage and rapid transmission of various media and information required, but also those that can automatically query and select media information, enabling the user to rapidly acquire the information needed without drowning in the vast ocean of multimedia. Identifying and managing visual information using computers, with the ultimate goal of automatic query retrieval, is an effective way of dealing with information expansion.

7.1.1 Content-Based Retrieval

Traditional visual information retrieval schemes often use text identifiers, for example, specific queries on images are carried out by means of image numbers or labels or tags. To achieve retrieval, first add to the image a text or numerical label describing it, and then retrieve the label during indexing. In this way, queries on images become label-based queries. Although simple, this approach suffers from several fundamental problems that affect the effective use of visual information.

First, since the richness of the image content is difficult to fully express with text labels, this approach often makes mistakes in querying images. Second, the textual description is a specific abstraction, and if the standard of the description changes, the label has to be remade to suit the new query. In other words, specific labels are only suitable for specific query requirements. Finally, at present, these text labels are selected and added by human operators, so they are greatly affected by subjective factors. Different people or the same person may give different descriptions of the same image under different conditions or at different times, so they are not sufficiently objective and lack a uniform standard.

To solve these problems, it is necessary to extract image content comprehensively, generically and objectively. Since people use images not only according to their visual quality but more importantly according to their visual content, it is only possible to obtain the required visual information effectively and accurately by retrieving it according to the content. **Content-based visual information retrieval** (CBVIR) methods are an effective way of acquiring and utilizing visual information (Zhang 2003). Relevant research has been carried out since the 1990s, and there have been many achievements in this century (Zhang and Lu 2002; Zhang 2015a), including **content-based image retrieval** (CBIR) and **content-based video retrieval** (CBVR). A special standard has also been developed internationally for this purpose, the **multimedia content description interface** (MPEG-7) (Zhang 1999).

7.1.2 Achieving and Retrieval Flowchart

The most common and basic content-based retrieval is done with the help of some visual features, which can be the picture (appearance) feature of the image, or its subject. From a visual point of view, it can be the color of the scene or object, the texture of the surface, the geometric shape of a specific target, the relationships of several objects in space, etc.

Figure 7.1 presents a block diagram of a content-based scheme for archiving, querying, and retrieval of images with features (Zhang and Liu 1997), consisting of two parts: image archiving and image retrieval. The extraction of image content can be performed when the image database is established. When the image is archived, the input image is first analyzed to extract the visual features of the image or object (common features include color,

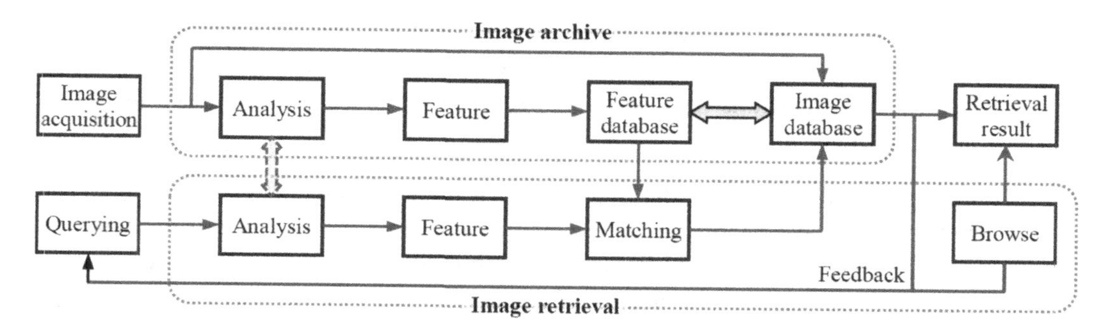

FIGURE 7.1 Principle block diagram of image archiving and image retrieval.

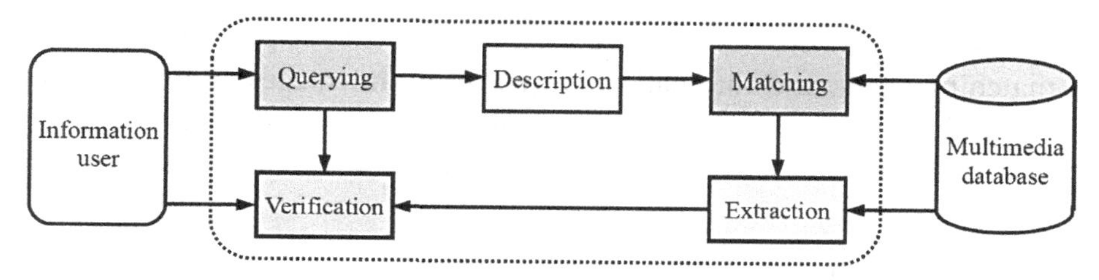

FIGURE 7.2 Five functional modules of image retrieval system.

texture, shape, spatial relationship, and motion, etc.; see Section 7.1.3). When the input image is stored in the image database, its corresponding feature representation is also stored in the feature database that is linked to the image database. In image query and retrieval, the example query method is generally used, the given query image is analyzed and its features are extracted first. The desired image can be extracted by matching the feature representation of the query image with the feature representation in the feature database (to determine the consistency and similarity of the image content) and searching the image database according to the matching result. The user can also browse the output retrieval results, select the desired image, or put forward opinions to modify the query (condition) for a new round of queries based on the preliminary feedback.

It can be seen from Figure 7.1 that there are three keys to feature-based image retrieval: selecting appropriate image features, adopting effective feature extraction methods, and having accurate feature-matching algorithms. When the image content is represented in other forms, the basic process is similar.

There are five functional modules in a content-based image retrieval system connecting information users with the image database (Figure 7.2).

(1) **Query module**. Provides users with various query means to support different applications. To query, users need to put forward query conditions, which are mainly based on the description of image content.

(2) **Description module**. Converts the user's query request into an abstract internal representation and description of the image content. The key here is how to acquire the

content of the image. It is necessary to analyze the image, establishing a description of the content with a specific data structure that can be easily represented by the computer. This module is also needed to represent and describe each image when building the image database.

(3) **Matching module**. Once the description of the queried image and the image in the image database have been established, the desired image content can be searched in the image database with the help of a search engine. Here, content matching and comparison between the description of the query image and the description of the queried image in the image database can determine their consistency and level of similarity in terms of content. The result of this match will be passed to the extraction module.

(4) **Extraction module**. Locates the image of interest in the image database according to the matching result, and automatically extracts all the images in the image database that meet the given query conditions on the basis of content matching for the user to select. If the image database is indexed in advance, it can improve efficiency when extracting images.

(5) **Verification module**. Allows the user to judge whether the extracted image meets the requirements. According to the current technical level and equipment conditions, on the basis of automatic query and extraction, users also need to have the means to finally verify the results. If the verification effect is not satisfactory, a new round of queries can be restarted by modifying the query criteria. For the verification feedback module, the appropriate user interface should be designed in combination with the query module.

7.1.3 Multi-Level Content Representation

Feature-based image retrieval is the primary stage of content-based image retrieval. The features mentioned here are mainly visual features, such as color, texture, shape, spatial relationship, motion information, etc. These features, which can be perceived intuitively, can easily be extracted by computer, and have been widely used.

From the semantic point of view, visual features are at a relatively low level. In traditional image description models based on low-level features, descriptions generally take the form of statistical data. In fact, these statistics are quite different from human understanding of the content of images. When humans describe image content, they often use semantic-level concepts and terms, such as concepts of object and scene (Zhang et al. 2004). These concepts and terms are at a higher semantic level than visual features (Figure 7.3). From the perspective of human cognition, people's description and understanding of images are mainly carried out at the semantic level. How to describe the image content so that it is as consistent as possible with human understanding of the image content is the key to image retrieval, but is also the main difficulty (Zhang 2015a).

Here, it is considered that the image content is represented by a collection of some basic visual features. The content that can be represented by these basic visual features can be

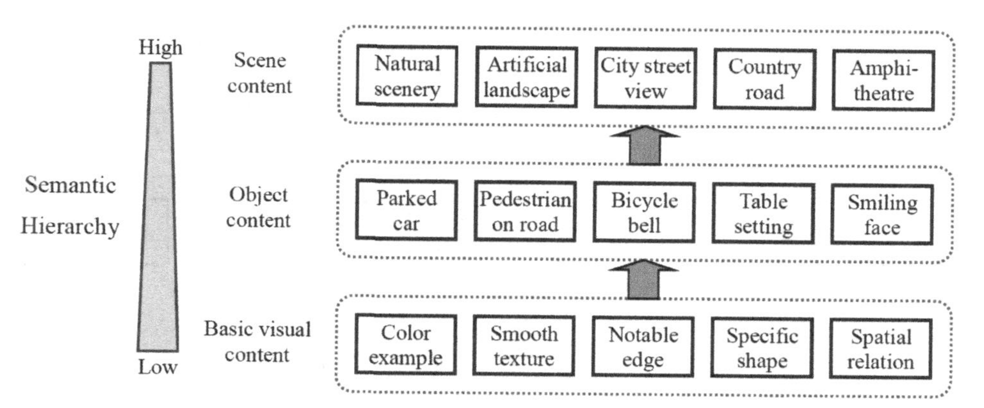

FIGURE 7.3 Hierarchical relationship of different contents.

called basic visual content. On this basis, higher-level image content can also be defined: object content and scene content. Object content and scene content are closely related to human knowledge about the semantic description of the scene (Zhang 2004). The object content is a set of semantic descriptions of basic visual content based on the object model. Scene content is a set of semantic descriptions of scene content based on a scene model. The object content describes the part of the scene that is of interest, while the scene content gives a global description of the image. In some cases, a layer of goal relationship is added between them, describing the connection between objects, such as "the man under the lamp", "the book on the table", "the sun in the sky", "the stool in the front row", "the tree in front of the building" and so on.

Semantic content is extracted on the basis of scene recognition and is the result of cognition of the objective world. In addition to describing objective things (such as images, cameras, etc.), semantics can also describe subjective feelings (such as beautiful, clear, etc.) and more abstract concepts (such as broad, rich, etc.). So from an abstract point of view, on the basis of cognition, an emotional level can also be defined. At the affective level, the goal of retrieval can be considered not to be strictly defined (or rather, broadly defined). In practice, sometimes the user's subjectivity plays an important role in retrieval, and this subjectivity is directly related to the user's emotional state or affection and the environment (Xu and Zhang 2005; Li et al. 2010; Li and Zhang 2011). In other words, it is the user's emotion that determines whether the retrieval results meet the requirements (like or dislike, interested or disinterested). In this way, a more general three-layer abstraction model for image retrieval can be given, as shown in Figure 7.4.

Incidentally, some people also call semantic features logical features and further divide them into objective features and subjective features (Djeraba 2002). Objective features are related to object recognition in image and object motion in video. They have been widely used in semantic retrieval based on object recognition. Subjective characteristics are related to the attributes of the object and scene, and represent the meaning and purpose of the object or scene. Subjective features can be further decomposed into event, activity type, affective meaning, belief, and other types. Users can query in the light of these more

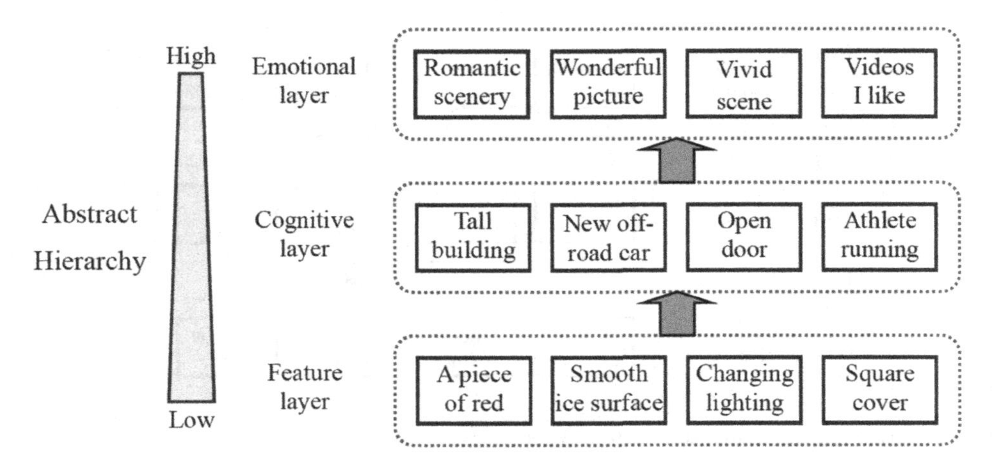

FIGURE 7.4 Three abstract levels of image retrieval.

abstract categories. According to this classification, the features of the cognitive layer in Figure 7.4 mainly correspond to objective features, while the features of the emotional layer mainly correspond to subjective features.

7.2 MATCHING AND RETRIEVAL OF VISUAL FEATURES

In content-based image retrieval, matching techniques play an important role. A typical image database query mode is **query by example**, that is, the user gives an example image and asks the system to retrieve and extract (all) similar images in the image database. A variation of this approach is to allow the user to combine multiple images or **draw sketches** to obtain example images (Zhang 2003). An important job in the query is to determine what method to use to match the query reference image with a large number of images in the image database, that is, to match the query description with the description of the queried information in the database to determine the consistency and similarity of their content. The basic approach is to (combine) match the visual features (mainly color, texture and shape, etc.) of images (Zhang et al. 2001; Zhang 2003).

The following is a brief summary of matching using color, texture, and shape features, respectively. Features such as the spatial relationship between objects' positions (Xu and Zhang 2005) and the structural information of the object itself (Zhang 2003) can also be used. In practice, various features are often used in combination.

7.2.1 Color Feature Matching

Color is an important feature to describe the content of an image, and many methods have been proposed to retrieve images with the help of color features, such as (Liu and Zhang 1999a, 2000; Niblack et al. 1998). The commonly used color spaces are mainly RGB and HSI spaces. Color information can be represented by means of a statistical histogram of colors. In this way, color feature matching between different images can be performed by calculating the distance between these histograms. Four basic methods are described below.

7.2.1.1 Histogram Intersection Method

Let $H_Q(k)$ and $H_D(k)$ be the feature statistical histograms of the query image Q and the database image D, respectively ($k = 0, 1, \ldots, L-1$), then the matching value between the two images can be calculated by the following method of **histogram intersection** (Swain and Ballard 1991):

$$P(Q,D) = \sum_{k=0}^{L-1} \min\left[H_Q(k), H_D(k)\right] \Big/ \sum_{k=0}^{L-1} H_Q(k) \tag{7.1}$$

7.2.1.2 Distance Method

In order to reduce the amount of calculation, the mean value of the histogram can be used to roughly represent the color information. For the three RGB components of the image (other color space components can be treated similarly), the combined feature vector is

$$f = \begin{bmatrix} \mu_R & \mu_G & \mu_B \end{bmatrix}^T \tag{7.2}$$

At this time, the matching value between the query image Q and the database image D is

$$P(Q,D) = \sqrt{(f_Q - f_D)^2} = \sqrt{\sum\nolimits_{R,G,B} (\mu_Q - \mu_D)^2} \tag{7.3}$$

7.2.1.3 Central Moment Method

For a histogram, the mean is its zero-order moment, and higher-order moments can also be used for more accurate descriptions. Let M^i_{QR}, M^i_{QG}, M^i_{QB} represent the i-th ($i \le 3$)-order central moment of the RGB three-component histogram of the query image Q, respectively; M^i_{DR}, M^i_{DG}, M^i_{DB} represent the i-th ($i \le 3$)-order central moment of the RGB three-component histogram of the database image D respectively, then the matching value between them is

$$P(Q,D) = \sqrt{W_R \sum_{i=1}^{3} \left(M^i_{QR} - M^i_{DR}\right)^2 + W_G \sum_{i=1}^{3} \left(M^i_{QG} - M^i_{DG}\right)^2 + W_B \sum_{i=1}^{3} \left(M^i_{QB} - M^i_{DB}\right)^2} \tag{7.4}$$

where W_R, W_G, W_B are weighting coefficients.

7.2.1.4 Reference Color Table Method

The distance method is too rough, and the histogram intersection method is too computationally expensive. A compromise method is to use a set of reference colors to represent the image color. This set of reference colors should cover the various colors that can be perceived visually (Mehtre et al. 1995). The number of reference colors is less than that in the original image, so that a simplified histogram can be obtained. The feature vector to be matched is

$$f = \begin{bmatrix} r_1 & r_2 & \cdots & r_N \end{bmatrix}^T \tag{7.5}$$

where r_i is the frequency of the i-th color, and N is the size of the reference color table. The matching value between the weighted query image Q and the database image D is

$$P(Q,D) = W\sqrt{(f_Q - f_D)^2} = \sqrt{\sum_{i=1}^{N} W_i (r_{iQ} - r_{iD})^2} \tag{7.6}$$

where

$$W_i = \begin{cases} r_{iQ} & r_{iQ} > 0 \quad \text{and} \quad r_{iD} > 0 \\ 1 & r_{iQ} = 0 \quad \text{or} \quad r_{iD} = 0 \end{cases} \tag{7.7}$$

The last three of the four methods presented above simplify the first one from the perspective of reducing the amount of calculation, but the histogram intersection method encounters another problem. When the features in the image do not take all possible values, there will be some zeros in the statistical histogram. The appearance of these zero values will affect the intersection of the histograms, so that the matching value calculated by Equation (7.1) cannot correctly reflect the color difference between the two images. To solve this problem, a cumulative histogram can be used. The cumulative histogram greatly reduces the likelihood of zero values appearing in the original statistical histogram. A further improvement can also utilize locally accumulated cumulative histograms (Zhang 1998).

7.2.2 Texture Feature Calculation

Texture is also an important feature to describe the content of an image. See Huang et al. (2003) for a method for JPEG image retrieval using texture features. Several texture descriptors are also specified in the international standard MPEG-7 (Xu and Zhang 2006b; Zhang 2000). An efficient method for texture feature extraction is based on the gray-level spatial correlation matrix, the co-occurrence matrix (Furht et al. 1995), because the joint frequency distribution of the co-occurrence of two grayscale pixels separated by $(\Delta x, \Delta y)$ in the image can be represented as gray-level co-occurrence matrix. If the number of gray levels in the image is N, the co-occurrence matrix is an $N \times N$ matrix, which can be written as $\boldsymbol{M}_{(\Delta x, \Delta y)}(h, k)$, where the element value m_{hk} at (h, k) represents the number of occurrences of two pixels (a pair) separated by $(\Delta x, \Delta y)$ with one pixel having gray-level h and another pixel having gray-level k. Various statistics on the gray-level co-occurrence matrix can be used as a measure of texture properties (Zhang 2017).

The following introduces a specific method for image retrieval using texture feature matching (Liu and Zhang 1999b). First divide the luminance component of the image into 64 gray levels, and construct co-occurrence matrices in four directions, namely $\boldsymbol{M}_{(1, 0)}$, $\boldsymbol{M}_{(0, 1)}$, $\boldsymbol{M}_{(1, 1)}$, $\boldsymbol{M}_{(1,-1)}$, and then calculate the following four texture parameters with the help of these four co-occurrence matrices:

(1) **Contrast** (or the moment of inertia of the main diagonal):

$$G = \sum_h \sum_k (h-k)^2 m_{hk} \tag{7.8}$$

For coarse textures, since the value of m_{hk} is more concentrated near the main diagonal, the value of $(h - k)$ is small at this time, so the corresponding G value is also small; for fine texture, on the other hand, the corresponding G value is relatively big.

(2) **Energy** (also called angular second moment):

$$J = \sum_h \sum_k (m_{hk})^2 \tag{7.9}$$

This is a measure of the uniformity of the grayscale distribution of an image. When the numerical distribution of m_{hk} is more concentrated near the main diagonal, the corresponding J value is larger; otherwise, the J value is smaller.

(3) **Entropy**:

$$S = -\sum_h \sum_k m_{hk} \log m_{hk} \tag{7.10}$$

When the values of m_{hk} in the gray-level co-occurrence matrix are not much different and scattered, the value of S is larger; when the values of m_{hk} are concentrated, on the other hand, the value of S is smaller.

(4) **Correlation**:

$$C = \frac{\sum_h \sum_k hkm_{hk} - \mu_h \mu_k}{\sigma_h \sigma_k} \tag{7.11}$$

Of these, μ_h, μ_k, σ_h, σ_k are the mean and standard deviation of m_h and m_k, respectively; $m_h = \sum_k m_{hk}$ is the sum of elements in each row of matrix $\boldsymbol{M}$; $m_k = \sum_h m_{hk}$ is the sum of elements in each column of matrix M. Correlation is used to describe the similarity of gray levels between row or column elements in a matrix.

After calculating the above four texture parameters of the image, their mean and standard deviation, namely μ_G, σ_G, μ_J, σ_J, μ_S, σ_S, μ_C, σ_C, can be used as the components of texture feature vector. Since the physical meanings and value ranges of these eight components are different, they need to be internally normalized. In this way, each component can have the same weight when calculating the similarity distance.

The Gaussian normalization method is a better normalization method, in that the influence of a small number of extreme large or extreme small element values on the distribution of the entire normalized element value is relatively small (Ortega et al. 1997). An N-D feature vector can be written as $\boldsymbol{F} = [f_1 \, f_2 \cdots f_N]^{\mathrm{T}}$. If I_1, I_2, $\cdots$, I_M are used to represent the images in the image database, then for any one of the images I_i, whose corresponding feature vector is $\boldsymbol{F}_i = [f_{i,1} \, f_{i,2} \cdots f_{i,N}]^{\mathrm{T}}$. Assuming that the eigen-component value series $[f_{1,j} \, f_{2,j} \cdots$

$f_{i,j} \cdots f_{M,j}]$ conforms to the Gaussian distribution, its mean m_j and standard deviation σ_j can be calculated, and then $f_{i,j}$ can be normalized to the interval [−1, 1] by the following formula:

$$f_{i,j}^{(N)} = \frac{f_{i,j} - m_j}{\sigma_j} \tag{7.12}$$

After normalization, according to Equation (7.12), each $f_{i,j}$ is transformed into $f_{i,j}^{(N)}$ with $N(0, 1)$ distribution. If normalized by $3\sigma_j$, the probability of the value of $f_{i,j}^{(N)}$ falling in the interval [−1, 1] can reach 99%. In practical applications, the value of $f_{i,j}$ outside the [−1, 1] interval can be set to −1 or 1 to ensure that all $f_{i,j}$ values fall within the [−1, 1] interval.

7.2.3 Multi-Scale Shape Features

Shape is also an important feature to describe the content of an image. There are three issues worth noting with retrieval by shape. First, shape is often associated with objects, so shape features can be seen as higher level than color or texture. To obtain the shape parameters of the object, it is often necessary to segment the image first, so the shape feature will be affected by the image segmentation effect. Second, the description of the object shape is a very complicated problem. In fact, no exact mathematical definition of the shape to make it completely consistent with the human feeling has yet been found. Human perception of shape is not only the result of a retinal physiological reflection, but also an integrated result of retinal perception and human knowledge about the real world. Finally, the shape of the object in images obtained from different perspectives may be very different. In order to accurately perform shape matching, it is necessary to solve the problem of translation, scale, and rotation invariance.

The shape of the target can often be represented by the contour of the object, and the contour is composed of a series of boundary points. It is generally believed that at larger scales, false detections can often be eliminated more reliably and real boundary points can be detected, but it is not easy to be accurate with the positioning of boundaries at larger scales. Conversely, the localization of true boundary points is often more accurate at smaller scales, but the proportion of false detections increases. Thus detecting the range of the real boundary points at a larger scale first, and then performing more accurate positioning of the real boundary points at a smaller scale can be considered. As a multi-scale and multi-channel analysis tool, wavelet transform and analysis are more suitable for multi-scale boundary detection of images.

The following briefly introduces a method of using wavelet transform modulo maxima to extract image multi-scale object edge information, and then using multi-scale-invariant moments as features to measure the similarity of object shapes in images (Yao and Zhang 2000).

(1) Calculate the maximum value of wavelet transform modulo

Generally, when discrete wavelet transform is performed by regular sampling, the obtained wavelet coefficients lack translation invariance, while the modulo maxima

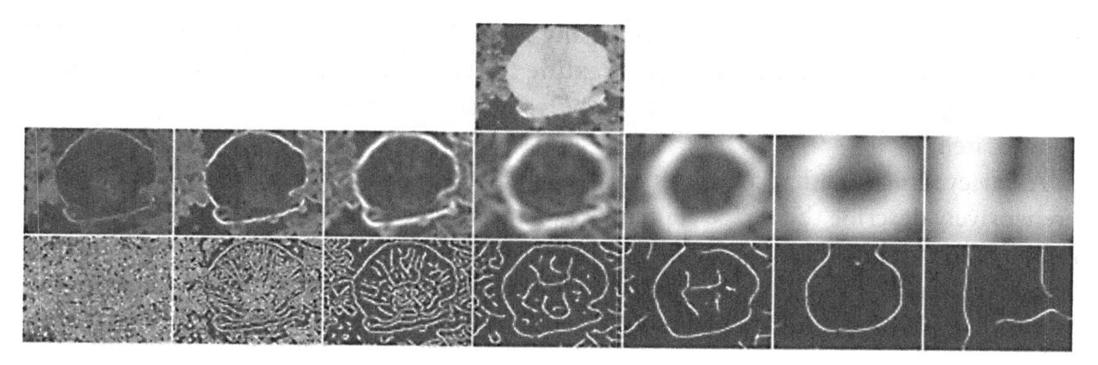

FIGURE 7.5 The modulus maximum points of image wavelet transform.

of wavelet transform are obtained on the basis of irregular sampling of multi-scale wavelet transform, which can overcome the above problems (Mallat and Hwang 1992). The wavelet transform modulo maxima can describe the singularity of the signal. For images, the wavelet transform modulo maxima describes the multi-scale (multi-layer) boundaries of objects in the image.

Figure 7.5 shows a set of examples of the modulo maxima of image wavelet transform. The single image in the first row is the original image. The second row is the modulo image of seven scales, and the scales increase from left to right. The third row is the modulo image for the location of the local maximum point. It can be seen from the figure that the maximum value of the wavelet transform modulus provides the object boundary at different levels.

(2) Calculate the invariant moment feature

It is difficult and inconvenient to directly measure the similarity of two images in the wavelet transform domain. Considering the expected translation, scale, and rotation invariance, the seven invariant moments used for the region representation can be taken to represent the features of multi-scale boundary after the wavelet transform. Retrieval is done with the help of eigenvectors consisting of invariant moments (details can be found in Yao and Zhang (2000); Zhang (2003)).

7.2.4 Retrieval with Composite Features

Various features describe different content attributes in images, so each has its own emphasis in retrieval. On the other hand, retrieval methods based on just a single feature can only represent part of the attributes of the image. Due to the relatively one-sided description of the image content and the lack of sufficient distinguishing information, suitable retrieval results can hardly be achieved, especially when the change of situation (such as scale or direction, etc.) is relatively large.

In fact, different features also have their own characteristics. For example, comparing color features and texture features, color features can describe the overall information of the image, while texture features are more focused on the description of local information; comparing color features and shape features, color features mostly have translation, rotation and scale differences. However, many shape features (such as edge direction) only have translation invariance; comparing texture features and shape features, it is generally easier to obtain texture features, while the calculation of shape features is often more complicated.

A retrieval method that combines use of color, texture, shape and other features to comprehensively describe the image content is called **composite feature retrieval**. The combination of features can not only achieve the effect of complementary advantages of different features, but also improve the flexibility of retrieval and the performance of the system to meet the needs of practical applications.

7.2.4.1 Combination of Color and Texture Features

Color features and texture features have certain complementarities in representing and describing objects in images. Figures 7.6–7.8 show the results of a group of experiments using more than 400 color flower images. Figure 7.6 shows the results obtained by using only color features. The first image from the left is the query image, and the rest are the retrieved images. From left to right, the relevant matching values decrease in turn. Since only the flower object color is used for the query, although the retrieved images are close to the query image from the perspective of color, the overall visual effect is not completely consistent with human perception. For example, although the color of the second image from the left is generally similar to the query image, the (color distribution) pattern is quite different (the query image is a large flower, while the second image from the left has many small flowers). It can also be said that the color distributions of the two images are very different. If it is arranged according to people's visual perception, it is more reasonable to move the 4th, 6th, and 8th pictures from the left (all of which are also a large yellow flower) to the front of the retrieval list. Figure 7.7 shows the result of retrieving using the same

FIGURE 7.6 Retrieval results using only color features.

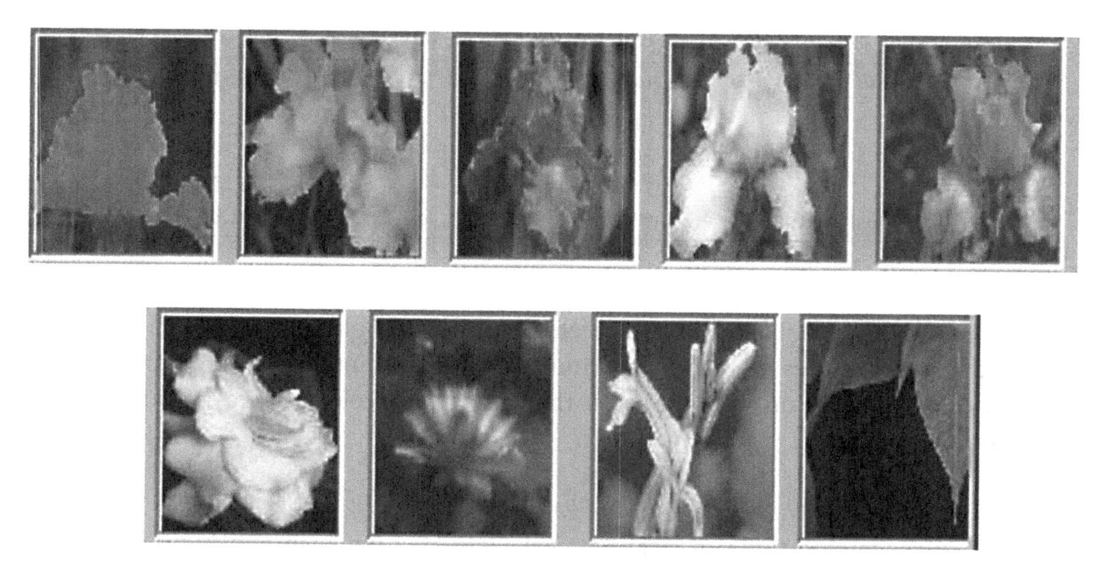

FIGURE 7.7 Retrieval results using only texture features.

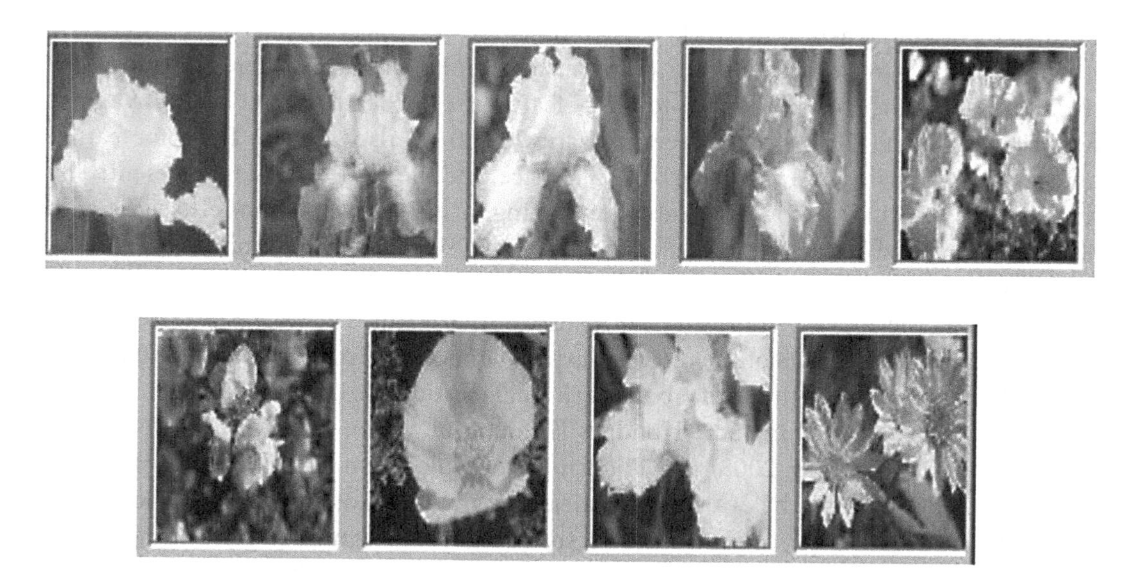

FIGURE 7.8 Retrieval results obtained by combining color and texture features.

query image but with only texture features (here, only luminance information is used for the calculation of texture features). From the perspective of flower patterns, it is better than Figure 7.6 (all are large flowers), but some pictures with completely different colors appear in the search results (such as the 6th to 9th pictures from the left). Finally, Figure 7.8 shows the results of retrieving the same query image using the composite feature method, where the weight ratio of the color feature and texture feature used in the retrieval is 3 to 2.

TABLE 7.1 Comparison of Composite Feature Results and Single Feature Results

Database	Holidays	Oxford	Paris	Flickr100K
HSV	61.95	–	–	53.38
SIFT	80.74	76.66	75.98	72.06
CNN	71.67	43.28	63.64	62.26
Composite Feature	**93.79**	**79.19**	**93.08**	**91.09**

From Figures 7.6–7.8, it can be clearly seen that the results obtained by combined use of color and texture features for retrieval are more in line with human visual requirements in terms of color distribution, and are better than using only color features or only texture features.

7.2.4.2 Combining Color, SIFT and CNN Features

In recent years, **convolutional neural networks** (CNN) have been widely used. Among them, with the help of local convolution, the locally extracted information is interconnected, and layer-by-layer feature extraction is used in the **deep convolutional neural network** (DCNN), and the pixel-level features are finally converted into higher-level CNN features through the neural network.

The following introduces a work (Fang et al. 2016) that fuses different features for retrieval. Among them, in addition to CNN features, color features and **scale-invariant feature transform** (SIFT) features (see Subsection 3.4.2) are also utilized. The color feature here is the color histogram feature of the HSV space, and the SIFT feature is represented by the SIFT description vector. For feature matching, a **bag-of-words model** (see Section 5.5.1) can be used to quantify and encode all SIFT vectors in an image to facilitate matching.

In retrieval, the above-mentioned HSV features, SIFT features and CNN features can be combined. Here, the results of the retrieval experiments comparing the method that fuses different features with methods that only use a single feature are listed in Table 7.1. Among them, the databases used include: Holidays (Jegou et al. 2010), Oxford5k (Philbin et al. 2007), Paris (Philbin et al. 2008) and Flickr100k (obtained from the Flickr website, containing 100,000 images); the evaluation indicator is the mean of **average precision** (AP), i.e., mAP, and the mean is based on the corresponding results of multiple query images. As can be seen from Table 7.1, the effect obtained by using the composite feature is improved compared to the effect of using a single feature (Fang et al. 2017).

7.3 VIDEO RETRIEVAL BASED ON MOTION FEATURES

Motion information represents the development and change of video content along the time axis, and it is important for understanding video content. While color, texture, and shape features are common to images and videos, motion features are unique to video data.

The matching of motion features is based on the extraction of motion information and features. A video sequence is composed of a series of temporally related image frames, and the difference between these image frames reflects the motion information contained in the video sequence. Motion information in video sequences can be divided into two categories: global motion information and local motion information.

7.3.1 Global Motion Features

The **global motion** is relative to the **background motion**, and is the overall movement of all points in the frame image caused by the motion of the camera itself (so it is also called **camera motion**). The global motion generally has the characteristics of strong integrity and relative regularity, and can often be represented by a unified feature or model for the whole image. It is generally represented by 2-D motion vectors here, and these 2-D motion vectors can be obtained by using low-level algorithms that do not require much prior knowledge, such as block matching methods, gradient-based optical flow algorithms, and so on.

To accurately extract motion information, it is necessary to consider image changes between adjacent frames (short-term motion analysis), which generally involves a relatively short time interval (usually tens of milliseconds). But people's overall understanding of a movement often requires a certain duration (often a second or more). In order to obtain meaningful motion content, the results of each short-term motion analysis need to be combined in chronological order.

In order to combine the results of short-term motion analysis, the concept of feature point sequence can be introduced (Yu and Zhang 2001a), and the motion information extracted from each pair of adjacent image frames after feature extraction is represented by a point in the motion feature space. A longer-term motion is represented as a sequence of points in the motion feature space. In this way, the motion information in the video sequence is represented by a set of feature points. This feature point sequence contains not only the motion information between each pair of adjacent frames, but also the temporal sequence relationship between the previous and subsequent frames, which more completely summarizes the motion content of the video sequence. The motion behavior over a longer duration can be understood by means of the matching of feature point sequences.

The string matching method (see Section 4.2.3) can be used to measure the similarity of feature point sequences. Suppose there are two video sequences, L_1 and L_2, and their lengths are N_1 and N_2, respectively, then the two feature point sequences can be represented as $\{f_1(i), i = 1, 2, \cdots, N_1\}$ and $\{f_2(j), j = 1, 2, \cdots, N_2\}$. If there is $N_1 = N_2 = N$, then it can be defined that the similarity between video sequences is equal to the sum of the similarities of their corresponding feature points, that is

$$S(L_1, L_2) = \sum_{i=1}^{N} S_f\left[f_1(i), f_2(i)\right] \tag{7.13}$$

where, $S_f[f_1(i), f_2(i)]$ is a function to calculate the similarity between two feature points, and various distance functions can be used.

If the lengths of the two sequences are different, let $N_1 < N_2$, then how to choose the starting point of the matching time must also be considered. Specifically, the sequence $L'_2(t)$ with the same length as L_1 is intercepted at different time starting points t in L_2. Since

L_1 and $L'_2(t)$ have the same length, the distance between them can be calculated using Equation (7.13). By moving the time starting point t, the similarity of the sub-sequences corresponding to all possible time starting points t can also be calculated. The similarity between the two sequences L_1 and L_2 can be selected as the maximum value, that is,

$$S(L_1, L_2) = \max_{0 \le t \le N_2 - N_1} \sum_{i=1}^{N} S_f\left[f_1(i), f_2(i+t)\right] \tag{7.14}$$

From the above formula, not only can the similarity between two sequences of different lengths be obtained, but also the matching position of the segment with the most similar motion in the short sequence and the long sequence can be obtained.

The calculation of the similarity of feature points depends on the representation for each feature point. Taking the common bilinear motion model as an example, the global motion of each adjacent frame is represented by eight model parameters. The global motion parameter model reflects the overall motion of all pixels in the video image caused by the motion of the camera. So the similarity between the parameters of the two motion models corresponding to the two images should be the similarity of all motion vectors contained within the range of the two motion models in the entire video image. Therefore, the distance between the two global motion models M_1 and M_2 can be defined as the sum of the squares of the distances between their motion vectors at various points, namely

$$D(M_1, M_2) = \sum_{x,y}\left[U_{M_1}(x,y) - U_{M_2}(x,y)\right]^2 + \left[V_{M_1}(x,y) - V_{M_2}(x,y)\right]^2 \tag{7.15}$$

where (U_{M1}, V_{M1}) and (U_{M2}, V_{M2}) are the global motion vectors obtained from M_1 and M_2, respectively. In this way, once the model parameters representing the global motion are obtained, matching can be performed and the similarity of the global motion in the two frames of images can be calculated to indicate the degree of matching (Yu and Zhang 2001b).

7.3.2 Local Motion Features

The **local motion** is relative to the **foreground motion**, which is caused by the own motion of the object in the scene. Therefore, the local motion is only reflected in the position of the corresponding moving object in the scene. Since there may be multiple objects in the scene, and each object may perform different motions, the local motion is often complex and irregular, and multiple features or models are often needed to express the local motion in a frame.

To extract the local motion information in the video sequence, it is necessary to find and determine the spatial position of each point on the object in the video in different image frames. Motion is linked to give a representation and description of the motion of that object or part of the object.

The motion representation and description of an object or part often uses a local motion vector field (Yu and Zhang 2001a), which gives the magnitude and direction of the local motion, where the direction of motion is important for distinguishing different motions. The local motion can be described using the **motion direction histogram** (MDH) as a

FIGURE 7.9 Retrieval results obtained by using the local motion features.

feature. On the basis of obtaining the local motion vector field, it can also be segmented to extract motion regions with different parameter models. A **motion region type histogram** (MRTH) can also be obtained by classifying the motion models. Since the parametric model representation of motion regions is a further abstraction of the local motion information in the video, the information obtained from MRTH often has higher-level meaning and more comprehensive description ability than the MDH. For the matching of the above two histogram features, methods such as the previous histogram intersection method can be used.

One set of retrieval experiment results using the above two histograms are shown in Figure 7.9. The experimental data are a video sequence of nine minutes for a basketball match, which comes from a MPEG-7 standard test database. The four images in Figure 7.9 are respectively the first frames of four clips retrieved with similar content (a penalty shot). It can be seen that although the sizes and positions of regions occupied by players are different, the actions performed by these players are all the same.

7.4 VIDEO PROGRAM RETRIEVAL

There are many kinds of video programs – news programs, sports competitions, movies, animation, advertising, etc. There are also non-professional videos such as home videos. The purpose of video program retrieval is to analyze and establish the (semantic) structure in video (Zhang and Lu 2000), and query or index according to this structure (Zhang and Lu 2002; Chen and Zhang 2008).

7.4.1 News Video Structuring

News video is a particularly widely used type of video program. In the structure analysis of news videos, the detection of announcer shots plays an important role in program structuring (Jiang and Zhang 2003a).

7.4.1.1 Features of News Video

The structural characteristics of news videos are relatively obvious; various types of **news programs** have good consistency, and their main content is a series of news story units. Each news story unit, that is, a **news item**, tells a relatively independent event in content, with clear semantics. It can be used as a basic unit for video analysis, indexing and querying (Jiang and Zhang 2003a).

The relatively fixed hierarchical structure of news videos provides a variety of information clues that are helpful for analyzing video content, building indexing structures, and performing content-based queries. These informative clues may exist at different structural levels and can provide relatively independent video features. For example, in many news programs, there is a repeated presentation of the host in the studio (that is, the announcer shot), which can be used as the start marker of the news item and the basis for the segmentation of the news item.

From a strategic point of view, there are two main types of announcer shot detection: direct detection and gradual exclusion. Since different news programs have different styles, and there are many other similar "speaker shots" in news programs (see below), it is difficult to directly and accurately detect announcer shots through a single detection method, which would generate many false detections and false alarms.

The following introduces a three-step, from coarse to fine, detection method (Jiang and Zhang 2005), which can be gradually refined, using different techniques in different detection scale ranges. The first step of the method is to detect all **main speaker close-ups** (MSCs) by analyzing the variation between image frames that contains real announcer shots and similar shots that are falsely detected. In the second step, all MSCs are divided into several character groups using the unsupervised clustering method, that is, a list of important speakers is established, and the results of news headline detection are used for post-processing. The third step is to distinguish the real announcer shot group from the list of important speakers through the statistical law of the time distribution of the shots.

7.4.1.2 Main Speaker Close-up Shot Detection

The main speaker close-up shot is the unique content in news video, and it is often of interest to the audience. Therefore, it can be used as an important annotation of video content and an important clue for video query (although it will also interfere with the detection of announcer lens). Therefore, it is worth detecting all main speaker close-up shots in news clips and using them to assist in the structuring of video content.

The main speaker close-up shot in a news video is a special kind of shot that often appears in general news programs: that is, a close-up shot of a single person-speaking

(a)

(b)

(c)

FIGURE 7.10 Comparison of various size changes in shot.

avatar that appears on the screen. In this case, the main object in the video screen is the character's avatar, and not background activities. Practical examples include live reporter coverage shots, interviewee close-up shots, speaker shots, etc. Unlike face detection, when detecting a speaker shot, the main concern is whether a dominant person's head (rather than other objects) appears in a specific position of the frame in the video. The detailed features of the character's face do not have to be extracted at this point, so a motion statistical model can be constructed, and the method of matching the head motion template of the character can be used to detect important speaker shots.

There are temporal, spatial, and motion visual features in main speaker close-up shots (see Figure 7.10).These include:

(1) For the general news video, the overall motion variation of the main speaker close-up shots (Figure 7.10(a) is much smaller than that of the normal shot (Figure 7.10(b)), but it is larger than the almost still-picture shot (Figure 7.10(c));

(2) The picture movement of main speaker close-up shots is mainly focused on a fixed speaker head, and the position of this head is concentrated in three relatively fixed positions in the news video: the middle, the left and the right, as shown in Figure 7.11. The body region in the three models of Figure 7.11 has the same area and form, only the horizontal position of the upper part corresponding to the head region of the character is different.

The change of the shot picture can be analyzed according to these two characteristics. Here, both the strength of the change between image frames (the former feature) and the spatial distribution of the change value (the latter feature) should be considered. A **map of**

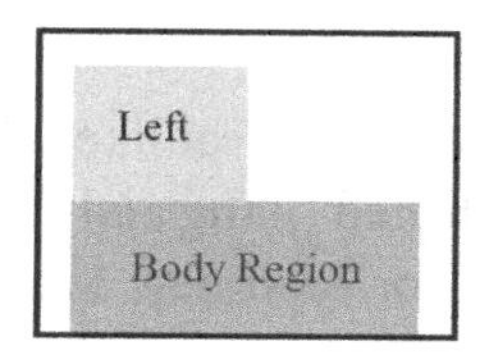

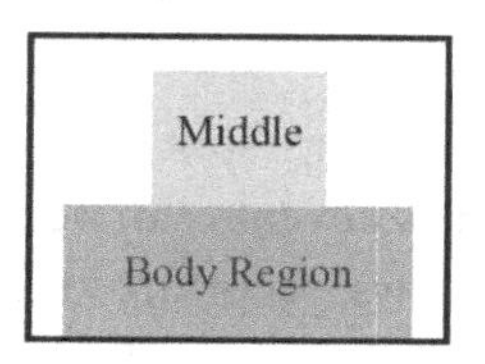

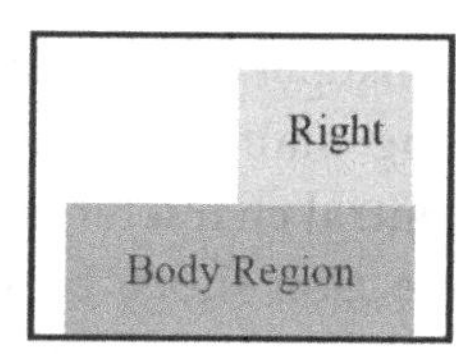

FIGURE 7.11 Three speaker position models.

average motion in shot (MAMS) (Jiang and Zhang 2005) can be used for this purpose. For the n-th shot, the calculation formula of the map of average motion in shot is as follows:

$$M_n(x,y) = \frac{1}{L}\sum_{i=1}^{L-v} | f_i(x,y) - f_{i+v}(x,y) | \qquad L = \frac{N}{v} \tag{7.16}$$

where N is the total number of frames in the n-th shot; i is the serial number of the frame; v is the number of intervals when measuring the difference between frames.

The map of average motion in shot is a 2-D distributed motion cumulative average map that preserves the spatial distribution information of the motion. Since the change intensity of the main speaker close-up shot is in the magnitude between the general shot and the still shot, the change intensity can be reflected by dividing the map of average motion in shot by the size of the map. Only shots that meet the following conditions are considered as possible main speaker close-up shots and proceed to the next step:

$$T_s < \frac{M_n(x,y)}{N_x \times N_y} < T_a \tag{7.17}$$

where T_s and T_a are the statistical averages of the image changes of still shots and general shots, respectively. Note that these two values are proportional to the number of intervals v that measures the difference between the frames.

7.4.1.3 Clustering of Main Speaker Close-up Shots

As mentioned earlier, the detected main speaker close-up shots are important annotations for news video content and important clues for video queries. However, just listing the detected main speaker close-up shots that occur at different times (and possibly multiple times) does not guarantee good results. In addition, it is more difficult to find further announcer shots among these disorganized main speaker close-up shots. Therefore, it is necessary to organize these shots, and the most direct method is to cluster according to the characters: according to the consistency of the external appearances of these characters in the same news content (that is, if the same character appears in the same news program many times, the visual characteristics such as his clothing and skin color generally have strong similarity), and the main speaker close-up shots of the same person that may appear at different times can be extracted to form shot groupings by people. Using this grouping list, you can easily query and retrieve related people.

Here, the color features of each detected main speaker close-up shot can be first extracted, and then the similarity of shots can be calculated using color histogram intersection. In addition, the three-position model shown in Figure 7.11 can be used for more accurately calculating the location model. Finally, all main speaker close-up shots can be clustered using unsupervised clustering.

After obtaining main speaker close-up shot clusters, news headlines describing the time, place, or central character of news events can be detected (Jiang and Zhang 2003b) to remove false detection shots. In a news program video, for each new appearance of the announcer, there is usually a news headline as a character annotation, otherwise the character will not be considered as an "important person" in the news program. Therefore, in the clustering result of person shots obtained above, shot classes that do not contain news headlines can be removed.

7.4.1.4 Announcer Shot Extraction

In order to achieve effective news video structuring, it is necessary to detect where the newscaster's footage appears as the starting point of each news item. Table 7.2 (Jiang and Zhang 2005) shows the statistical results of four features of length, number, interval and temporal coverage of shots in news videos of about six hours selected from the MPEG-7 test dataset.

It can be seen from Table 7.2 that the number of shots, interval, and coverage have a good ability to distinguish between announcer shots and other shots. Compared with other shot types, the announcer shot type has more repetitions, more dispersed time, and covers a much longer timespan. According to these characteristics, the detection of announcer shots can be achieved by further filtering the clustering results of main speaker close-up shots using the following conditions:

(1) The total number of such shots is greater than a threshold;

(2) The average time interval between the occurrence of such adjacent shots is greater than a threshold;

(3) The interval (number of shots) between the earliest and the latest in this type of shot, that is, the coverage of this type of shot is greater than a threshold.

In this way, according to the temporal distribution characteristics of the shots, the main speaker close-up shots can be divided into two cases: the announcer shot class (represented by A_1, A_2, …); and the other person-speaking shot class (represented by P_1, P_2, … indicate). For each news item, starting with the location of the newscaster's footage, followed by the relevant reporting footage and person-speaking footage, a structured result of the news video can be obtained. In order to give users a concept of time, a streaming structure in units of shots can be used, as shown in Figure 7.12. Using this video structured representation, users can locate and replay any news item and any interesting news shots, which can further realize high-level non-linear browsing and query (Zhang 2003).

TABLE 7.2 Average Statistics of Anchor Shots in Contrast to Other MSC Shots

	Shot Length	# of Shots in 20 Minutes	Shot Interval	Range of All Shots
Anchor group	15.72 sec	≈ 9 shots	≈ 12 shots	> 60 shots
Any other MSC group	10.38 sec	≈ 2 shots	≈ 5 shots	< 10 shots

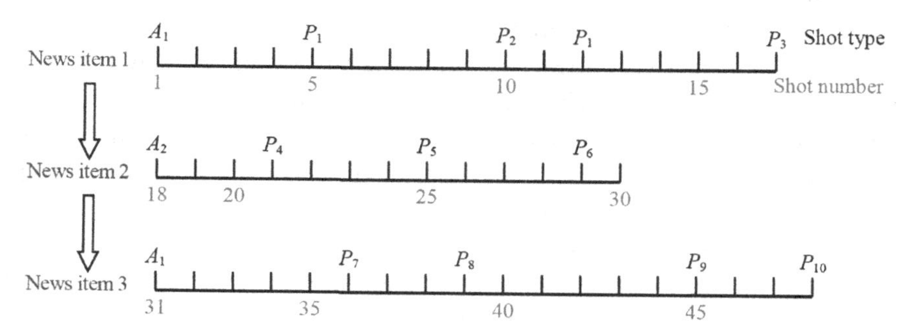

FIGURE 7.12 News item and shot structure.

7.4.2 Video Ranking of Sports Games

Highlight shots of special moments in sports games are a big draw. Ranking shots according to their degree of brilliance is beneficial for news reporting and prioritizing the viewing of shots of interest.

7.4.2.1 Features of Sports Video

Sports-related programs generally have a strong structure. For example, a football game is divided into first and second halves, and an NBA basketball game is divided into two halves, with each half further divided into two sections. These features provide temporal cues and constraints for sports video analysis. In addition, there are always some climax events in sports games, such as shooting in football games, dunks and good passes in basketball games, and so on. Video footage is dynamically organized around the event. The environment of sports competitions is mostly specific, but there are many uncertain factors in the competition. The time and location of events cannot be determined in advance, so the video generation process cannot be controlled during the competition. The shooting methods of sports games also have many characteristics. For example, dunks in basketball games are not only shot in the air, but also shot from the basket up.

The peculiarities of sporting events provide the basis for using specific analysis techniques to extract segments of interest from the game, or the possibility of viewing the same action from multiple different angles. In sports competitions, a specific scene often has its fixed color, moving object, and spatial object distribution characteristics. Since there are only a few fixed cameras on the playing field, the occurrence of an event often corresponds to a specific scene change. This allows people to judge specific events based on the extraction and recognition of these features. For example, for a sequence with athletes appearing, the outline of the athlete, the color of the sportswear, the movement track of the athlete, etc. can be extracted as an index; for the audience sequence, the action posture of the audience can be extracted as an index; for important segment sequences, such as basketball sports, the position of the basketball basket and the movement trajectory of the basketball can be extracted, while for track and field throwing events, the shooting angle and landing position can be extracted, and both can be used as indexes.

Most existing sports game highlight analysis systems use prior knowledge to define highlight events for specific sports game types, and complete the highlight scene detection by detecting specific highlights in sports games. Highlights have different meanings, contents, and video presentations for different sports. For example, shots in football games are undoubtedly the focus of everyone's attention; while dunks, good passes, and fast breaks in basketball games are always very attractive. From the query point of view, the query can be made according to these excellent shots, and can also be queried according to the elements constituting them, such as football, goal, basketball, rebound, etc. In addition, according to the different characteristics of these shots, queries such as penalty kicks, free kicks, free throws, and three-pointers can be made.

Some people divide the features in sports games into three levels (Duan et al. 2003), namely low-level, middle-level and high-level semantic features. The low-level features include motion vectors, color distribution, and other features that can be directly obtained from the image. The middle-level features include camera motion and the resulting field-of-view changes, object motion and attitude transformation, etc. These features can be analyzed from the underlying features, for example, camera motion is estimated from the motion vector histogram. The high-level semantic features correspond to an event, and the relationship between the high-level features and the middle- and/or low-level features is generally defined in advance through prior knowledge.

7.4.2.2 Structure of Table Tennis Competition Program

Unlike football matches, which have a fixed time, table tennis matches are score based and have a relatively definite structure, and a match consists of relatively fixed, typically structured, and constantly repeated scenes. The table tennis game program can be divided into several competition scenes: serving, rest, audience, and replay. Each scene has its relatively definite characteristics. For example, in the game scene, the camera shooting range will cover the global shot including the table, the players of both sides and a part of the field. In the serve scenes, there are mostly close-up shots of players or rackets. A table tennis match generally consists of three to seven games, and each game consists of multiple play rounds. Table tennis is made up of these repeated structural units, and the occurrence of each event has a relatively definite time sequence relationship. For example, the serve scene is followed by the game scene, and the replay scene is followed by the highlight game scene. Figure 7.13 shows a schematic diagram of the structure.

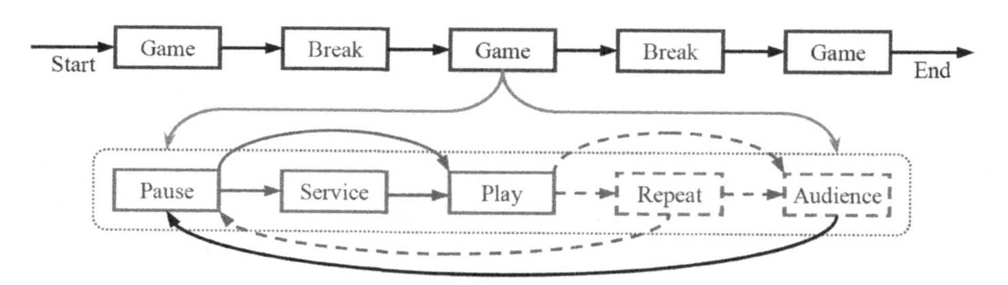

FIGURE 7.13 The structure of table tennis matches.

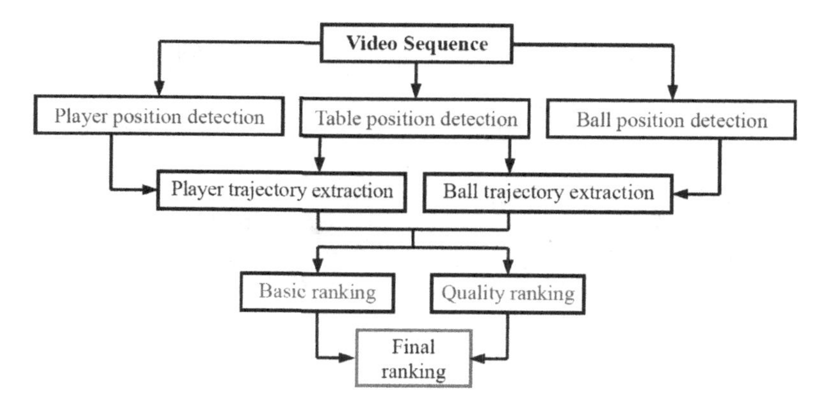

FIGURE 7.14 Flowchart of object detection, tracking, and shot ranking.

7.4.2.3 Object Detection and Tracking

In table tennis, brilliant shots are often associated with specific strokes. In order to determine the brilliance of a table tennis match, it is necessary to identify the types of technical actions included in each round of the match, perform statistics and analysis on the technical actions that occur in each round, and then determine which are the most consistent with people's subjective feelings. The brilliance of a table tennis match is judged by two main methods: objective indicators, such as the duration of the game or the number of attacking rounds; or assisted by other relevant information and events, such as detecting applause and picture playback.

In order to count objective indicators, some objects in the scene need to be detected first. This includes players, tables, and table tennis balls. On this basis, it is also necessary to track the moving objects in the scene, including the tracking of players and ball. By tracking and obtaining the movement trajectory of the player and the movement trajectory of the ball, it is possible to further make the **brilliance ranking** of shots. A flow diagram of this process can be seen in Figure 7.13 (Chen and Zhang 2006).

Specifically, the detection and tracking of athletes is similar to the detection and tracking of pedestrians (Li and Zhang, 2010). An example of ping-pong tracking is shown in Figure 7.15, in which Figures 7.15(a) to 7.15(d) are a sequence of four frames of images with equal intervals, and Figure 7.15(e) shows a segment of the ping-pong trajectory tracked sequentially (superimposed on the last frame of the image).

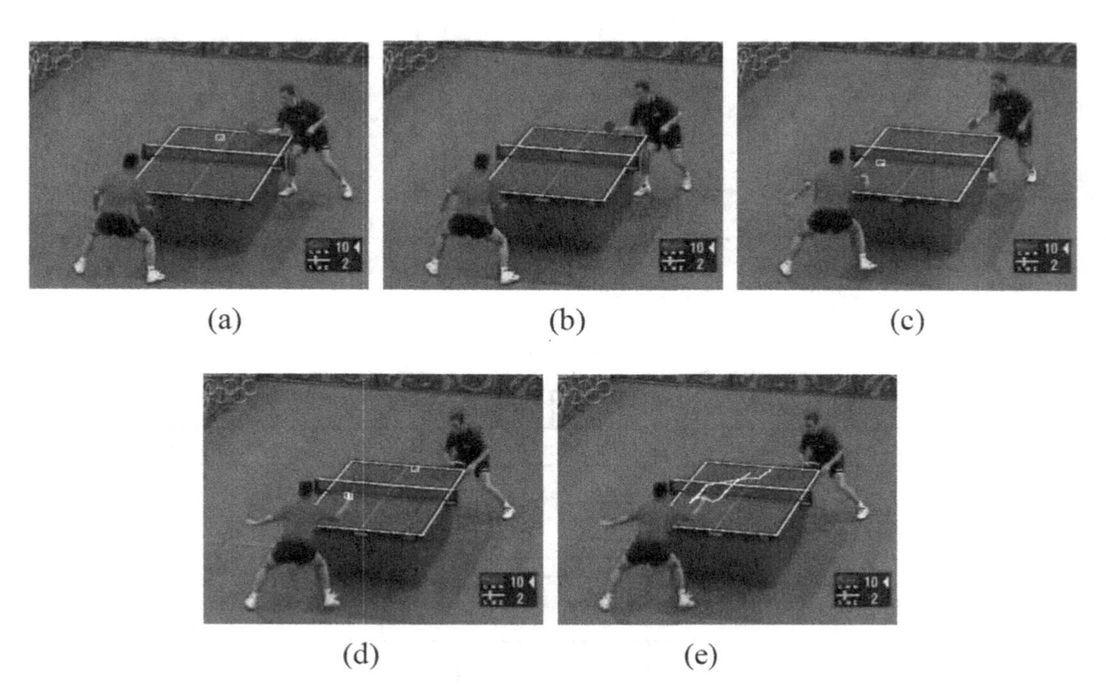

FIGURE 7.15 Example of table tennis tracking.

7.4.2.4 Make the Brilliance Ranking

To rank video clips for their brilliance and make them as relevant as possible to human perception and viewing habits, the criteria and opinions used by those watching the game when evaluating the brilliance of the game can be used. The relevant content can be divided into three levels: (i) the basic level, such as how many times a ball has been hit back and forth, the trajectory and speed of the ball movement, etc.; (ii) the quality layer, such as various hitting styles, the speed of the player's movement etc.; and (iii) the sensory layer, such as the perception of the referee based on the evaluation of a large number of games. The last layer is more subjective, and only some work in the first two layers will be discussed below.

(1) Ranking of the basic layer

The main consideration here is how to rank the brilliance of clips with the help of features directly detected from the video. A ranking metric is defined as follows (Chen and Zhang 2006):

$$R = N\left(w_v h_v + w_b h_b + w_p h_p\right) \tag{7.18}$$

N is the total number of hits in a round (one point) in the game; w_v, w_b, w_p are the weights. The content of the brilliance is determined by three parts. The first is the average speed of the ball movement:

$$h_v = f\left(\sum_{i=1}^{N} | v(i) | / n\right) \tag{7.19}$$

where $v(i)$ is the speed of the i-th shot. This is followed by the average distance of ball movement between consecutive shots:

$$h_b = f\left[\frac{1}{N_1}\sum_{i=1}^{N_1} | b_1(i+1) - b_1(i) | + \frac{1}{N_2}\sum_{i=1}^{N_2} | b_2(i+1) - b_2(i) |\right] \tag{7.20}$$

where N_1 and N_2 are the total hits of the first player and the second player, respectively; b_1 and b_2 are the positions of the ball when the first player and the second player hit the ball for the i-th time, respectively. Finally, the average distance traveled by the players between two consecutive hits:

$$h_p = f\left[\frac{1}{N_1}\sum_{i=1}^{N_1} | p_1(i+1) - p_1(i) | + \frac{1}{N_2}\sum_{i=1}^{N_2} | p_2(i+1) - p_2(i) |\right] \tag{7.21}$$

where p_1 and p_2 are the positions of the first player and the second player on their i-th shot, respectively.

In the above three equations, $f(\cdot)$ is the Sigmoid function:

$$f(x)=\frac{1}{1+\exp\left[-(x-\bar{x})\right]} \tag{7.22}$$

It is used to convert each variable value to brilliance.

(2) Ranking of quality layers

The ranking of quality layers relies on some relatively high-level concepts such as movement before hitting, hitting action, ball trajectory and speed, similarity and consistency between two adjacent hits, etc. These concepts are all based on a shot as the time unit, and can more suitably be described by fuzzy sets. For example, the intensity of an athlete's movement can be expressed as follows:

$$m(i)=w_{p}f\left[|p(i)-p(i-2)|\right]+w_{s}f\left[|s(i)-s(i-2)|\right] \tag{7.23}$$

where $p(i)$ and $s(i)$ are the position and shape of the players at the i-th hit, respectively; w_p and w_s are the corresponding weights, respectively.

The quality of the ball trajectory can be represented as:

$$t(i)=w_{l}f\left[l(i)\right]+w_{v}f\left[v(i)\right] \tag{7.24}$$

where $l(i)$ and $v(i)$ are the trajectory length and movement speed of the ball between the i-th and i–1-th hits, respectively; w_l and w_v are the corresponding weights, respectively.

The change in the hits can be represented as:

$$u(i)=w_{l}f\left[l(i)-l(1-i)\right]+w_{v}f\left[v(i)-v(1-i)\right]+w_{d}f\left[d(i)-d(1-i)\right] \tag{7.25}$$

where $l(i)$, $v(i)$ and $d(i)$ are the trajectory length, speed and direction of the ball between the i-th and i–1th hits, respectively, w_l, w_v and w_d, are the corresponding weight, respectively.

According to the prior knowledge, the corresponding fuzzy membership function can be designed for the above quality variables, and the final quality-level ranking will be the sum of the quality rankings of each hit. Video clips with different degrees of brilliance can be selected with the ranking by quality.

7.5 SEMANTIC CLASSIFICATION RETRIEVAL

Content-based retrieval has focused on semantic retrieval in recent years (Zhang 2004; Zhang 2007a). Unlike low-level visual features such as color, texture, shape, and motion, semantic features, also known as logical features, are high-level features that can be further divided into objective and subjective features (Djeraba 2002). Objective features are related to the identification of objects in images and the motion of objects in videos, and have been widely used in semantic retrieval based on object identification; see the literature (Gao and Zhang 2001; Gao et al. 2003; Zhang 2004, 2007a). Subjective features are related to the attributes of the object and the scene, and represent the meaning and purpose of the object or scene. Subjective features can be further decomposed into types such as events, activity categories, emotional meanings, beliefs, etc., and users can query these more abstract categories.

7.5.1 Image Classification Based on Visual Keywords

With the enlargement of image databases and the expansion of network applications, the number of objects retrieved in images is often in the order of hundreds of thousands or even millions. Searching in such a large space is time-consuming and performance suffers. To solve this problem, before formal retrieval, images can be classified, the large space can be divided into small spaces, and the search can be more targeted.

Image classification can be done according to different criteria and requirements, such as separating images and graphics (Dai and Zhang 2002); separating images containing different objects (Li and Zhang 2002; Li et al. 2002; Zhang 2006b); separating images with different environments or situations, such as indoor/outdoor, beach/mountain (Mitko et al. 2009). Here we mainly discuss classification according to the scene. The detection of specific regions in the image is often combined to obtain the semantic description of the image. The following is an example (Xu and Zhang 2007a, 2007b).

7.5.1.1 Feature Selection

In describing image content, local **salient features** often provide more accurate information than global features. Using local salient features requires, first, detecting local regions with salient features (salient patches), which can be done with the help of the SIFT operator introduced in Subsection 3.4.2.

Feature selection identifies the most informative and discriminatory features in the image class. According to information theory, this can be determined and selected by computing the mutual information between features and related classes (Xu and Zhang 2006a). Further, the descriptor of each salient patch is regarded as a visual keyword, and a keyword set can be constructed by clustering to represent the main content of the whole image.

Consider that there are N objects in the image, which can be represented and described by a set of vectors $\boldsymbol{V} = \{\boldsymbol{v}_1, \cdots, \boldsymbol{v}_N\}$. The density of each object in each cluster needs to be calculated. Based on the density of the entire object set, the individual clusters can be further calculated using the Parzen estimation algorithm (Bian and Zhang 2000). Assuming that the density function of the object $\boldsymbol{v}_i$ in the cluster C_j, $j = 1, 2, \ldots, J$ is $D(\boldsymbol{v}_i|C_j)$, the goal function is:

$$D(v_i \mid C_j) = \max_l \left[D(v_i \mid C_l) \right] \tag{7.26}$$

The Parzen estimation algorithm performs three steps:

(1) Initialize clustering first;

(2) For the object $\boldsymbol{v}_i$, calculate its conditional density in each cluster, and then mark $\boldsymbol{v}_i$ according to Equation (7.26);

(3) If there is a change in the mark of the object $\boldsymbol{v}_i$, go back to step (2).

By following the above steps, the object can be clustered according to its density distribution: the higher the density, the more compact the clustering. The clustering of salient patches is equivalent to giving a combination of visual keywords. The mutual information between visual keywords and clusters can be used to measure the impact and contribution of the visual keywords to a specific cluster, and the greater the mutual information, the greater the impact and contribution.

The mutual information between visual keywords and clusters reflects the representation of visual keywords in image categories, so that for each image category, salient patches with large mutual information can be selected as the feature of the image category. In practical applications, the salient patches can be sorted according to their mutual information from large to small, and the first M salient patches are selected to form a feature vector. Here M is determined in advance considering the computational complexity and the description ability. Too many saliency patches will lead to too much computation, while too few saliency patches will not be able to describe the image content effectively.

7.5.1.2 Image Classification

Once the feature descriptors are selected, the image classification problem is reduced to a **multi-class supervised learning** problem. During the training phase, multiple classifiers are trained using the annotated images to distinguish multiple types.

An experimental result using the above image classification method on the Corel image database is as follows: 25 types of images in the database (most of which have more prominent objects) are selected; each type has 100 images, and five tests are carried out; each time 80 of them are tested and the remaining 20 of them are used for testing. Figure 7.16 presents the accuracy results (Xu and Zhang 2006a) for 25 classes of images.

It can be seen from the above classification results that it is easier to obtain more accurate classifications from images with clear objects and more consistent backgrounds.

7.5.2 High-Level Semantics and Atmosphere

From the perspective of human cognition, people's description and understanding of images are mainly carried out at the semantic level. In addition to describing objective things (such as images, cameras, etc.), semantics can also describe subjective feelings (such as beautiful, clear, etc.) and more abstract concepts (such as broad, rich, etc.).

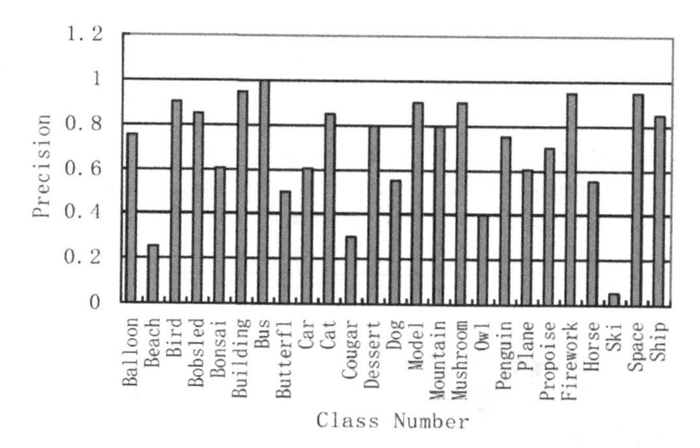

FIGURE 7.16 Accuracy results for classification of 25 categories of images.

In other words, in addition to **cognitive-level** object semantics, scene semantics, etc., there are more subjective **abstract attribute** semantics (such as atmosphere, emotion, etc.).

Atmosphere plays an important role in filming and can convey information other than the episodic story. The semantic information provided by atmospheres is generally relatively abstract, although the atmosphere itself can often be defined in terms of ambient brightness and object illumination conditions (Aner-Wolf 2004). Work on image classification with the help of atmosphere semantics (Xu and Zhang 2005) is presented below.

7.5.2.1 Five Atmospheric Semantics

Illumination (embodied as the brightness of the image) and color are the two main ways of representing the **atmospheric semantics** in the image. General light intensities can be divided into two groups: well-lit scenes convey a happy atmosphere, and dim scenes are depressing or convey some sense of mystery. In addition, the contrast of lighting can also convey some motivational information (large contrast is shocking and small contrast is more peaceful). Color is another factor that represents the atmosphere of an image. In the visual perception-oriented color model HSI, the luminance component and the chrominance component are independent of each other. The chrominance components include H (hue) and S (saturation), where H is closely related to atmosphere semantics. General tones can also be divided into two categories: warm-toned scenes reflect vitality and success, while cool-toned scenes give a sense of peace or desolation.

According to human experience and the theory of visual psychology, five typical atmospheres can be defined using different combinations of the two characteristics of global illuminance (intensity and distribution of illuminance) and main hue; see Table 7.3 (Xu and Zhang 2005).

Figure 7.17 shows an example of a set of images with different atmospheres. The first row (numbered according to the five atmospheres in Table 7.3) gives one image of each of the five typical atmospheres, the second row gives the corresponding illuminance component map. and the third row gives the corresponding hue component map.

TABLE 7.3 Illuminance and Color Tone Characteristics of Five Atmospheres

#	Atmosphere	Illuminance and Contrast	Hue
1	Vigor and strength	High illumination, high contrast	Colorful
2	Mystery or horror	High contrast	Dim/serene
3	Victory and brightness	High illumination, low contrast	Warm-toned
4	Peace or desolation	Low contrast	Cool-toned
5	Lack unity and disjointed	Messy distribution	—

FIGURE 7.17 Examples of different atmosphere images.

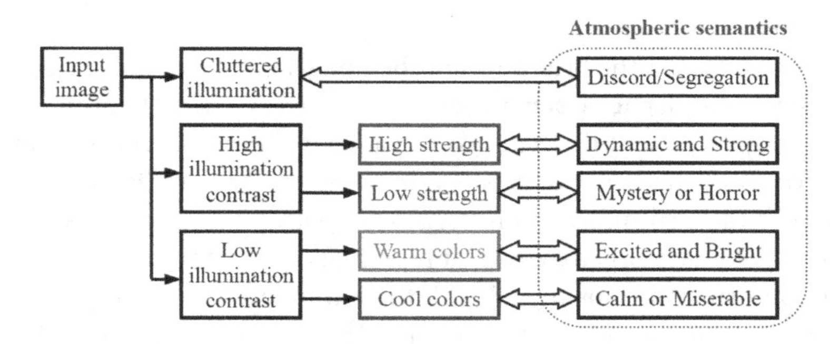

FIGURE 7.18 Hierarchical classification flow chart.

7.5.2.2 Classification of Atmosphere

According to Table 7.3, images of different atmospheres can be classified with the help of layered SVM, and the process is shown in Figure 7.18. The first step starts by distinguishing the case of the scattered illuminance distribution from the other cases. At this time, only the standard variance matrix value of the illuminance is used (the hue is not used), and a large value indicates that the distribution of the illuminance component is disordered. The second step distinguishes high luminance contrast from low luminance contrast. The luminance contrast of an image region can be represented by the ratio of the highest luminance to the lowest luminance in the region. The third step uses two SVMs. For the case of high illumination contrast, it is divided into high intensity and low intensity according to whether the average illumination intensity in the image block is greater than a given threshold; for the case of low illumination contrast, it is divided into warm tone and cool tone according to whether the average hue in the image block is greater than a given threshold.

After the various images are classified according to the atmosphere, they can be marked with the help of text. For example, combining the text describing the atmospheric semantics in the image with the corresponding features expressed in XML (using the XML in MPEG-7 for text markup) can be used for retrieval and indexing of various atmosphere images.

7.6 SOME RECENT DEVELOPMENTS AND FURTHER RESEARCH

In the following sections, technical developments and promising research directions of the last few years are briefly reviewed.

7.6.1 Deep Learning-Based Cross-Modal Retrieval

The purpose of (content-based) **cross-modal retrieval** is to use the data of one modality as query input to retrieve and output related data from another modality. A cross-modal relationship model needs to be built so that users can retrieve the modal data they expect by submitting the modal data they have. The key here is how to measure the content similarity between different modal data.

Modeling of cross-modal correlations with deep learning requires multimodal common representation learning. Based on multimodal common representation, cross-modal similarity can be effectively measured. General common representations include **real-valued representation** (representing the learned different modalities as real values, usually vectors) and **binary representation** (representing the learned different modalities as a code consisting of −1 and 1). Methods based on the latter are also known as **cross-modal hashing**.

According to the cross-modal information provided when learning common representations, deep cross-modal retrieval can be divided into three categories (Yin et al. 2021): (i) based on one-to-one correspondence between cross-modal data; (ii) based on similarity between cross-modal data; and (iii) based on semantic annotations of cross-modal data. On the other side, different learning techniques can also be used for treating different cross-modal data information. They mainly include seven categories: **canonical correlation analysis** (CCA), one-to-one correspondence maintenance, metric learning, likelihood analysis, learning ranking, semantic prediction, and adversarial learning. The division of these techniques is mainly reflected in the difference in achieving the optimization goal of common representation learning.

Not all techniques are suited to treating all types of information. The existing combinations are indicated in Table 7.4 (Yin et al. 2021).

TABLE 7.4 Classification of Deep Cross-modal Retrieval Techniques

	Based on one-to-one correspondence between cross-modal data	Based on the similarity between cross-modal data	Based on semantic annotation of cross-modal data
Canonical correlation analysis	√		√
One-to-one correspondence maintenance	√		
Metric learning		√	√
Likelihood analysis	√	√	√
Learning ranking	√		√
Semantic prediction			√
Adversarial learning	√	√	√

More descriptions on the cross-modal information provided are as follows:

(1) *Based on one-to-one correspondence between cross-modal data.* The descriptions of different modalities of the same data samples coexist and correspond one to one, that is, the representation of one modality for a certain sample is the same as that of another modality. In other words, there is a corresponding relationship between the representations of the samples. In multimodal representation learning, only this one-to-one correspondence between cross-modal data is used.

(2) *Based on the similarity between cross-modal data.* There is a relationship of similarity (or dissimilarity) between cross-modal data, that is, the similarity (or dissimilarity) between two samples can be determined. Generally, such similarity information between the cross-modal data covers the one-to-one correspondence information between the cross-modal data.

(3) *Semantic annotation based on cross-modal data.* Cross-modal data has single- or multi-label semantic annotation of samples, that is, for any sample in the database, it can be determined whether it has a certain type of semantic existence. Generally, while providing data semantic information, one-to-one correspondence information between cross-modal data is also provided, and the similarity information between cross-modal data can be deduced or calculated.

More descriptions on the learning techniques are as follows:

(1) *Canonical correlation analysis.* The two modal data are projected into a low-dimensional space through linear projection, and the above projection is learned by maximizing the correlation between the modalities. The introduction of deep neural networks generally replaces the above linear projection and is conducive to the optimization of the correlation maximization objective function.

(2) *One-to-one correspondence maintenance.* A one-to-one correspondence between cross-modal data is constructed in the common presentation layer, minimizing the

distance between corresponding cross-modal data. This type of techniques has modeling versatility and is widely used in unsupervised cross-modal retrieval.

(3) *Metric learning.* A metric function or a deep neural network is introduced to make similar samples have a smaller distance in the common representation space while the dissimilar samples have a larger distance. The technical modeling has a certain generality.

(4) *Likelihood analysis.* The generative model performs the generative modeling of the observation data through the maximum likelihood optimization objective function. In the cross-modal data, the observation data can be multimodal features, the correspondence between the data, and the similarity between the data. This type of technique can generate observations such as data features, similarity, etc. by using cross-modal co-representation, and can perform effective learning of co-representation based on training datasets.

(5) *Learning ranking.* Building a ranking model ensures that the ranking relationship between data is maintained in a common representation space. In cross-modal data, this ranking information is generally constructed in the form of triples through intra-modal or inter-modal similarity relationships. Because similarity can provide more accurate relationships between data than ranking, this type of techniques is less used given the similarity information between data.

(6) *Semantic prediction.* Through the general classification task model, the similarity structure of intra-modal data is maintained, that is, with the same semantic annotation, there is a similar common representation. According to this, the cross-modal data relationship construction is indirectly realized, in which the cross-modal common representation is consistent under the same semantics. Because of the need to provide semantic annotation of cross-modal data, this type of technique is generally used when semantic annotation information can be provided.

(7) *Adversarial learning.* With the introduction of the idea of a generative-adversarial network, a generative-adversarial task to learn multimodal common representation completed by construction. The modeling process forces similar cross-modal data to be jointly represented with statistical inseparability, and then realizes the similarity calculation between modalities. This type of technique can realize the statistical inseparability of common representation by easily constructing common representation learning and discrimination, similarity generation and discrimination, and modal raw data generation and discrimination.

Finally, there are many types of multimodal data, such as 3-D model, audio, image, text, video, etc. Existing methods consider the retrieval of two modalities the most. Of the more than 70 techniques mentioned in Yin et al. (2021), all but one consider only two modalities; the exception treats the five modalities of 3-D model, audio, image, text, and video. The strongest pair is (image, text), as shown by the statistics in Table 7.5.

TABLE 7.5 Dual Modality Treated by Existing Techniques

	3D model	Audio	Image	Text	Video
3D model		∅	∅	∅	∅
Audio	—		2	1	∅
Image	—	—		70	∅
Text	—	—	—		2
Video	—	—	—	—	

7.6.2 Hashing in Image Retrieval

With the rapid increase in image data, the cost of optimal query and search is getting higher and higher. In big data applications, the **approximate nearest neighbor** (ANN) search is widely used, in which hashing has become one of the most popular and effective techniques due to its fast query speed and low memory cost. With the help of hashing, image data can be transformed from the original high-dimensional space to a more compact hamming space, while maintaining the similarity of the data. This can not only significantly reduce storage cost and achieve constant or sub-linear time complexity in information search, but it also preserves the semantic structure existing in the original space.

7.6.2.1 Supervised Hashing

Existing hashing methods can be roughly divided into two categories: data-independent and data-dependent hashing methods. As the most typical data-independent hashing methods, **local sensitive hashing** (LSH) and its extension obtain the hash function by random projection. However, they require longer binary code to achieve high accuracy.

Data-dependent hashing methods learn binary codes from available training data, or learning hashes. Existing data-dependent hashing methods can be further divided into unsupervised and supervised hashing according to whether supervised information is used for learning. Unsupervised hashing just tries to use data structures to learn compact binary codes to improve performance, while supervised hashing uses supervised information to learn hash functions.

Specifically, given a training set with N images: $\boldsymbol{X} = \{x_i\}_{i=1}^{N} \in \mathbb{R}^{D \times N}$, where D is the dimension of the sample. The purpose of hash learning is to learn a set of K-bit binary codes $\boldsymbol{B} \in \{-1, 1\}^{k \times N}$, where the i-th column $b_i \in \{-1, 1\}^N$ represents the K-bit binary code of the i-th sample x_i. Generally, $b_i = h(x_i) =[h_1(x_i), h_2(x_i), \ldots, h_c(x_i)]$, where $h(x_i)$ represents the hash function to be learned. Consider common supervised hashing based on supervised information as pairwise labels. The label information is expressed as $\boldsymbol{Y} = \{y_i\}_{i=1}^{N} \in \mathbb{R}^{C \times N}$, $y_i \in \{0, 1\}^N$ corresponds to the sample x_i, and C is the number of dataset categories.

Paired images are associated with similarity labels $s_{i,j}$. Here $\boldsymbol{S} = \{s_{i,j}\}$, $s_{i,j} \in \{0, 1\}$ is used to represent the similarity between two images: $s_{i,j} = 1$ means x_i is similar to x_j, $s_{i,j} = 0$ means x_i and x_j not similar. The hash function to be learned in supervised hashing can map data points from the original space to the binary code space and preserve the semantic similarity of $\boldsymbol{S}$ in the binary code space. For two binary codes b_i and b_j, the hamming distance between them is defined as $\text{dis}(b_i, b_j) = (K - \langle b_i, b_j \rangle)/2$. Therefore, the similarity of hash codes can be measured using the inner product. In order to maintain the similarity between data points, when the data points x_i and x_j are similar (i.e., $s_{i,j} = 1$), the hamming distance between the binary codes b_i and b_j should be relatively small. Conversely, when the data points x_i and x_j are not similar (i.e., $s_{i,j} = 0$), the hamming distance between the binary codes b_i and b_j should be relatively large.

7.6.2.2 Asymmetric Supervised Deep Discrete Hashing

In recent years, some hash methods based on deep learning have been proposed to learn image representation and hash coding at the same time. However, existing deep supervised hash methods mainly use pairwise supervision for hash learning, and the semantic information is not fully utilized, while the information helps to improve the semantic recognition ability of hash codes. Moreover, for most data sets, each item is annotated with multi-label information. Therefore, it is necessary not only to ensure high correlation between multiple different item pairs, but also to maintain multi-label semantics in a framework to generate high-quality hash codes.

The following describes an **asymmetric supervised deep discrete hashing** (ASDDH) method (Gu et al. 2021). This method uses multi-label binary code mapping to make the hash code have multi-label semantic information, so as to generate a hash code that can completely retain the multi-label semantics of all items. In the optimization process, in order to reduce the quantization error, the discrete cyclic coordinate descent method is used to optimize the objective function to maintain the discreteness of hash code.

The process framework of this method is shown in Figure 7.19. There are two main modules: feature learning and loss function. The two modules are integrated into the same end-to-end framework. During training, each module can give feedback to the other. Here, the AlexNet network is used as the backbone network, which contains five convolutional layers and three fully connected layers, and the first seven layers use ReLU as the activation function. In order to get the final binary code, the last layer uses a fully connected hash layer (activation function is tanh), which can project the outputs of the first seven layers into the $\mathbb{R}^K$ space. The binary code used is $b_i = \text{sign}(h_i)$, and the final output is $\boldsymbol{H} = \{h_i\}_{i=1}^N \in \mathbb{R}^{K \times N}$.

7.6.2.3 Hashing in Cross-Modal Image Retrieval

In cross-modal retrieval, hash methods can map high-dimensional different modal data to a unified hamming space, and then measure the similarity of different modal data, so as to solve the problem of heterogeneous gap. The depth hashing method based on depth

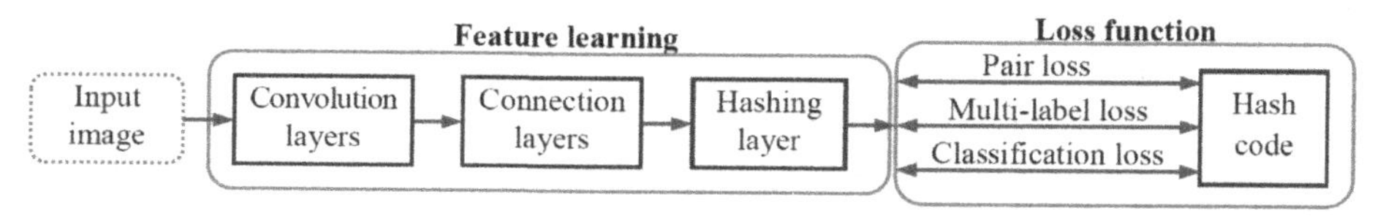

FIGURE 7.19 Asymmetric supervised depth discrete hashing flow chart.

learning maps the data of different modes into a unified hash code through convolutional neural network, and measures the similarity through hamming distance in hamming space. However, data with different modes may be very different in the expression of features or feature dimensions. If these data with different modes are directly mapped to hamming space, it is difficult to obtain a consistent hash code, which will affect retrieval accuracy. In addition, many methods combine the category label information to improve the discrimination of hash code, but the robustness will be affected when the category label information is missing or wrong.

Structure-preserving hashing methods with coupled projections are proposed for treating these two problems (Min et al. 2021). Taking image and text bimodal as an example, the retrieval problem for them can be described as follows. Suppose there are N training samples $\{\boldsymbol{x}_1, \boldsymbol{x}_2, \ldots, \boldsymbol{x}_N\}$, each sample point $\boldsymbol{x}_i$ consists of a pair of images and texts with the same semantics, namely $\boldsymbol{x}_i = (\boldsymbol{v}_i, \boldsymbol{t}_i)$, where $\boldsymbol{v}_i \in \mathbb{R}^{1\times Mv}$ represents the feature vector of the image, $\boldsymbol{t}_i \in \mathbb{R}^{1\times Mt}$ represents the feature vector of the text, M_v and M_t are the image feature dimension and the text feature dimension, respectively. Re-define the class label of the sample as $\boldsymbol{y}_i \in \mathbb{R}^{1\times C}$, where C represents the number of sample classes. When the sample point $\boldsymbol{x}_i$ belongs to the k-th class, the corresponding column $y_{ik} = 1$ in $\boldsymbol{y}_i$, otherwise $y_{ik} = 0$. In addition, suppose the input image sample is $\boldsymbol{V}$, $\boldsymbol{V} = \{\boldsymbol{v}_i\}_{i=1}^{N} \in \mathbb{R}^{N\times Mv}$; suppose the input text sample is $\boldsymbol{T}$, $\boldsymbol{T} = \{\boldsymbol{t}_i\}_{i=1}^{N} \in \mathbb{R}^{N\times Mt}$.

A flowchart of this method is shown in Figure 7.20. First, the input image and input text are projected into their respective subspaces $\boldsymbol{M}_v$ and $\boldsymbol{M}_t$ with projection matrices $\boldsymbol{P}_{v\text{-}M}$ and $\boldsymbol{P}_{t\text{-}M}$ to narrow the difference between the two modal data. This can be expressed as:

$$\boldsymbol{P}_{v-M}^{\mathrm{T}}\boldsymbol{V} \rightarrow \boldsymbol{M}_v \qquad \boldsymbol{P}_{t-M}^{\mathrm{T}}\boldsymbol{T} \rightarrow \boldsymbol{M}_t \tag{7.27}$$

where $\boldsymbol{M}_v \in \mathbb{R}^{N\times Dv}$, $\boldsymbol{M}_t \in \mathbb{R}^{N\times Dt}$, D_v and D_t represent the image embedding space dimension and text embedding space dimension, respectively.

In order to maintain the original structure information of image and text data in their respective subspaces, a graph model is introduced in each of the image and text subspaces. Define the weight matrix $\boldsymbol{W}$ as

$$\mathbf{W}_{ij} = \begin{cases} 1/N_C & y_i = y_j \in C \\ 0 & \text{otherwise} \end{cases} \tag{7.28}$$

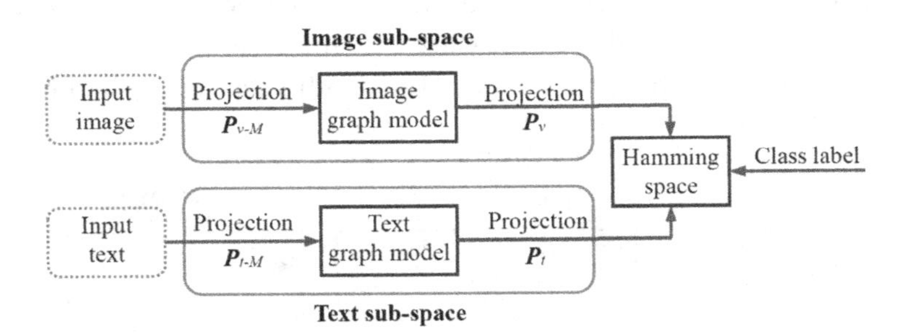

FIGURE 7.20 Structure-preserving hashing flow diagram with coupled projection.

where N_C is the number of class C samples, and y_i and y_j are the labels of images and texts.

To preserve the structural information of the data in the image and text subspaces, the two graph models are:

$$\min \frac{1}{2} \sum_{i,j} W_{ij} \left(P_{v-M}^{\mathrm{T}} v_i - P_{v-M}^{\mathrm{T}} v_j \right)^2 = \mathrm{tr}\left(P_{v-M}^{\mathrm{T}} V L_v V^{\mathrm{T}} P_{v-M} \right) \tag{7.29}$$

$$\min \frac{1}{2} \sum_{i,j} W_{ij} \left(P_{t-M}^{\mathrm{T}} t_i - P_{t-M}^{\mathrm{T}} t_j \right)^2 = \mathrm{tr}\left(P_{t-M}^{\mathrm{T}} T L_t T^{\mathrm{T}} P_{t-M} \right) \tag{7.30}$$

Next, the data in the subspace is re-projected to hamming space $\boldsymbol{H}$, with the help of projection matrices $\boldsymbol{P}_v$ and $\boldsymbol{P}_t$, which can be expressed as:

$$P_v^{\mathrm{T}} M_v \to H \qquad P_t^{\mathrm{T}} M_t \to H \tag{7.31}$$

Finally, in order to improve the discrimination of the hash code, a class label is introduced to dynamically learn the classifier W, that is, the class learned by the improved hash code is consistent with the class of the original data:

$$W^{\mathrm{T}} H_i \to y_i \tag{7.32}$$

To sum up, the objective function of the whole method includes eight items:

$$\begin{aligned} \min_{\substack{W,H,M_v,M_t \\ P_{v-M},P_{t-M},P_v,P_t}} \quad & \left\| Y - W^{\mathrm{T}} H \right\|_F^2 + a_v \left\| M_v - P_{v-M}^{\mathrm{T}} V \right\|_F^2 + a_t \left\| M_t - P_{t-M}^{\mathrm{T}} T \right\|_F^2 + b_v \left\| H - P_v^{\mathrm{T}} M_v \right\|_F^2 + \\ & b_t \left\| H - P_t^{\mathrm{T}} M_t \right\|_F^2 + c_v \left(P_{v-M}^{\mathrm{T}} V L_v V^{\mathrm{T}} P_{v-M} \right) + c_t \left(P_{t-M}^{\mathrm{T}} T L_t T^{\mathrm{T}} P_{t-M} \right) + \\ & d \left(\left\| W \right\|_F^2 + \left\| P_{v-M} \right\|_F^2 + \left\| P_{t-M} \right\|_F^2 + \left\| P_v \right\|_F^2 + \left\| P_t \right\|_F^2 \right) \\ & \text{s.t.} \quad h_i \in \{-1, 1\}, \quad P_{v-M}^{\mathrm{T}} P_{v-M} = I, \quad P_{t-M}^{\mathrm{T}} P_{t-M} = I \end{aligned} \tag{7.33}$$

Among them, a_v, a_t, b_v, b_t, c_v, c_t are the penalty parameters, d is the regular parameter, $\|\cdot\|_F^2$ represents the F norm, and the projection matrix $\boldsymbol{P}_{v\text{-}M}$, $\boldsymbol{P}_{t\text{-}M}$, $\boldsymbol{P}_v$, $\boldsymbol{P}_t$, the unified hash code matrix $\boldsymbol{H}$, the embedding spaces $\boldsymbol{M}_v$ and $\boldsymbol{M}_t$, and the linear classifier W, can be obtained after the optimization.

Items in the objective function of Equation (7.33) have different functions. The first item introduces label information, which improves the discrimination of the hash code; the second and third items project the original modal data into their corresponding subspaces, to avoid the problem of directly mapping heterogeneous data to a common hamming space. The fourth and fifth items remap the embedding space to the hamming space, so that the learned hash codes are not only built a common bridge for the different modal data with heterogeneous characteristics, and the similarity between the data can be measured by the hamming distance. The sixth and seventh items are the graph models of the image subspace and the text subspace, respectively. They maintain the structure information between the data in the respective subspaces. The eighth item is used to suppress the overfitting problem that may occur when the projection matrix is updated.

REFERENCES

Aner-Wolf, A. 2004. Extracting semantic information through illumination classification. *Proceedings of the CVPR*, 1: 269–274.

Bian, Z.Q. and X.G. Zhang. 2000. *Pattern Recognition*. Beijing: Tsinghua University Press.

Chen, W. and Y.-J. Zhang. 2006. Tracking ball and players with applications to highlight ranking of broadcasting table tennis video. *Proceedings of the Multi-conference on Computational Engineering in Systems Applications*, 2: 1896–1903.

Chen, W. and Y.-J. Zhang. 2008. Parametric model for video content analysis. *Pattern Recognition Letters*, 29(3): 181–191.

Dai, S.Y. and Y.-J. Zhang. 2002. Image and graphic filtering in GIF format on the web. *Application of Electronic Technique*, 28(1): 48–49.

Djeraba C. 2002. Content-based multimedia indexing and retrieval. *IEEE Transaction on Multimedia*, 9(2): 18–22.

Duan, L.Y., M. Xu and Q. Tian. 2003. Semantic shot classification in sports video. *SPIE*, 5021: 300–313.

Fang, M.Y., Y.-J. Zhang, X. Li, et al. 2016. Normalized similarity measurement and query adaptive fusion for graph based visual reranking. *Proceedings of the ICSP*, 607–612.

Fang, M.Y., Y.-J. Zhang, X. Li, et al. 2017. Query adaptive fusion for graph based visual reranking. *IEEE Journal of Selected Topics in Signal Processing*, 11(6): 908–917.

Furht, B., S.W. Smoliar and H.J. Zhang. 1995. *Video and image processing in multimedia systems*. The Netherlands: Kluwer Academic Publishers, 226–270.

Gao, Y.Y. and Y.-J. Zhang. 2001. Progressive image content understanding based on multi-level image description model. *Acta Electronica Sinica*, 29(10): 1376–1380.

Gao, Y.Y., Y.-J. Zhang, and Y. Luo. 2003. Image retrieval system based on semantic features of objects. *Journal of Electronics and Information Technology*, 25(10): 1341–1348.

Gu, G.H., W.H. Huo, M.Y. Su, et al. 2021. Asymmetric supervised deep discrete Hashing based image retrieval. *Journal of Electronics & Information Technology*, 43(12): 3530–3537.

Huang, X.Y., Y.-J. Zhang and D. Hu. 2003 Image retrieval based on weighted texture features using DCT coefficients of JPEG images. *Proceedings of the 4th PCM*, 3: 1571–1575.

Jegou, H., M. Douze and C. Schmid. 2010. Improving bag-of-features for large-scale image search. *International Journal of Computer Vision*, 87(3): 316–336.

Jiang, F. and Y.-J. Zhang. 2003a. News video indexing with scene sectioning and summary generation. *Chinese Journal of Computers*, 26(7): 859–865.

Jiang, F. and Y.-J. Zhang. 2003b. A caption detection algorithm based on morphological operation. *Journal of Electronics and Information Technology*, 25(12): 1647–1652.

Jiang, F. and Y.-J. Zhang. 2005. News video indexing and abstraction by specific visual cues: MSC and news caption. In: *Video Data Management and Information Retrieval*, IRM Press, Chapter 11 (254–281).

Li, Q. and Y.-J. Zhang. 2002. Image classification based on feature element and association-rule. *Acta Electronica Sinica*, 30(9): 1262–1265.

Li, Q., Y.-J. Zhang and S.Y. Dai. 2002. Image search engine with selective filtering and feature element based classification. *SPIE*, 4672: 190–197.

Li, S. and Y.-J. Zhang. 2010. A novel system for video retrieval of frontal-view indoor moving pedestrians. *Proceedings of the 5th ICIG*, 266–271.

Li, S., Y.-J. Zhang and H.C. Tan. 2010. Discovering latent semantic factors for emotional picture categorization. *Proceedings of the 17th ICIP*, 1065–1068.

Li, S. and Y.-J. Zhang. 2011. Semi-supervised classification of emotional pictures based on feature combination. *SPIE*, 7881A, 78810X-1~78810X-8.

Liu, Z.W. and Y.-J. Zhang. 1999a. Image retrieval using color features. *Application of Electronic Technique*, 25(2): 19–20.

Liu, Z.W. and Y.-J. Zhang. 1999b. Image retrieval using both color and texture features. *Journal of China Institute of Communications*, 20(5): 36–40.

Liu, Z.W. and Y.-J. Zhang. 2000. A comparative and analysis study of ten color feature-based image retrieval algorithms. *Signal Processing*, 16(1): 79–84.

Mallat, S. and W.L. Hwang. 1992. Singularity detection and processing with wavelets, *IEEE Transaction on Information Technology*, 38(2): 617–643.

Mehtre, B.M., M.S. Kankanhalli, A.D. Narasimhalu, et al. 1995. Color matching for image retrieval. *Pattern Recognition Letters*, 16: 325–331.

Min, K.L., G.B. Zhang, L. Wang, et al. 2021. Structure-preserving hashing with coupled projections for cross-modal retrieval. *2018 IEEE International Conference on Acoustics, Speech and Signal Processing (ICASSP)*, 26(07): 1558–1567.

Mitko, V., K. Tomislav and I. Zoran. 2009. Content-based indoor/outdoor video classification system for a mobile platform. *World Academy of Science, Engineering and Technology*, 57: 91–96.

Niblack, W., J.L. Hafner, T. Breuel, et al. 1998. Updates to the QBIC system. *SPIE*, 3312: 150–161.

Ortega, M., Y. Rui, K. Chakrabarti, et al. 1997. Supporting similarity queries in MARS, *Proceedings of the ACM Multimedia*, 403–413.

Philbin, J., O. Chum, M. Isard, et al. 2007. Object retrieval with large vocabularies and fast spatial matching. *Proceedings of the CVPR*, 1–8.

Philbin, J., O. Chum, M. Isard, et al. 2008. Lost in quantization: Improving particular object retrieval in large scale image databases. *Proceedings of the CVPR*, 1–8.

Swain, M.J. and D.H. Ballard. 1991. Color indexing. *International Journal of Computer Vision*, 7: 11–32.

Xu, F. and Y.-J. Zhang. 2005. Atmosphere-based image classification through illumination and hue. *SPIE*, 5960: 596–603.

Xu, F. and Y.-J. Zhang. 2006a. Feature selection for image categorization. *Proceedings of the 7th ACCV*, 2: 653–662.

Xu, F. and Y.-J. Zhang. 2006b. Comparison and evaluation of texture descriptors proposed in MPEG-7. *International Journal of Visual Communication and Image Representation*, 17: 701–716.

Xu, F. and Y.-J. Zhang. 2007a. A novel framework for image categorization and automatic annotation. In: *Semantic-Based Visual Information Retrieval*, Hershey, PA: IRM Press, Chapter 5 (90–111).

Xu, F. and Y.-J. Zhang. 2007b. Integrated patch model: A generative model for image categorization based on feature selection. *Pattern Recognition Letters*, 28(14): 1581–1591.

Yao, Y.R. and Y.-J. Zhang. 2000. Shape-based image retrieval using wavelet and moment. *Journal of Image and Graphics*, 5A(3): 206–210.

Yin, Q.Y., Y. Huang, J.G. Zhang, et al. 2021. Survey on deep learning based cross-modal retrieval. *Journal of Image and Graphics*, 26(6): 1368–1388.

Yu, T.L. and Y.-J. Zhang. 2001a. Motion feature extraction for content-based video sequence retrieval. *SPIE*, 4311: 378–388.

Yu, T.L. and Y.-J. Zhang. 2001b. Retrieval of video clips using global motion information. *IEE Electronics Letters*, 37(14): 893–895.

Zhang, Y.-J. 1998. Color-based image retrieval using sub-range cumulative histogram. *High Technology Letters*, 4(2): 71–75.

Zhang, Y.-J. 1999. The international standard being developed --- MPEG-7. *Electronic Technology Herald (Electronic Commerce)*, 11: 15–18.

Zhang, Y.-J. 2000. MPEG-21 -- An international standard just started. *Journal of Image and Graphics*, 6B(11–12): 7–10.

Zhang, Y.-J. 2003. *Content-Based Visual Information Retrieval*. Beijing: Science Press.

Zhang, Y.-J. 2004. Progresses in semantic-based visual information retrieval. *Science Technology and Engineering*, 4(4): 321–324.

Zhang, Y.-J. 2005. New advancements in image segmentation for CBIR. *Encyclopedia of Information Science and Technology, Idea Group Reference*, 4, Chapter 371 (2105–2109).

Zhang, Y.-J. (ed.). 2006a. *Advances in Image and Video Segmentation*. Hershey, PA: Idea Group, Inc.

Zhang, Y.-J. 2006b. Mining for image classification based on feature elements. *Encyclopedia of Data Warehousing and Mining, Idea Group Reference*, 1: 773–778.

Zhang, Y.-J. (ed.). 2007a. *Semantic-Based Visual Information Retrieval*. Hershey, PA: IRM Press.

Zhang, Y.-J. 2007b. Toward high level visual information retrieval. In: *Semantic-Based Visual Information Retrieval*, Hershey, PA: IRM Press, Chapter 1 (1–21).

Zhang, Y.-J. 2008. Image classification and retrieval with mining technologies. *Handbook of Research on Text and Web Mining Technologies*, Chapter VI (96–110).

Zhang, Y.-J. 2009a. Advanced techniques for object-based image retrieval. *Encyclopedia of Information Science and Technology*, 2nd Ed., 59–64.

Zhang, Y.-J. 2009b. Information fusion of multi-sensor images. *Encyclopedia of Information Science and Technology*, 2nd Ed., 1950–1956.

Zhang, Y.-J. 2009c. Organization of home video. *Encyclopedia of Information Science and Technology*, 2nd Ed., VI: 2917–2922.

Zhang, Y.-J. 2009d. Recent progress in image and video segmentation for CBVIR. *Encyclopedia of Information Science and Technology*, 2nd Ed., VII: 3224–3228.

Zhang, Y.-J. 2015a. Up-to-date summary of semantic-based visual information retrieval. *Encyclopedia of Information Science and Technology*, 3rd Ed., Chapter 123 (1294–1303).

Zhang, Y.-J. 2015b. *A hierarchical organization of home video*. *Encyclopedia of Information Science and Technology*, 3rd Ed., Chapter 210 (2168–2177).

Zhang, Y.-J. 2017. *Image Engineering, Vol. 2: Image Analysis*. Germany: De Gruyter.

Zhang, Y.-J., Y.Y. Gao and Y. Luo. 2004. Object-based techniques for image retrieval. *Multimedia Systems and Content-based Image Retrieval*, Idea Group Publishing, Chapter 7 (156–181).

Zhang, Y.-J. and Z.W. Liu. 1997. Color image retrieval based on HSI model and cumulative histogram. *Proceedings of the Eighth National Signal Processing Committee Joint Conference*, 256–260.

Zhang, Y.-J. and H.B. Lu. 2000. Scheme and techniques for hierarchical organization of video. *Engineering Science*, 2(3): 18–22.

Zhang, Y.-J. and H.B. Lu. 2002. A hierarchical organization scheme for video data. *Pattern Recognition*, 35(11): 2381–2387.

Zhang, Y.-J., Y. Xu, Z.W. Liu, et al. 2001. A test-bed for retrieving images with extracted features. *Journal of Image and Graphics*, 6A(5): 439–443.

CHAPTER 8

Understanding Spatial-Temporal Behavior

AN IMPORTANT TASK IN image understanding is to interpret the scene and guide the action by processing the image obtained from the scene. To do this, it is necessary to determine which objects are in the scene, and how their position, attitude, speed, relationship, etc. in space and in time change. In short, it is necessary to grasp the action of the scene in time and space, determine the purpose of the action, and then understand the semantic information they convey.

Image/video-based automatic object behavior understanding is a challenging research problem. It includes acquiring objective information (acquiring image sequences), processing relevant visual information, analyzing (representing and describing) the image to extract information content, and interpreting the information of images/videos on this basis to realize learning and recognition behavior.

Much research attention has been devoted to action detection and recognition, where significant progress has been made. Relatively speaking, research on behavior recognition and interpretation (related to semantics and intelligence) at a high level of abstraction has not yet been fully carried out, many concepts are not yet clearly defined, and many technologies are constantly being developed and updated.

This chapter is organized as follows. Section 8.1 defines spatial-temporal techniques in image understanding and reviews their development. Section 8.2 introduces the detection of key spatial-temporal points of interest that reflect the concentration and variation of spatial-temporal motion and change information. Section 8.3 discusses the dynamic trajectories and activity paths of the subject that are formed by connecting points of interest. Section 8.4 discusses strengths and weaknesses of typical action classification and recognition techniques currently under investigation. Section 8.5 presents a classification of techniques for modeling and identifying activities and behaviors. Section 8.6 further introduces

DOI: 10.1201/9781003362388-8

techniques for joint modeling and recognition of actions and activities, including single-label subject-action recognition, multi-label subject-action recognition, and subject-action semantic segmentation. Section 8.7 reviews technique developments and promising research directions of the last year.

8.1 SPATIAL-TEMPORAL TECHNOLOGY

The relatively new research field of **spatial-temporal technology** is a technology oriented to **understanding spatial-temporal behavior**. The main objects studied are moving/changing people or objects, and sceneries (especially people) in the objective world. According to the abstraction level of its representation and description, it can be divided into multiple levels from bottom to top (Zhang 2018):

(1) **Action primitive**. The atomic unit used to construct an action, which generally corresponds to the short-term specific motion information in the scene.

(2) **Action**. A meaningful aggregate (ordered combination) composed of a series of action primitives of the subject/initiator. Typically, action represents simple patterns of movement, often performed by one person, typically only lasting the order of seconds. The results of human actions often lead to changes in human posture.

(3) **Activity**. The combination (mainly emphasizing logical combination) of a series of actions performed by the subject/initiator in order to complete a certain work or achieve a certain goal. Activities are relatively large-scale movements that generally depend on the environment and interacting people. Activities often represent complex sequences of (possibly interacting) actions performed by multiple people, often lasting for an extended period of time.

(4) **Event**. A specific (irregular) activity that occurs in a specific time period and a specific spatial location. The actions in it are usually performed by multiple subjects/initiators (group activities). Detection of specific events is often associated with anomalous (abnormal) activity.

(5) **Behavior**. The subject/initiator mainly refers to people or animals, emphasizing that the subject/initiator is dominated by thoughts and changes actions in a specific environment/context, continues activities, and describes events.

Taking table tennis as an example, some typical instances of various levels mentioned above are shown in Figure 8.1. A player's step, swing, etc. can be regarded as typical action primitives. Typical actions by a player include completing a serve (including tossing the ball, swinging arms, shaking the wrist, hitting the ball, etc.) or returning the ball (including stepping, extending the arm, turning the wrist, and drawing the ball, etc.). Going to the paddle and picking up the ball is often seen as an activity. In addition, two players hitting the ball back and forth to win points is also a typical activity scene. The competition between sports teams is generally regarded as an event, and the awarding of certificates

FIGURE 8.1 Several pictures in the table tennis match.

after the competition is also a typical event. Although a player's self-motivation by making a fist after winning the game can be regarded as an action, it is more often regarded as a behavioral performance. When the player hits a beautiful shot, the audience's applause, shouting, cheering, etc. are also attributed to the behavior of the audience.

It should be pointed out that the concepts of the last three levels are often used loosely in many studies. For example, when an activity is called an event, it generally refers to some anomalous activities (such as a dispute between two people, an old man walking and falling, etc.); when an activity is called an action, the meaning (behavior) and nature of the activity (such as stealing or the act of breaking into a house or over a wall) are emphasized. In the following discussion, unless otherwise emphasized, activities (in a broad sense) will be used to collectively represent the latter three levels.

8.2 SPATIAL-TEMPORAL POINTS OF INTEREST

The change of the scene stems from the movement of the scene, especially the accelerated movement. The accelerated motion of the local structure of the video image corresponds to the accelerated motion of the scenery in the scene. They are in the position of the image with unconventional motion values. It can be expected that these positions (image points) contain the force information causing the scene movement and changing the scene structure in the physical world, which is very helpful for understanding the scene.

In spatial-temporal scenes, the detection of **points of interest** (POIs) tends to expand from space to space-time (Laptev 2005).

8.2.1 Detection of Spatial Points of Interest

In the image space, the **linear scale-space representation** can be used to model the image, namely $L^{sp}: \mathbb{R}^2 \times \mathbb{R}_+ \to \mathbb{R}, f^{sp}: \mathbb{R}^2 \to \mathbb{R}$. For example,

$$L^{sp}\left(x, y; \sigma_z^2\right) = g^{sp}\left(x, y; \sigma_z^2\right) \otimes f^{sp}(x, y) \tag{8.1}$$

That is, convolving f^{sp} with a Gaussian kernel with variance σ_z^2:

$$g^{sp}\left(x, y; \sigma_z^2\right) = \frac{1}{2\pi\sigma_z^2} \exp\left[-\left(x^2 + y^2\right) / 2\sigma_z^2\right] \tag{8.2}$$

Next, use the **Harris interest point detector** to detect points of interest. The idea of detection is to determine the spatial position of the f^{sp} where there are significant changes in both the horizontal and vertical directions. For a given observation scale σ_z^2, these points can be computed by means of a matrix of second moments summed in a Gaussian window with variance σ_z^2:

$$\begin{aligned} \mu^{sp}\left(\cdot; \sigma_z^2, \sigma_i^2\right) &= g^{sp}\left(\cdot; \sigma_i^2\right) \otimes \left\{\left[\nabla L\left(\cdot; \sigma_z^2\right)\right]\left[\nabla L\left(\cdot; \sigma_z^2\right)\right]^{\mathrm{T}}\right\} \\ &= g^{sp}\left(\cdot; \sigma_i^2\right) \otimes \begin{bmatrix} \left(L_x^{sp}\right)^2 & L_x^{sp} L_y^{sp} \\ L_x^{sp} L_y^{sp} & \left(L_y^{sp}\right)^2 \end{bmatrix} \end{aligned} \tag{8.3}$$

where L_x^{sp} and L_y^{sp} are Gaussian differentials calculated at the local scale σ_z^2 according to $L_x^{sp} = \partial_x[g^{sp}(\cdot; \sigma_z^2) \otimes f^{sp}(\cdot)]$ and $L_y^{sp} = \partial_y[g^{sp}(\cdot; \sigma_z^2) \otimes f^{sp}(\cdot)]$.

The second-order moment descriptor in Equation (8.3) can be regarded as the orientation distribution covariance matrix of a 2-D image in the local neighborhood of a point. So the eigenvalues λ_1 and λ_2 ($\lambda_1 \leq \lambda_2$) of μ^{sp} constitute the descriptors of the variation of f^{sp} along the two image directions. If the values of λ_1 and λ_2 are both large, there is a point of interest. To detect such points, the positive maxima of the corner function can be detected:

$$H^{sp} = \det\left(\mu^{sp}\right) - k \cdot \operatorname{trace}^2\left(\mu^{sp}\right) = \lambda_1\lambda_2 - k\left(\lambda_1 + \lambda_2\right)^2 \tag{8.4}$$

At the point of interest, the ratio of eigenvalues $a = \lambda_2 / \lambda_1$ should be large. According to Equation (8.4), for the positive local extrema of H^{sp}, a should satisfy $k \leq a/(1+a)^2$. So, if $k = 0.25$, the positive maximum value of H would correspond to an ideal isotropic point of interest (where $a = 1$, i.e., $\lambda_1 = \lambda_2$). Smaller values of k are more suitable for detection of sharper points of interest (corresponding to larger values of a). A commonly used value of k in the literature is $k = 0.04$, which corresponds to detecting points of interest with $a < 23$.

8.2.2 Detection of Spatial-Temporal Points of Interest

The detection of points of interest in space is extended to space-time, that is, to detect the positions in the local space-time volume with significant changes in image values along time and space. A point with this property will correspond to a spatial point of interest with a specific position in time, which is located in a spatial-temporal neighborhood with unconventional values. Detecting spatial-temporal points of interest is a method to extract the underlying motion features without background modeling. Here, the given video can be convoluted with a 3-D Gaussian kernel at different spatial-temporal scales. Then the spatial-temporal gradients are calculated at each layer of the scale-space representation, and their neighborhoods at each point are combined to obtain the stability estimation of the spatial-temporal second-order moment matrix. The local features can be extracted from the matrix.

Figure 8.2 shows a segment of a player's swing and stroke in a table tennis game, from which several spatial-temporal points of interest are detected. The density of spatial-temporal points of interest along the time axis is related to the frequency of action, and the position of spatial-temporal points of interest in space corresponds to the motion trajectory and action amplitude of the racket.

To model a spatial-temporal image sequence, the function f: $\mathbb{R}^2 \times \mathbb{R} \rightarrow \mathbb{R}$ can be used, and its linear scale-space representation can be constructed by convolving f with an isotropic Gaussian kernel (uncorrelated spatial variance σ_z^2 and temporal variance τ_z^2) to construct the spatial representation L: $\mathbb{R}^2 \times \mathbb{R} \times \mathbb{R}_+^2 \rightarrow$ R:

$$L\left(\cdot;\sigma_z^2,\tau_z^2\right) = g\left(\cdot;\sigma_z^2,\tau_z^2\right) \otimes f\left(\cdot\right) \tag{8.5}$$

where the Gaussian kernel of space-time separation is

$$g\left(x,y,t;\sigma_z^2,\tau_z^2\right) = \frac{1}{\sqrt{(2\pi)^3 \sigma_z^4 \tau_z^2}} \exp\left[-\frac{x^2+y^2}{2\sigma_z^2} - \frac{t^2}{2\tau_z^2}\right] \tag{8.6}$$

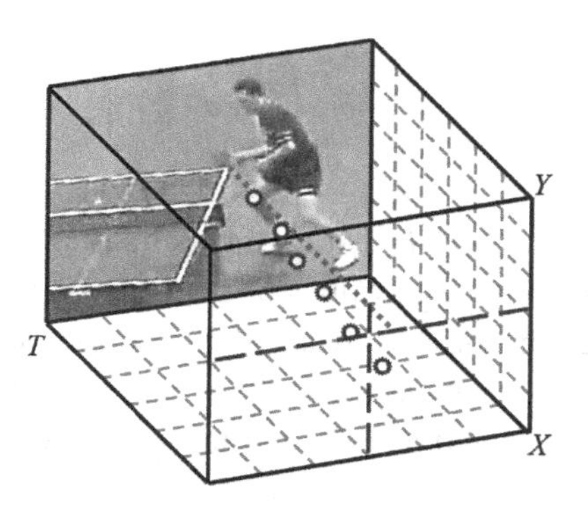

FIGURE 8.2 Example of spatial-temporal points of interest.

Using a separate scale parameter for the temporal domain is critical because events on the temporal and spatial scales are generally independent. In addition, the events detected by the points of interest operator depend on both the spatial and temporal observation scales, so the scale parameters σ_z^2 and τ_z^2 need to be treated separately.

Similar to the spatial domain, consider a matrix of second-order moments in the spatial-temporal domain, which is a 3 × 3 matrix, including the first-order space and the first-order time differentiation convolved with the Gaussian weight function $g(\cdot;\sigma_i^2, \tau_i^2)$:

$$\mu = g\left(\cdot;\sigma_i^2,\tau_i^2\right)\otimes\begin{bmatrix} L_x^2 & L_xL_y & L_xL_t \\ L_xL_y & L_y^2 & L_yL_t \\ L_xL_t & L_yL_t & L_t^2 \end{bmatrix} \tag{8.7}$$

The integral scales σ_i^2 and τ_i^2 are related to the local scales τ_i^2 and τ_z^2 according to $\sigma_i^2 = s\sigma_z^2$ and $\sigma_i^2 = s\sigma_z^2$. The first-order differential is defined as

$$\begin{aligned} L_x\left(\cdot;\sigma_z^2,\tau_z^2\right) &= \partial_x\left(g\otimes f\right) \\ L_y\left(\cdot;\sigma_z^2,\tau_z^2\right) &= \partial_y\left(g\otimes f\right) \\ L_t\left(\cdot;\sigma_z^2,\tau_z^2\right) &= \partial_t\left(g\otimes f\right) \end{aligned} \tag{8.8}$$

To detect points of interest, the search for regions with significant eigenvalues λ_1, λ_2, λ_3 in f is performed. This extends the Harris corner detection function defined in space, that is, Equation (8.4), to the spatial-temporal domain by combining the determinant and rank of μ:

$$H = \det\left(\mu\right) - k\cdot\text{trace}^3\left(\mu\right) = \lambda_1\lambda_2\lambda_3 - k\left(\lambda_1+\lambda_2+\lambda_3\right)^3 \tag{8.9}$$

To prove that the positive local extrema of H correspond to points with large λ_1, λ_2 and λ_3 ($\lambda_1 \le \lambda_2 \le \lambda_3$) values, define the ratios $a = \lambda_2 / \lambda_1$ and $b = \lambda_3 / \lambda_1$, and rewrite H as

$$H = \lambda_1^3\left[ab - k\left(1+a+b\right)^3\right] \tag{8.10}$$

Since $H \ge 0$, there is $k \le ab/(1+a+b)^3$, and k takes its maximum possible value $k = 1/27$ when $a = b = 1$. For significantly large values of k, positive local extrema of H correspond to points with large changes in image values along both time and space directions. In particular, if a and b are assumed to have a maximum value of 23 as in space, the value of k used in Equation (8.9) will be $k \approx 0.005$. So spatial-temporal points of interest in f can be obtained by detecting positive local spatio-temporal maxima in H.

8.3 DYNAMIC TRAJECTORY LEARNING AND ANALYSIS

Dynamic trajectory learning and analysis attempts to provide a grasp of the state of the monitored scene by understanding and characterizing the behavior of various moving objects in it (Morris and Trivedi 2008).

8.3.1 Overall Process

A flow diagram of dynamic trajectory learning and analysis for video is shown in Figure 8.3. First, the object is detected (such as pedestrian detection from a camera in a car; see Jia and Zhang 2007) and tracked, and then the obtained trajectory is used to automatically construct the scenario model, which is finally used to describe monitored conditions and provide annotations for activities.

In scene modeling, the image region where the event occurs is first defined as the **point of interest** (POI), and then in the next learning step, the **activity path** (AP) is defined, which describes how the object moves/travel between the points of interest. The model thus constructed may be referred to as the POI/AP model.

The main work in POI/AP learning includes:

(1) **Activity learning**: The learning of activities can be performed by comparing trajectories; although the length of the trajectories may be different, the key is to maintain an intuitive understanding of the similarity.

(2) **Adaptation**: Research on techniques for managing POI/AP models. These technologies need to be able to adapt online to how to add new activities, remove discontinued activities, and validate models.

(3) **Feature selection**: Determine the correct kinetic representation level for a specific task. For example, a car's route can be determined using only spatial information, but speed information is often also required to detect an accident.

8.3.2 Automatic Scene Modeling

Automatic scene modeling with dynamic trajectories includes the following three points (Makris and Ellis 2005).

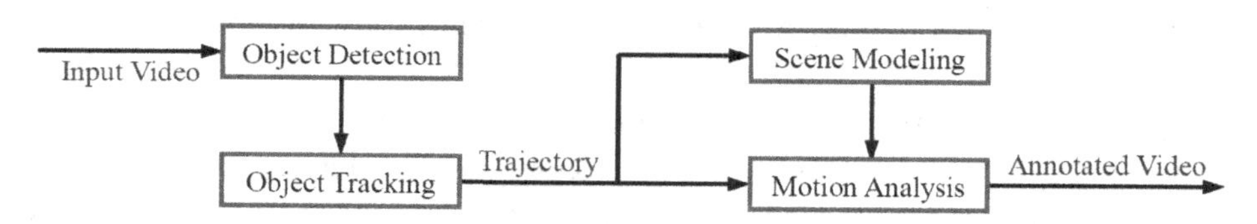

FIGURE 8.3 Flowchart for dynamic trajectory learning and analysis.

8.3.2.1 Object Tracking

Tracking objects requires identity maintenance for each observable object in each frame. For example, an object being tracked in T-frame video generates a series of inferred tracking states:

$$S_T = \{s_1, s_2, \cdots, s_T\} \tag{8.11}$$

where each s_t can describe the object characteristics such as position, speed, appearance, shape, etc. This trajectory information forms the cornerstone of further analysis. Activities can be identified and understood by careful analysis of this information.

8.3.2.2 Interest Point Detection

The first task of scene modeling is to find regions of interest in an image. In a topographic map indicating tracking objects, these regions correspond to nodes in the map. Two types of nodes that are often considered include entry/exit (in/out) regions and stop regions. Take the example of a professor going to the classroom to teach: the former corresponds to the classroom door and the latter corresponds to the podium.

The entry/exit region is where the object enters or leaves the **field of view** (FOV) or where the tracked object appears or disappears. These regions can often be modeled with a 2-D **Gaussian mixture model** (GMM), $Z \sim \sum^{W}_{i=1} w_i N(\mu_i, \sigma_i)$, with W components. This can be solved using the EM algorithm (see Subsection 5.5.2). Entry point data includes the position determined in the first tracking state, and exit point data includes the position determined in the last tracking state. They can be distinguished by a density criterion; the mixing density at state i is defined as

$$d_i = \frac{w_i}{\pi\sqrt{|\sigma_i|}} > T_d \tag{8.12}$$

It measures how compact a Gaussian mixture is, where the threshold

$$T_d = \frac{w}{\pi\sqrt{|\mathbf{C}|}} \tag{8.13}$$

indicates the average density of signal clusters. Here, $0 < w < 1$ are user-defined weights and $\mathbf{C}$ is the covariance matrix of all points in the regional dataset. A compact mixing indicates the correct region while a loose mixing indicates tracking noise due to tracking interruptions.

The stop region comes from the scene landmarks, that is, the position where the object tends to be fixed in a period of time. These stop regions can be determined by two different methods: (i) the speed of the tracked point in this region is lower than a very low threshold determined in advance; (ii) all tracked points remain in a limited distance loop for at least a certain period of time. By defining a radius and a time constant, the second method can ensure that the object is indeed maintained in a specific range, while the first method may still include slow moving objects. For activity analysis, in addition to determining the location, the time spent in each stop region also needs to be established.

8.3.2.3 Activity Path Learning

To understand behavior, the **activity path** needs to be identified. False alarms or track-disrupted noise can be filtered from the training set using POI, keeping only trajectories that start after entering the active region and end before the active terminating region. The tracking trajectory through the active region is divided into two sections corresponding to entering the active region and leaving the active region, and an activity should be defined between the two points of interest of the object start action and end action.

In order to distinguish moving objects that change over time (such as pedestrians walking or running along a sidewalk), temporal dynamics information needs to be incorporated into path learning. Figure 8.4 shows the three basic structures of the path learning algorithm. Their main differences include the type of input, motion vectors, trajectories (or video clips), and the way the motion is abstracted. In Figure 8.4(a), the input is a single trajectory at time t, and the points in the path are implicitly ordered in time. In Figure 8.4(b), a complete trajectory is used as input to the learning algorithm to directly build the path to the output. In Figure 8.4(c), the decomposition of the path by video timing is shown. A video clip (VC) is decomposed into a set of action words to describe the activity, or the video clip is given a certain activity label based on the occurrence of the action word.

8.3.3 Automated Activity Analysis

Once the scenario model is established, the behavior and activities of the object can be analyzed. A basic function of surveillance video is to verify events of interest. Generally speaking, it is only good to define interest in certain circumstances. For example, a parking management system will focus on whether there is still space for parking, while in a smart conference room system it is concerned with the communication between people. In addition to only identifying specific behaviors, all atypical events also need to be checked. By observing a scene over a long period of time, the system can perform a series of activity analyses to learn which events are of interest.

Some typical activity analyses are as follows:

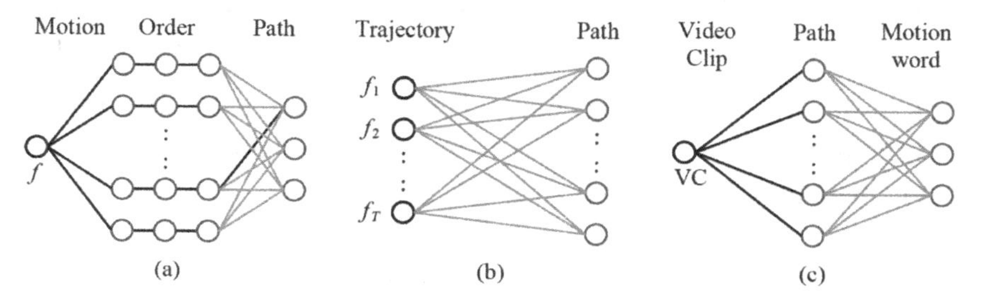

FIGURE 8.4 Scheme for trajectory and path learning.

(1) **Virtual fence**: Any monitoring system has a monitoring range, and setting up sentries on the boundary of the range can give early warnings to events that occur within the range. This is equivalent to establishing a virtual fence at the boundary of the monitoring range, and triggering analysis once there is an intrusion, such as controlling a high-resolution **PTZ camera** to obtain the details of the intrusion, and starting to count the number of intrusions.

(2) **Speed analysis**: The virtual fence only uses location information, while with the help of tracking technology, dynamic information can also be obtained to realize speed-based early warning, such as vehicle speeding or road congestion.

(3) **Path classification**: Velocity analysis only uses the current tracked data, and in practice, it can also use the activity path (AP) obtained from the historical motion mode. The behavior of emerging objects can be described by means of maximum a posteriori MAP path:

$$L^* = \arg\max_k p(l_k \mid G) = \arg\max_k p(G,\ l_k) p(l_k) \tag{8.14}$$

This can help to determine which activity path best interprets the new data. Because the prior path distribution $p(l_k)$ can be estimated by training set, the problem is simplified to maximum likelihood estimation with HMM.

(4) **Anomaly detection**: Anomaly detection is an important task of the monitoring system. Because the activity path can indicate typical activities, exceptions can be found if a new track does not match the existing one. Anomalous modes can be detected by intelligent thresholding:

$$p(l^* \mid G) < L_l \tag{8.15}$$

Among them, the value of the active path l^* most similar to the new trajectory G is still smaller than the threshold value L_l.

(5) **Online activity analysis**: Being able to analyze, identify, and evaluate activities online is more important than using the entire trajectory to describe movement. A real-time system needs to be able to quickly reason about what is happening based on incomplete data (often based on graph models). Two cases are considered here:

(i) *Path prediction*. The tracking data to date can be used to predict future behavior, and the prediction can be refined as more data is collected. Predicting activity using incomplete trajectories can be represented as

$$\hat{L} = \arg\max_j p(l_j \mid W_t G_{t+k}) \tag{8.16}$$

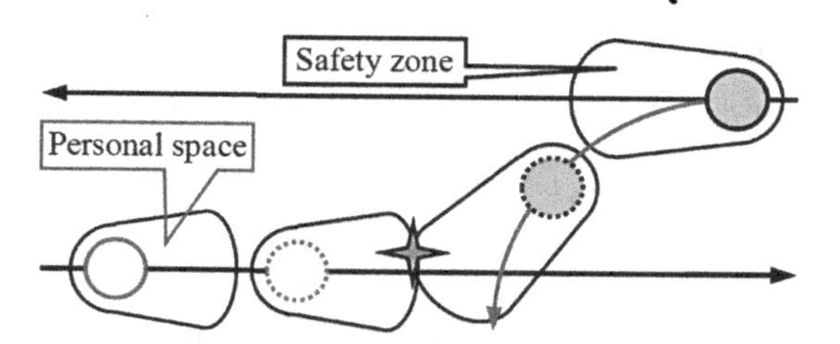

FIGURE 8.5 Use paths for collision assessment.

where W_t represents the window function, G_{t+k} is the trajectory up to the current time t, and the k predicted future tracking states.

(ii) *Tracking anomalies.* In addition to classifying entire trajectories as anomalies, it is also necessary to *detect* anomalous events as soon as they occur. This can be achieved by substituting W_tG_{t+k} for G in Equation (8.15). The window function W_t does not have to be the same as in the prediction, and the threshold may need to be adjusted according to the amount of data.

(6) **Object interaction characterization**: Higher-level analysis is expected to further describe the interaction between objects. Similar to anomalous events, strictly defining object interactions is difficult. There are different types of interactions between different objects in different contexts. Taking car crashes as an example, each car has its own space dimension, which can be regarded as its personal space. When the car is driving, its personal space needs to increase a minimum safe distance (minimum safety zone) around the car, so the spatial-temporal personal space will change with the movement: the faster the speed, the more the minimum safe distance increases (especially in the direction of travel). A schematic diagram is shown in Figure 8.5, where the personal space is represented by a circle, and the safety zone changes with speed (both size and direction). If the safety zones of two vehicles meet, there is potential for a collision, which can help plan driving routes.

Finally, it should be pointed out that for simple activities, the analysis can be carried out only by relying on the object position and velocity, but for more complex activities, more measurements may be required, such as adding the curvature of the profile to identify strange motion trajectories. To provide more comprehensive coverage of activities and behaviors, multi-camera networks are often required. Activity trajectories can also be derived from objects composed of interconnected parts (e.g., the human body), where activity needs to be defined relative to a set of trajectories.

8.4 ACTION CLASSIFICATION AND RECOGNITION

Vision-based human action recognition is a process of labeling image sequences (videos) with action (class) labels. Human action recognition can be turned into a classification problem based on the representation of observed images or videos.

8.4.1 Action Classification

Techniques for the classification of actions can take many forms (Poppe 2010).

8.4.1.1 Direct Classification

In direct classification methods, no special attention is paid to the temporal domain. These methods add information from all frames in an observation sequence into a single representation or perform action recognition and classification for each frame separately.

In many cases, the representation of images is high-dimensional. This results in very computationally expensive matching. In addition, features such as noise may also be included in the representation. Therefore, a compact and robust feature representation in a low-dimensional space is required for classification. Dimensionality reduction techniques can use both linear and nonlinear methods. For example, PCA is a typical linear method, while **local linear embedding** (LLE) is a typical nonlinear method.

The classifiers used for direct classification can also be different. Discriminative classifiers focus on how to distinguish between different classes, rather than modeling individual classes; SVM is a typical one. Under the bootstrapping framework, a series of weak classifiers (each often using only 1-D representations) are used to build a strong classifier. Except for AdaBoost, LPBoost can obtain sparse coefficients and converge quickly.

8.4.1.2 Time-State Model

The **generative model** learns a joint distribution between observations and actions, modeling each action class (considering all variations). Discriminative models learn the probabilities of action classes under observation conditions. They do not model classes but focus on differences between classes.

The most typical of the generative models is the hidden Markov model (HMM), in which the hidden states correspond to the various steps of the action. Hidden states model state transition probabilities and observation probabilities. There are two independent assumptions here. One is that the state transition only depends on the previous state, and the other is that the observation only depends on the current state. Variations of HMM include **maximum entropy Markov model** (MEMM), **factored-state hierarchical HMM** (FS-HHMM), and **hierarchical variable transition hidden Markov model** (HVT-HMM).

On the other hand, discriminative models model the conditional distribution given an observation, and combining multiple observations to distinguish different action classes. This model is beneficial for distinguishing related actions. **Conditional random field** (CRF) is a typical discriminant model, and its improvement includes **decomposition of conditional random field** (FCRF), **generalization of conditional random field**, etc.

8.4.1.3 Action Detection

Action detection-based methods do not explicitly model object representations nor actions in images. They link observation sequences to numbered video sequences to directly detect (defined) actions. For example, a video segment can be described as a bag of words encoded at different temporal scales, each word corresponding to the gradient orientation of a local patch. Local patches with slow temporal changes can be ignored, so that the representation will mainly focus on the motion region.

When the movement is periodic (such as walking or running), the action is circular, that is, **circular action**. At this time, the time domain segmentation can be carried out by analyzing the self-similarity matrix. Furthermore, the person in move can be labeled, and the self-similarity matrix can be constructed by tracking the labels and using the affine distance function. By frequency transformation of the self-similarity matrix, the peak in the spectrum corresponds to the frequency of motion (if you want to distinguish the walking person or the running person, you can calculate the gait cycle). The types of actions can be determined by analyzing the matrix structure.

The main methods of human action representation and description can be divided into two categories: (i) appearance-based methods, which directly use the description of foreground, background, contour, optical flow and change of image; and (ii) manikin-based methods, which use a manikin to represent the structural characteristics of actors, such as describing actions with human joint point sequences. Whichever method is adopted, detecting and tracking the human body and important parts of the human body (such as head, hands, feet, etc.) will play an important role.

Many public databases have been established for the purpose of experimenting with action detection and verifying the effect of action detection methods. Figure 8.6 shows some example pictures of actions in the Weizmann action recognition database (Blank et al. 2005). From top to bottom, the left column is Jack, side move, bend, walk and run, and the right column is wave1 (wave one hand), wave2 (wave two hands), skip (one-foot forward jump), jump (jump forward with both feet), pjump (both feet jump in place).

FIGURE 8.6 Example pictures of actions in Weizmann action recognition database.

8.4.2 Action Recognition

The representation and recognition of actions and activities is a relatively new but immature field (Moeslund et al. 2006). Most of the methods used depend on the researcher's purpose. In scene interpretation, the representation can be independent of the subject that leads to the activity (such as a person or car). Monitoring applications generally concern people's activities and interactions. In the holistic method, global information is better than component information, for example, when it is necessary to determine human gender. For simple actions such as walking or running, local methods can also be considered, in which more attention is paid to detailed actions or action primitives.

8.4.2.1 Holistic Recognition

Holistic recognition emphasizes the recognition of the whole human body or each part of a single human body. For example, the walking and walking gait of persons can be recognized based on the structure and dynamic information of the whole body. Most of the methods here are based on the silhouette or outline of the human body without distinguishing between various parts of the body. For example, a human body-based recognition technology uses human silhouettes and uniformly samples their contours, and then processes the decomposed contours with PCA. In order to calculate the time-space correlation, the trajectories can be compared in the eigen-space. On the other hand, using dynamic information can not only recognize identity, but also determine what people are doing. Based on the recognition of body parts, the action is recognized through the position and dynamic information of body parts.

8.4.2.2 Pose Modeling

The recognition of human actions is closely linked to the estimation of human poses. Human posture can be divided into action posture and body posture. The former corresponds to the action behavior of a person at a certain moment, and the latter corresponds to the orientation of the human body in 3-D space.

The representation and calculation methods of human body posture can be mainly divided into three types:

(1) *Appearance-based method.* The human pose is analyzed using information such as color, texture, and contour, instead of modeling the physical structure of the human directly. Since only the apparent information in the 2-D image is exploited, it is difficult to estimate the human pose.

(2) *Human body model-based methods.* The human body is first modeled using a line-graph model, 2-D or 3-D model, and then the human pose is estimated by analyzing these parameterized human models. Such methods usually require high image resolution and accurate object detection.

(3) *Method based on 3-D reconstruction.* First, the 2-D moving objects obtained by multiple cameras at different positions are reconstructed into 3-D moving objects through corresponding point matching, and then the camera parameters and imaging formulas are used to estimate the human poses in 3-D space.

Pose can be modeled based on spatial-temporal points of interest. If only the spatial-temporal **Harris interest point detector** is used, the obtained spatial-temporal points of interest are mostly in the region of sudden movement. The number of such points is small, which belongs to the sparse type, and it is easy to lose important motion information in the video, resulting in detection failure. To overcome this problem, the dense spatial-temporal points of interest can also be extracted with the help of motion intensity to fully capture motion-induced changes. Here, the motion intensity can be calculated by convolving the image with a spatial Gaussian filter and a temporal Gaber filter. After the spatial-temporal points of interest are extracted, a descriptor is first established for each point, and then each pose is modeled. A specific method is to first extract the spatial-temporal feature points of the pose in the training sample library as the underlying feature, so that one pose corresponds to a set of spatial-temporal feature points. The pose samples are then classified using unsupervised classification methods to obtain clustering results of typical poses. Finally, each typical pose category is modeled using an EM-based Gaussian mixture model.

A recent trend in pose estimation in natural scenes is to use a single frame for pose detection in order to overcome the problem of tracking with a single view in unstructured scenes. For example, robust part detection and probabilistic combination of parts have resulted in better estimates of 2-D poses in complex movies.

8.4.2.3 Active Reconstruction

Actions lead to changes in posture. If each static posture of the human body is defined as a state, then by means of the state space method (also called the probability network method), the states are switched through the transition probability, and an activity sequence can be constructed by performing a traversal between the states of the corresponding pose.

Significant progress has also been made in automatically reconstructing human activity from videos based on pose estimation. The original model-based analysis-synthesis scheme leverages multi-view video acquisition to efficiently search the pose space. Many current methods focus more on capturing the overall body motion and less on precisely building the details.

There have also been many advances in single-view human activity reconstruction with the help of **statistical sampling techniques**. The current focus is on using the learned model to constrain activity-based reconstruction. Research has shown that using a strong prior model is helpful for tracking specific activities in a single view.

8.4.2.4 Interactive Activities

Interactive activities are more complex activities that can be divided into two categories: (i) interaction between human and environment, such as driving and taking a book; (ii) interpersonal interaction, often referring to the communication activities or contact behaviors of two (or more) people. Single-person activities can be described with the help of a probability graph model, which is a powerful tool for modeling continuous dynamic feature sequence, and has a relatively mature theoretical basis. Its disadvantage is that the topology of its model depends on the structural information of the activity itself, so it needs a lot of training data to learn the topology of a graph model for complex interactive activities.

In order to combine single-person activities, the **statistical relationship learning** (SRL) method can be used. SRL is a machine learning method that integrates relational/logical representation, probabilistic reasoning, machine learning and data mining to obtain the likelihood model of relational data.

8.4.2.5 Group Activities

Quantitative changes lead to qualitative changes, and a substantial increase in the number of subjects involved in activities will bring new problems and new research. For example, group subject movement analysis mainly takes people flow, traffic flow and dense biological groups in nature as the objects, studies the representation and description method of a group subject movement, and analyzes the movement characteristics of the group subject and the influence of boundary constraints on group subject movement. At this time, the grasp of the unique behavior of a special individual is weakened, and more attention is paid to the abstraction of the individual and the description of the entire collective activity. For example, some researchers use the macro kinematics theory to explore the motion law of particle flow and establish the motion theory of particle flow. On this basis, dynamic evolution phenomena such as aggregation, dissipation, differentiation and merger in group subject activities are semantically analyzed, in order to explain the trend of the whole scene.

In the analysis of group activity, the statistics of the number of individuals participating in the activity represent basic data. For example, in many public places, such as squares, stadium entrances and exits, it is necessary to have certain statistics on the number of people. Figure 8.7 shows a picture of people counting in a surveillance scenario (Jia and Zhang 2009). Although there are many people in the scene, with different movement patterns, the concern here is the number of people in a certain region (the region enclosed by the box).

8.4.2.6 Scene Interpretation

Unlike the recognition of objects in a scene, **scene interpretation** mainly considers the entire image without verifying a specific object or person. Many methods in practice only consider the results captured by the camera, from which they learn and recognize activity by observing object motion without necessarily determining the object's identity. This strategy is effective when the target is small enough to be represented as a point in 2-D space.

For example, a system for detecting anomalous conditions includes the following modules. The first is to extract objects such as 2-D position and velocity, size and binary silhouette, and use vector quantization to generate an example codebook. To account for the temporal relationship between each other, co-occurrence statistics can be used. Iteratively defining the probability function between the examples in the two codebooks and determining a binary tree structure, where the leaf nodes correspond to the probability distributions in the co-occurrence statistics matrix, and the higher-level nodes correspond to simple scene activities (such as movement of pedestrians or car motion) can further contribute to scene interpretation.

FIGURE 8.7 Statistics on the number of people in the monitoring of people flow.

8.5 ACTIVITY AND BEHAVIOR MODELING

A general action/activity recognition system consists of several working steps from an image sequence to high-level interpretation (Turaga et al. 2008):

(1) Obtain the input video or sequence image

(2) Extract the refined underlying image features

(3) Obtain middle-level action descriptions based on underlying features

(4) Make high-level semantic interpretation by starting from basic actions.

Generally practical activity recognition systems are hierarchical. The bottom layer includes foreground-background segmentation module, tracking module and object detection module, etc. The middle layer is mainly the action recognition module. The most important of the high layers is the inference engine, which encodes the semantics of the activity in terms of the action primitives of the lower layers, and makes an overall understanding in terms of the learned model.

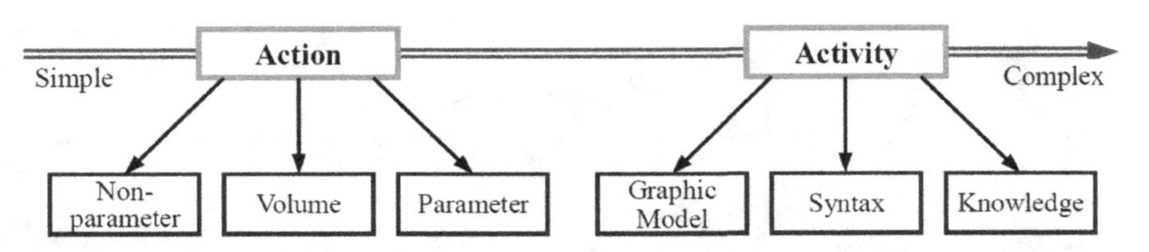

FIGURE 8.8 Classification of approaches for action and activity recognition.

As pointed out in Section 8.1, in terms of abstraction, activities are higher than actions. From a technical point of view, the modeling and recognition of actions and activities often use different techniques, and they range from simple to complex. Many commonly used action and activity modeling and recognition techniques can be classified, as shown in Figure 8.8 (Turaga et al. 2008).

8.5.1 Action Modeling

Action modeling methods can be divided into three categories: **non-parametric modeling**, **3-D modeling**, and **parametric time-series modeling**. Non-parametric methods extract a set of features from each frame of the video and match these features to a stored template. 3-D approaches do not extract features frame by frame, but treat the video as a 3-D volume of pixel intensities and extend standard image features (e.g., scale-space extrema, spatial filter responses) to 3-D. Parametric time-series methods model the temporal dynamics of motion, estimating parameters specific to a set of actions from a training set.

8.5.1.1 Non-Parametric Modeling Methods

Common non-parametric modeling methods are as follows.

(1) 2-D template

This type of method involves the steps of performing motion detection and then tracking objects in the scene. After tracking, build a cropped sequence containing the object. Changes in scale can be compensated for by normalizing the object size. A periodic index for a given action is calculated, and if the periodicity is strong, perform action recognition. For identification, an estimate of the period is used to segment the sequence of periods into individual periods. The average period is decomposed into several temporal segments and flow-based features are computed for each spatial point in each segment. The flow features in each segment are averaged into a single frame. The average flow frame in this activity cycle constitutes the template for each action group.

A typical approach is to construct temporal templates as action models. The background is first extracted, and the background patches extracted from a sequence are combined into a still image. There are two ways of combining: one is to assign the same weight to all frames in the sequence, and the resulting representation can be called a **motion energy image** (MEI); the other is to assign different weights to different frames in the sequence, with new frames generally given larger weights and older frames given smaller weights. The resulting representation can be called a **motion history image** (MHI). For a given action, use the combined images to form a template. Calculate its region invariant moments on the template and recognize it.

(2) 3-D object model

The 3-D object model is a model established for a spatial-temporal object, such as the generalized cylinder model, the 2-D contour stacking model and so on. The motion and shape information of the object is included in the 2-D contour stacking model, from which the geometric features of the object surface, such as peaks, pits, valleys, ridges, etc., can be extracted. If you replace the 2-D contour with blobs in the background, you get a **binary space-time volume**.

(3) Manifold learning method

A lot of action recognition involves data in high-dimensional space. Since the feature space becomes exponentially sparse with dimensionality, a large number of samples are required to construct an effective model. The inherent dimension of the data can be determined by using the manifold where the learning data is located, which has a relatively small degree of freedom and can help design an effective model in a low-dimensional space. The easiest way to reduce dimensionality is **principal component analysis** (PCA), in which the data are assumed to be in a linear subspace. In practice, except in very special cases, the data are not in a linear subspace, so methods that can learn the eigen-geometry of the manifold from a large number of samples are needed. Nonlinear dimensionality reduction techniques allow data points to be represented in terms of how close they are to each other in a nonlinear manifold. Typical methods include **local linear embedding** (LLE) and **Laplace eigen-maps**.

8.5.1.2 3-D Modeling Methods

Common 3-D modeling methods are as follows.

(1) Spatial-temporal filtering

Spatial-temporal filtering is a generalization of spatial filtering, which uses a set of spatial and temporal filters to filter the data of the **video volume**. Specific features are further deduced from the response of the filter bank. It has been hypothesized that the spatial-temporal properties of cells in the visual cortex can be described by spatial-temporal filter structures, such as **oriented Gaussian kernels and their derivatives** as well as **oriented Gaber filter banks**. For example, a video segment can be considered as a spatial-temporal volume defined in XYT space, and a Gaber filter bank is used for each voxel (x, y, t) to compute local appearance models for different orientations and spatial scales as well as a single temporal scale. Actions are recognized using the average spatial probability of each pixel in a frame image. Because the action is analyzed at a single time scale, this method cannot be applied when the frame rate varies. To do this, locally normalized spatial-temporal gradient histograms can be extracted at several time scales, and χ^2 between the histograms can be used to match the input video to the stored samples. Another method is to use a

Gaussian kernel to filter in the spatial domain, use a Gaussian differential to filter in the time domain, and combine them into the histogram after thresholding the response. This method provides simple and effective features for far-field (non-close-up) video.

The filtering method can be implemented simply and quickly with the help of efficient convolutions. In most applications, however, the bandwidth of the filter is not known in advance, so large filter banks at multiple temporal and spatial scales are required to efficiently capture action. The use of large filter banks with multiple temporal and spatial scales is also limited by the requirement that the response of each filter output has the same dimensionality as the input data.

(2) Part-based approach

A 3-D video volume can be seen as a collection of many local parts, each part having a special movement pattern. A typical approach is to use the spatial-temporal points of interest in Section 8.2 to represent. In addition to using the **Harris interest point detector**, the spatial-temporal gradients extracted from the training set can also be clustered. Actions can also be represented using a bag-of-words model, which can be obtained by extracting spatial-temporal points of interest and clustering features.

Because **points of interest** are local in nature, long-term correlations are ignored. To solve this problem, a **correlogram** can be used. Think of a video as consisting of a series of sets, each set containing parts in a small sliding window of time. This approach does not directly model the global geometry of the local components, but treats them as a package of features. Different actions may contain similar spatial-temporal components but may have different geometric relationships. If global geometric information is incorporated in the part-based video representation, this constitutes the parts of a **constellation**. This model becomes more complex when there are many parts. It is also possible to combine the constellation model and the bag-of-words model into a hierarchical structure, with only a smaller number of components in the high-level constellation model, and each component is contained in the lower-level bag of features. Thus, the advantages of both models are combined.

In most part-based methods, the detection of parts is often based on some linear operations, such as filtering, spatial-temporal gradients, etc., so the descriptors are sensitive to appearance changes, noise, occlusions, etc. But on the other hand, these methods are more robust to non-stationary backgrounds due to their inherent locality.

(3) Sub-volume matching

Sub-volume matching refers to matching between the sub-volumes in the video and the template. For example, actions can be matched to templates by means of spatial-temporal action correlation. The main difference between this method and the part-based method is that it does not need to extract action descriptors from extreme points in the scale space but instead checks the similarity between two local spatial-temporal patches (by comparing the motion between two patches). However, it can be time-consuming to perform the relevant calculations for the entire video volume. One way to solve this problem is to generalize the fast Haar features (box features) that have been successful in object detection to 3-D. A 3-D Haar feature is the output of a 3-D filter bank with coefficients 1 and -1. Combining the outputs of these filters with the bootstrapping method results in robust performance. Another approach is to view a video volume as a collection of sub-volumes of any shape, with each sub-volume being a spatially consistent volumetric region obtained by clustering pixels that are both apparently similar and spatially close. The given video is then over-segmented into many sub-volumes or **super-voxels**. Action templates are matched by searching for the smallest set of regions in these sub-volumes that maximizes the overlap ratio between the set of sub-volumes and the templates.

The advantage of sub-volume matching is that it is robust to noise and occlusion, and if combined with optical flow features, it is also robust to apparent changes. The disadvantage of sub-volume matching is that it is susceptible to background changes.

(4) Tensor-based method

Tensors are a generalization of 2-D matrices in multidimensional spaces. A 3-D space-time volume can naturally be viewed as a tensor with three independent dimensions. For example, human action, human identity and joint trajectories can be viewed as three independent dimensions of a tensor. By decomposing the total data tensor into dominant patterns (similar to the generalization of PCA), it is possible to extract the signatures corresponding to the actions and identities of people (the person performing the action). Of course, the 3-D of the tensor can also be directly taken as the 3-D of the space-time domain, namely (x, y, t).

The tensor-based approach provides a straightforward way of matching videos as a whole, which does not need to consider the mid-level representations used by the previous methods. In addition, other kinds of features (such as optical flow, spatial-temporal filter responses, etc.) are easily incorporated by increasing the tensor dimension.

8.5.1.3 Parametric Time-Series Modeling Methods

The first two modeling methods are more suitable for simpler movements, and the modeling method described below is more suitable for complex movements that span the time domain, such as complex dance steps in ballet videos, and special gestures of musical instrument players.

(1) Hidden Markov model

Hidden Markov model (HMM) is a typical model of state space, which is very effective for modeling time-series data, has good generalization and discrimination, and is suitable for tasks that require recursive probability estimation. In the process of constructing discrete hidden Markov models, the state space is regarded as a finite set of discrete points. Evolution over time is modeled as a series of probabilistic steps transitioning from one state to another. The three key issues of hidden Markov models are inference, decoding and learning. The hidden Markov model was first used to identify the action of tennis shots, such as forehand, forehand volley, backhand, backhand volley, smash and so on. A series of background-subtracted images are modeled as hidden Markov models corresponding to specific categories. Hidden Markov models can also be used to model actions that vary over time, such as gait.

Using a single hidden Markov model can model single-person actions. For multi-person actions or interactive actions, a pair of hidden Markov models can be used to represent alternate actions. In addition, domain knowledge can be incorporated into the construction of hidden Markov models, or the hidden Markov models can be combined with object detection to exploit the connection between actions and (action) objects. For example, prior knowledge of state duration can be incorporated into the framework of a hidden Markov model, and the resulting model is called a **semi-hidden Markov model** (semi-HMM). If a discrete label for modeling high-level behavior is added to the state space, a mixed-state hidden Markov model can be used to model non-stationary behavior.

(2) Linear dynamic system

Linear dynamic systems (LDS) are more general than hidden Markov models in that the state space is not restricted to a set of finite symbols but can be continuous values in $\mathbb{R}^k$ space, where k is the dimension of the state space. The simplest linear dynamic system is the first-order time-invariant Gauss–Markov process, which can be represented as

$$x(t) = Ax(t-1) + w(t) \quad w \sim N(0, P) \tag{8.17}$$

$$y(t) = Cx(t) + v(t) \quad v \sim N(0, Q) \tag{8.18}$$

where $\boldsymbol{x} \in \mathbb{R}^d$ is the d-D state space, $\boldsymbol{y} \in \mathbb{R}^n$ is the n-D observation vector, $d << n$, $\boldsymbol{w}$ and $\boldsymbol{v}$ are the process and observation noise, respectively, both of which are Gaussian distributed with zero mean and covariance matrices $\boldsymbol{P}$ and $\boldsymbol{Q}$, respectively. Linear dynamic systems can be regarded as an extension of the hidden Markov model with Gaussian observation model in continuous state space, which is more suitable for processing high-dimensional time-series data, but still less suitable for non-stationary actions.

(3) Nonlinear dynamic system

Consider the following sequence of actions. A person first bends over to pick up an item, then walks to a table and places the item on the table, and finally sits on a chair. There are a series of short steps, each of which can be modeled with LDS. The whole process can be seen as the conversions between different LDSs. The most general form of time-varying LDS is

$$x(t) = A(t)x(t-1) + w(t) \quad w \sim N(0,P) \tag{8.19}$$

$$y(t) = C(t)x(t) + v(t) \quad v \sim N(0,Q) \tag{8.20}$$

In contrast to Equations (8.17) and (8.18), here both A and C can vary with time. The method commonly used to solve such complex dynamic problems is **switched linear dynamic systems** (SLDS) or **jump linear systems** (JLS). A switched linear dynamic system consists of a set of linear dynamic systems and a switching function that changes model parameters by switching between models. To identify complex motion, a multi-layered approach can be employed that involves several different levels of abstraction. At the lowest level is a series of input images, and the upper level includes regions of consistent motion, called blobs. The further upper layer combines the trajectories of the blobs temporally, and the highest layer includes a hidden Markov model that represents complex behavior.

Although switched linear dynamic systems are more capable of modeling and describing than hidden Markov models and linear dynamic systems, learning and reasoning are much more complex in switched linear dynamic systems, so approximation methods are generally required. In practice, determining an appropriate number of switching states is difficult, often requiring a large amount of training data or cumbersome manual adjustments.

8.5.2 Activity Modeling and Recognition

Activity is not only long-lasting compared to action, but most applications of activity that people focus on, such as monitoring and content-based indexing, include multiple people in action. Their activities not only interact with each other but also with **contextual entities**. To model complex scenes, high-level representation and reasoning of the eigen-structure and semantics of complex behaviors are required.

8.5.2.1 Graph Model

Common graph models are as follows.

(1) Belief network

A Bayesian network is a simple **belief network**. It first encodes a set of random variables as a **local conditional probability density** (LCPD) and then encodes the complex conditional dependencies between them. **Dynamic belief networks** (DBNs, also known as dynamic Bayesian networks) are a generalization of simple Bayesian networks that incorporate time dependencies between random variables. Compared with traditional HMMs that can only encode one latent variable, DBN can encode complex conditional dependencies among several random variables.

Interactions between two people, such as pointing, squeezing, pushing, hugging, etc., need to be modeled using a two-step process. The pose is first estimated by a Bayesian network, and then the temporal evolution of the pose is modeled with a DBN. Actions can be recognized based on contextual information derived from other objects in the scene, and human–human or human–object interactions can be explained using Bayesian networks.

DBN is more general than HMM if the dependencies among multiple random variables are considered. But in DBN, the time model is also a Markov model, as in HMM, so the basic DBN model can only deal with the behavior of the sequence. The development of graph models for learning and reasoning allows them to model structured behavior. However, learning local CPD for large networks often requires a large amount of training data or complicated manual adjustments by experts, both of which impose certain limitations on the use of DBNs in large-scale environments.

(2) Petri net

A **Petri net** is a mathematical tool for describing the connections between conditions and events. It is particularly suitable for modeling and visualizing behaviors such as ordering, concurrency, synchronization, and resource sharing. A Petri net is a bilateral graph that contains two kinds of nodes – positions and transitions – where positions refer to the state of an entity and transitions refer to changes in the state of the entity.

Consider an example of a car pick-up activity represented by a probabilistic Petri net, as shown in Figure 8.9. In the figure, the positions are marked as p_1, p_2, p_3, p_4, p_5 and the transitions are marked as t_1, t_2, t_3, t_4, t_5, t_6. In this Petri net, p_1 and p_3 are the starting nodes, and p_5 is the ending node. A car enters the scene and places a token at position p_5. Transition t_1 can be activated at this time, but it will not officially start until the conditions related to this (that is, the car must be parked in a nearby parking space) are met. At this point the token at p_1 is eliminated and placed at p_2. Similarly, when a person enters a parking space, the token is placed at p_3, and the transition starts after the person leaves the parked car. The token is then removed from p_3 and placed at p_4.

Now, a token is placed in each allowed position of transition t_6, so that when the relevant condition (here the car leaves the parking space) is met, the fire can be fired. Once the car leaves, t_6 fires, the tokens are all removed and one token is placed at the final position p_5. In this example, ordering, concurrency, and synchronization all happen.

Petri nets have been used to develop systems for high-level interpretation of image sequences. Here the structure of the Petri net needs to be determined in advance, which is a very complicated task for large networks that represent complex activities. This work can be semi-automated by automatically mapping a small set of logical,

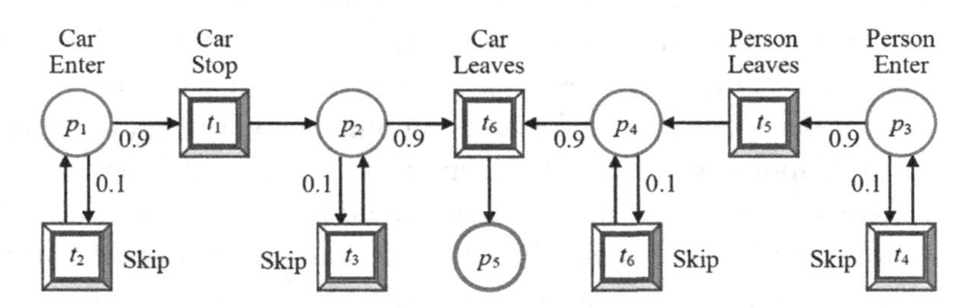

FIGURE 8.9 The probability Petri nets representing a car pick-up activity.

spatial and temporal operations onto the graph structure. With such an approach, interactive tools for querying video surveillance by mapping user query requirements to Petri nets can be developed. However, the method is based on deterministic Petri nets, so it cannot handle uncertainty in the low-level modules (trackers, object detectors, etc.).

Further, real human activities are not exactly consistent with rigorous models, which need to allow for differences from the expected sequence and penalize significant differences. To this end, the concept of **probabilistic Petri net** (PPN) is proposed. In PPN, transitions are associated with weights, which record the probability of transition initiation. Robustness to missing observations in the input stream is achieved by exploiting skip transitions and penalizing them with low probability. In addition, the uncertainty of identifying the target or the uncertainty of the unfolding activity can be effectively combined into the token of the Petri net.

Although the Petri net is a relatively intuitive tool for describing complex activities, its disadvantage is that it requires manual description of the model structure. The problem of learning structure from training data is also not formally addressed.

(3) Other graph models

In view of the shortcomings of DBN, especially the limitation of sequence activity description, some other graph models have also been proposed. Under the framework of DBN, some graph models specially used to model complex time connections, such as sequence, time period, parallelism, synchronization, etc., are constructed. A typical example is the **past-now-future** (PNF) structure, which can be used to model complex temporal ordering situations. In addition, a propagation net can be used to represent activities that use partially ordered time intervals. One of the activities is constrained by time, logical order, and the length of the activity interval. The new approach treats a time-expanding activity as a series of event labels. With the help of context and activity-specific constraints, sequence labels can be found to have some inherent partial ordering properties. For example, you need to open your mailbox before you can view your mail. With these constraints, the activity model can be viewed as a set of sub-sequences that represent partial ordering constraints of different lengths.

8.5.2.2 Synthesis Methods

Synthetic methods are mainly implemented with the help of grammatical concepts and rules.

(1) Grammar

Grammar uses a set of production rules to describe the structure of processing. Similar to the grammar in the language model, production rules indicate how to construct sentences (activities) from words (activity primitives) and how to recognize the rules that sentences (videos) meet in a given grammar (activity model). The early grammar for recognizing visual activities was used to recognize the work of disassembling objects. At that time, there was no probability model in the grammar. Subsequently, **context-free grammar** (CFG) was applied, to model and recognize human motion and multi-human interaction. A hierarchical process is used here. At the lower level, HMM and BN are combined, and at the higher level, the interaction is modeled with CFG. The context-free grammar method has a strong theoretical basis and can model structured processes. In the synthetic method, we only need to enumerate the **primitive events** to be detected and define the production rules of high-level activities. Once the rules of CFG are constructed, the existing analytical algorithms can be used.

Because the deterministic syntax expects very good accuracy at the low level, it is not suitable for errors caused by tracking and missing observations at low level. In complex scenarios with multiple time connections (such as parallelism, coverage, synchronization, etc.), it is often difficult to build grammar rules manually. Learning grammar rules from training data is a promising alternative, but it has proved to be very difficult in general situations.

(2) Stochastic grammar

The algorithms used to detect low-level primitives are often probabilistic in nature. Therefore, **stochastic context-free grammar** (SCFG) extends the probability of context-free grammar, which is more suitable for combining practical visual models. SCFG can be used to model the semantics of activities whose structure assumptions are known. HMM is used in the detection of low-level primitives. The production rules of grammar are supplemented by probability, and a skip transition is introduced. In this way, the robustness to the insertion error in the input stream can be improved, and the robustness in the low-level module can also be improved. SCFG is also used to model multi-tasking activities (including multiple independent execution threads, intermittent related interaction activities, etc.).

In many cases, it is necessary to associate some additional properties or characteristics with event primitives. For example, the exact location where an event primitive occurs may be important to describe an event, but this may not be recorded in the set of event primitives beforehand. In these cases, attribute grammars are more descriptive than traditional grammars. The probabilistic attribute grammar has been used to handle multi-agent activity in monitoring.

An example of attribute syntax, passenger boarding, is shown in Figure 8.10. Production rules and event primitives such as "appear", "disappear", "move-close"

$S \rightarrow \text{BOARDING}_{\text{N}}$

$\text{BOARDING} \rightarrow \text{appear}_0\ \text{CHECK}_1\ \text{disappear}_1$

(isPerson (appear, class) ∧ isInside (appear.loc, Gate) ∧ isInside (disappear.loc, Plane))

$\text{CHECK} \rightarrow \text{moveclose}_0\ \text{CHECK}_1$

$\text{CHECK} \rightarrow \text{moveaway}_0\ \text{CHECK}_1$

$\text{CHECK} \rightarrow \text{moveclose}_0\ \text{moveaway}_1\ \text{CHECK}_1$

(isPerson (moveclose, class) ∧ moveclose.idr = moveaway.idr

FIGURE 8.10 An example of attribute grammar for passenger boarding.

and "move-away" are used to describe activities. Event primitives are further associated with attributes such as where the event appears and disappears (loc), classifying a set of objects (class), and identifying related entities (idr).

While SCFGs are more robust to errors and missed detections in the input stream than CFGs, they also have the same limitations as CFGs in modeling temporal connections.

8.5.2.3 Knowledge- and Logic-Based Methods

Knowledge and logic are closely linked.

(1) Logic-based approach

Logic-based methods rely on strict logical rules to describe domain knowledge in a general sense to describe activities. Logical rules are useful for describing user-entered domain knowledge or for representing high-level reasoning results in an intuitive and user-readable form. The **declarative model** describes all expected activities in terms of scene structures, events, etc. The activity model includes interactions between objects in the scene. A hierarchical structure can be used to identify a series of actions performed by an agent. Symbolic descriptors for actions can be extracted from low-level features through some intermediate layers. Next, a rule-based approach is used to approximate the probability of a particular activity by matching the properties of the agent with the expected distribution (expressed in terms of mean and variance). This method considers that an activity is composed of several action threads, and each action thread can be modeled as a finite random state automaton. Constraints between different threads are propagated in a temporal logical network.

For example, in a system based on logic programming, when it expresses and identifies high-level activities, it first uses low-level modules to detect event primitives, and then uses a Prolog-based high-level inference engine to identify activities expressed by logical rules between event primitives. These methods do not directly address the problem of uncertainty in observing the input stream. To deal with these problems, logical and probabilistic models can be combined, where logical rules are represented in terms of first-order logical predicates. Each rule is also associated with a weight that indicates the accuracy of the rule. Further reasoning can be done with the help of Markov logic networks.

While logic-based methods provide a natural way to incorporate domain knowledge, they often involve time-consuming audits of constraints being met. Also, it is unclear how much domain knowledge needs to be incorporated. It can be expected that more knowledge of the results will make the model more rigorous and not easy to generalize to other situations. Finally, logical rules require domain experts to perform time-consuming traversal of each configuration pair.

(2) Ontological approach

In most practical configurations using the aforementioned methods, the definition of symbolic activity is constructed empirically. Rules such as grammar or a set of logical rules are specified manually. Although empirically constructed designs are fast and

```
PROCESS (cruise-parking-lot (vehicle v, parking-lot lot),
Sequence (enter (v, lot),
    Set-to-zero (i),
    Repeat-Until (
        AND (inside (v, lot), move-in-circuit (v), increment (i) ),
        Equal (i, n) ),
    Exit (v, lot) ) )
```

FIGURE 8.11 Example of ontology used to describe car cruising in parking lot.

work well in most cases, they are less generalizable and limited to the specific cases in which they are designed. Therefore, a centralized representation of activity definitions or an algorithm-independent activity ontology is also required. Ontologies can standardize the definition of activities, allow porting to specific registrations, enhance interoperability between different systems, and easily replicate and compare system performance. Typical practical examples include analyzing social interactions in nursing rooms, classifying meeting videos, setting up bank interactions, etc. However, while ontologies provide concise high-level activity definitions, they do not guarantee the correct "hardware" to "parse" the ontologies used to identify tasks.

Figure 8.11 shows an example of using the ontology concept to describe a car cruising activity. This ontology records the number of times the car turns around on the road in the parking lot without stopping. When this number exceeds a threshold, a cruising activity is detected.

Incidentally, since 2003, an international **video event competition work conference** has been held to integrate various capabilities to build a domain ontology based on general knowledge. The conference has defined six areas of video surveillance: (i) perimeter and interior security; (ii) railway crossing surveillance; (iii) visual bank surveillance; (iv) visual subway surveillance; (v) warehouse security; and (vi) airport apron security. The meeting has also guided the formulation of two formal languages: **video event representation language** (VERL), which helps to complete the ontology representation of complex events based on simple sub-events; and **video event markup language** (VEML), which is used to annotate VERL events in the video.

8.6 JOINT MODELING OF ACTOR AND ACTION

With the deepening of research, the categories of actors (subjects) and actions that need to be considered for **understanding spatial-temporal behavior** are increasing. To do this, the actor and action need to be jointly modeled (Xu et al. 2015). In fact, jointly detecting an ensemble of several objects in an image is more robust than detecting individual objects individually. Therefore, joint modeling is necessary when considering multiple different types of actors performing multiple different types of actions.

Consider the video as a 3-D image $f(x, y, t)$ and represent the video using the graph structure $G = (N, A)$. The node set $N = (n_1, \ldots, n_M)$, which represents M voxels (or M super-voxels), and the arc set $A(n)$ represents the voxel set in the neighborhood of a certain voxel n in N. Assume that the actor label set is denoted by X, and the action label set is denoted by Y.

Consider a set of random variables $\{x\}$ representing actors and a set of random variables $\{y\}$ representing actions. The actor-action understanding problem of interest can be viewed as a maximum a posteriori problem:

$$(x^*, y^*) = \underset{x,y}{\arg\max} P(x, y \mid M) \tag{8.21}$$

The general actor-action understanding problem includes three cases: single-label actor-action recognition, multi-label actor-action recognition, and actor-action semantic segmentation. They correspond to the three stages of successive granularity refinement.

8.6.1 Single-Label Actor-Action Recognition

Single-label actor-action recognition is the coarsest-grained case, which corresponds to the general action recognition problem. Here x and y are both scalars, and Equation (8.21) represents that given a video, a single action y is initiated by a single actor x. There are three models available at this time:

(1) Naïve Bayes model

In the **naïve Bayes model**, it is assumed that the actor and the action are independent of each other, that is, any actor can initiate any action. At this point, a set of classifiers need to be trained in the action space to classify different actions. This is the simplest method, but does not emphasize the existence of actor-action tuples, that is, some actors may not initiate all actions, or some actors can only initiate certain actions. In this way, when there are many different actors and different actions, sometimes unreasonable combinations (such as people can fly, birds can swim, etc.) occur when using the naive Bayes model.

(2) Joint product space model

The **joint product space model** utilizes the actor space X and the action space Y to generate a new label space Z. Here, the product relationship is used: $Z = X \times Y$. In the joint product space, a classifier can be learned directly for each actor-action tuple. Obviously, this method emphasizes the existence of actor-action tuples, which can eliminate the appearance of unreasonable combinations; and it is possible to use more cross-actor-action features to learn more discriminative classifiers. However, this approach may not take advantage of commonalities across different actors or different actions, such as steps and arm swings for both adults and children to walk.

(3) Three-level model

The three-level model unifies the naive Bayes model and the joint product space model. It simultaneously learns a classifier in actor space X, action space Y and joint actor-action space Z. At inference time, it infers Bayesian terms and joint product space terms separately, and then combines them linearly to get the final result. It not only models actor-action intersection, but also models different actions initiated by the same actor and the same action initiated by different actors.

8.6.2 Multi-Label Actor-Action Recognition

In practice, many videos have multiple actors and/or initiate multiple actions, which is the case for multi-labels. At this point, both $\boldsymbol{x}$ and $\boldsymbol{y}$ are binary vectors with dimensions $|X|$ and $|Y|$. The value of $\boldsymbol{x}_i$ is 1 if the i-th actor type exists in the video, and 0 otherwise. Similarly, the value of $\boldsymbol{y}_j$ is 1 if the j-th action type exists in the video, and 0 otherwise. This generalized definition does not confine specific elements in $\boldsymbol{x}$ to specific elements in $\boldsymbol{y}$. This facilitates an independent comparison of the multi-label performance of actors and actions with the multi-label performance of actor-action tuples.

For example, to study the situation where multiple actors initiate multiple actions, a corresponding video database has been constructed (Xu et al. 2015). This database is called the **actor-action database** (A2D). Among them, a total of seven actor categories are considered: adults, babies, cats, dogs, birds, cars, balls; and a total of nine action categories are considered: walking, running, jumping, rolling, climbing, crawling, flying, eating, and none (not the first eight categories of action). The main actor includes both articulated, such as adults, babies, cats, dogs, birds, and rigid bodies, such as cars, balls. Many actors can initiate the same action, but no actor can initiate all of them. So, while there are 63 combinations of them, some of them are unreasonable (or hardly ever), resulting in a total of 43 reasonable actor-action tuples. Using the text of these 43 actor-action tuples, 3782 video clips were collected on YouTube, ranging in length from 24 to 332 frames (136 frames per clip on average). The number of video clips corresponding to each actor-action tuple is shown in Table 8.1. The blanks in the table correspond to unreasonable actor-action tuples, so no video is collected. It can be seen from Table 8.1 that the number of video clips corresponding to each actor-action tuple is about a hundred.

Among these 3782 video clips, the number of video clips containing different numbers ($1 \sim 5$) of actors, the number of video clips containing different numbers ($1 \sim 5$) of actions, and the number of video clips containing different numbers of actors-actions, respectively, are shown in Table 8.2. It can be seen from Table 8.2 that in more than one-third of the video clips, the number of actors or actions is greater than 1 (the last four columns of the bottom row in the table, including one actor initiated more than two actions or more than two actors initiated one action).

TABLE 8.1 Number of Video Clips Corresponding to Actor-action Labels in the Database

	Walk	Run	Jump	Roll	Climb	Crawl	Fly	Eat	None
Adult	282	175	174	105	101	105		105	761
Baby	113			107	104	106			36
Cat	113	99	105	103	106			110	53
Dog	176	110	104	104		109		107	46
Bird	112		107	107	99		106	105	26
Car		120	107	104			102		99
Ball			105	117			109		87

TABLE 8.2 Number of Video Clips Corresponding to Actor, Action, and Actor-action Labels in the Database

	1	2	3	4	5
Actor	2794	936	49	3	0
Action	2639	1037	99	6	1
Actor-Action	2503	1051	194	31	3

For the case of **multi-label actor-action recognition**, three classifiers can still be considered, similar to single-label actor-action recognition: a multi-label actor-action classifier using naive Bayes model; a multi-label actor-action classifier in the joint product space; and an actor-action classifier based on a three-level model that combines the first two classifiers.

Multi-label actor-action recognition can be viewed as a retrieval problem. Experiments on the previously introduced database (with 3036 clips as training set and 746 clips as test set, with basically similar ratios for various combinations) show that the multi-label actor-action classifier in the joint product space performs better than naive Bayes, and the performance of the multi-label actor-action classifier based on the three-level model can still be improved somewhat (Xu et al. 2015).

8.6.3 Actor-Action Semantic Segmentation

Actor-action semantic segmentation is the most fine-grained case of action behavior understanding, and it also encompasses other coarser-grained problems such as detection and localization. Here, the task is to find labels for the actor-action on each voxel throughout the video. Still define two sets of random variables $\{x\}$ and $\{y\}$ whose dimensions will be determined by the number of voxels or super-voxels, and $x_i \in X$ and $y_j \in Y$. The objective function of Equation (8.21) is unchanged, but the way in which the graph model of $P(x, y|M)$ is implemented requires radically different assumptions about the relationship between actor and action variables.

This relationship is discussed in detail below. We first introduce the method based on the naive Bayes model, which handles the labels of the two classes separately. We then introduce a method based on the joint product space model, which utilizes the tuple $[x, y]$ to jointly consider the actor and action. Next consider a two-level model that considers the association of actor and action variables. Finally, a three-level model is introduced, which considers both intra-category linkages as well as linkages between categories.

(1) Naive Bayes model

Similar to the case in single-label actor-action recognition, the naive Bayes model can be represented as

$$\begin{aligned} P(x,y \mid M) = P(x \mid M)P(y \mid M) &= \prod_{i \in M} P(x_i)P(y_i) \prod_{i \in M} \prod_{j \in A(i)} P(x_i, x_j)P(y_i, y_j) \\ &\propto \prod_{i \in M} q_i(x_i) r_i(y_i) \prod_{i \in M} \prod_{j \in A(i)} q_{ij}(x_i, x_j) r_{ij}(y_i, y_j) \end{aligned} \tag{8.22}$$

Among them, q_i and r_i encode the potential functions defined in the actor and action models, respectively, and q_{ij} and r_{ij} encode the potential functions in the actor node set and the action node set, respectively.

Now train the classifier $\{f_c|c \in X\}$ on the actor and use the features on the action set to train the classifier $\{g_c|c \in Y\}$. The paired edge potential functions have the form of the following contrast-sensitive Potts model:

$$q_{ij} = \begin{cases} 1 & x_i = x_j \\ \exp\left[-k / \left(1 + \chi_{ij}^2\right)\right] & \text{Otherwise} \end{cases} \tag{8.23}$$

$$r_{ij} = \begin{cases} 1 & x_i = x_j \\ \exp\left[-k / \left(1 + \chi_{ij}^2\right)\right] & \text{Otherwise} \end{cases} \tag{8.24}$$

where χ^2_{ij} is the χ^2 distance between the feature histograms of nodes i and j, and k is the parameter to be learned from the training data. Actor-action semantic segmentation can be obtained by solving these two flat conditional random fields independently.

(2) Joint product space model

Consider a new set of random variables $\boldsymbol{z} = \{z_1, \ldots, z_M\}$, which are also defined over all super-voxels in a video and pick labels from the actor and action product space $Z = X \times Y$. This way jointly captures the actor-action tuple as the only element, but cannot model the common factor of actor and action in different tuples (the model introduced below will solve this problem). Thus, it has a single-layer graph model:

$$\begin{aligned} P(x, y \mid M) = P(z \mid M) = \prod_{i \in M} P(z_i) \prod_{i \in M} \prod_{j \in A(i)} P(z_i, z_j) \\ \propto \prod_{i \in M} s_i(z_i) \prod_{i \in M} \prod_{j \in A(i)} s_{ij}(z_i, z_j) = \prod_{i \in M} s_i([x_i, y_i]) \prod_{i \in M} \prod_{j \in A(i)} s_{ij}([x_i, y_i], [x_j, y_j]) \end{aligned} \tag{8.25}$$

where s_i is the potential function of the joint actor-action product space label, and s_{ij} is the internal node potential function between the two nodes of the corresponding tuple $[x, y]$. Specifically, s_i contains the classification score obtained for node i with the trained actor-action classifier $\{h_c|c \in Z\}$, and s_{ij} has the same form as in Equation (8.23) or Equation (8.24). For illustration, see Figure 8.12(a) and Figure 8.12(b).

(3) Two-level model

Given the actor node x and the action node y, the two-level model uses edges that encode the potential function of the tuple to connect each random variable pair $\{(x_i, y_i)^M{}_{i=1}\}$, and directly obtains the covariance cross-actor and action labels.

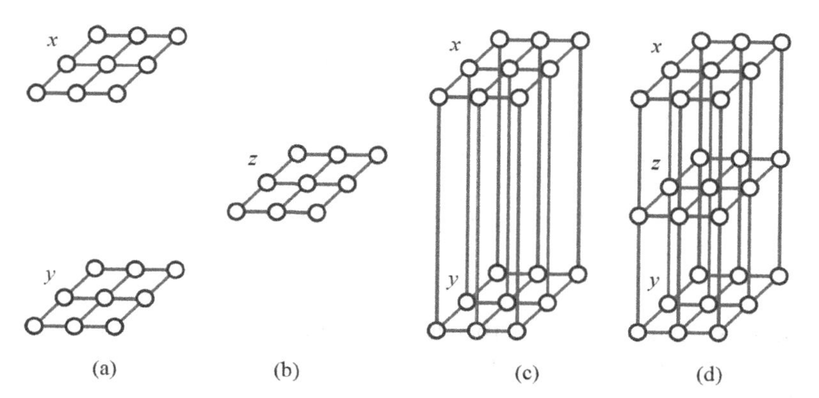

FIGURE 8.12 Schematic representation of different graph models.

$$
\begin{aligned}
P(x,y \mid M) &= \prod_{i\in M} P(x_i,y_i) \prod_{i\in M}\prod_{j\in A(i)} P(x_i,x_j)P(y_i,y_j) \\
&\propto \prod_{i\in M} q_i(x_i)r_i(y_i)t_i(x_i,y_i) \prod_{i\in M}\prod_{j\in A(i)} q_{ij}(x_i,x_j)r_{ij}(y_i,y_j)
\end{aligned}
\tag{8.26}
$$

where $t_i(x_i, y_i)$ is the potential function learned for the labels of the entire product space, which can be obtained as s_i in Figure 8.12; see Figure 8.12(c). Here, the connecting edges across layers are added.

(4) Three-level model

The naive Bayes model represented by Equation (8.22) does not consider the connection between the actor variable x and the action variable y. The joint product space model of Equation (8.25) combines features across actors and actions as well as interaction features within the neighborhood of an actor-action node. The two-layer model of Equation (8.26) adds actor-action interactions between separate actor nodes and action nodes, but does not account for the spatial-temporal variation of these interactions.

A three-level model is given below that explicitly models the spatial-temporal variation of Figure 8.12(d). It combines the nodes of the joint product space with all the actor nodes and action nodes:

$$
\begin{aligned}
P(x,y,z \mid M) &= P(x \mid M)P(y \mid M)P(z \mid M)\prod_{i\in M} P(x_i,z_i)P(y_i,z_i) \\
&\propto \prod_{i\in M} q_i(x_i)r_i(y_i)s_i(z_i)u_i(x_i,z_i)v_i(y_i,z_i) \\
&\quad \prod_{i\in M}\prod_{j\in A(i)} q_{ij}(x_i,x_j)r_{ij}(y_i,y_j)s_{ij}(z_i,z_j)
\end{aligned}
\tag{8.27}
$$

where

$$u_i(x_i, z_i) = \begin{cases} w(y_i' \mid x_i) & \text{For } z_i = [x_i', y_i'] \text{ has } x_i = x_i' \\ 0 & \text{Otherwise} \end{cases} \tag{8.28}$$

$$v_i(y_i, z_i) = \begin{cases} w(x_i' \mid y_i) & \text{For } z_i = [x_i', y_i'] \text{ has } y_i = y_i' \\ 0 & \text{Otherwise} \end{cases} \tag{8.29}$$

where $w(y_i'|x_i)$ and $w(x_i'|y_i)$ are the classification scores of conditional classifiers specially trained for this three-level model.

These conditional classifiers are the main reason for the performance improvement. Based on the condition of actor type, action oriented and separated classifier can take advantage of the unique characteristics of actor-action tuple. For example, when training a conditional classifier on the action "eat" given an actor adult, all other actions of the actor adult can be treated as negative training samples. In this way, this three-level model takes into account all connections in each actor space and each action space, as well as in the joint product space. In other words, the first three basic models are all special cases of the three-level model. It can be shown that maximizing (x^*, y^*, z^*) of Equation (8.27) also maximizes Equation (8.21) (Xu et al. 2015).

8.7 SOME RECENT DEVELOPMENTS AND FURTHER RESEARCH

In the following sections, technical developments and promising research directions from the last few years are briefly overviewed.

8.7.1 Behavior Recognition Using Joints

Behavior recognition is based on action recognition and activity recognition, and is considered at a higher level. Compared with RGB data and depth data, skeleton joint point data corresponds to higher-level features of the human body and is not easily affected by object appearance. In addition, it can better avoid the noise impact caused by background occlusion, illumination change and viewing angle change. At the same time, it is also effective in calculation and storage.

Joint point data is usually represented as the coordinate vector of a series of points (4-D points in spatial-temporal space). That is, a joint point is represented by a 5-D function $J(l, x, y, z, t)$, where l is the label, (x, y, z) represent the space coordinates, t represents the time coordinate. In different deep learning networks and algorithms, joint point data are often represented in different forms (such as pseudo image, vector sequence and topology graph).

The research based on deep learning method mainly involves three aspects: the data processing method, network architecture and data fusion method (Liu et al. 2021). At present, there are three main commonly used network architectures: convolutional neural network (CNN), recurrent neural network (RNN) and graph convolution network (GCN). The representation methods of their corresponding joint point data are pseudo image, vector sequence and topological graph, respectively.

8.7.1.1 Using CNN as Backbone

CNN is an effective network architecture for extracting human behavior features, which can be recognized by local convolution filter or kernel learned from data. Behavior recognition methods based on CNN encode the temporal and spatial position coordinates of the joint into rows and columns, respectively, and then feed the data to CNN for recognition. Generally, in order to facilitate the use of a CNN-based network for feature extraction, the joint point data will be transposed and mapped into the image format, in which the row represents different joints l, the column represents different time t, and the 3D spatial coordinates (x, y, z) are regarded as three channels of the image, and then the convolution operation is carried out.

Some recent techniques treating the problem that the accuracy of complex interactive behavior recognition is not high enough are briefly described in Table 8.3.

8.7.1.2 Using RNN as Backbone

RNN can process sequence data of variable length. The behavior recognition methods based on RNN first represent the joint point data as a vector sequence that contains the position information of all joint points in a time (state) sequence; then the vector sequence is sent into the behavior recognition network with RNN as the backbone. The long/short-term memory (LSTM) model is a variant RNN. Because its cell state can determine which time states should be left and which should be forgotten, it has greater advantages in processing timing data such as joint point video.

Some recent techniques taking LSTM as network structure for behavior recognition are briefly described in Table 8.4.

TABLE 8.3 Some Recent Techniques Using CNN

Techniques	Feature and Description	Relative Merits
(Ji et al. 2019)	Designing a dual stream fusion of RGB information and joint point information for improving accuracy. The key frame is extracted before the RGB video information is sent to the CNN to reduce training time.	Fast training time.
(Yan et al. 2019)	Using pose-based mode for behavior recognition. The CNN framework includes three sematic modules: CNN for spatial posture, CNN for timing posture, and CNN for action. It can be used as another semantic stream complementary to RGB stream and optical flow stream.	The network structure is simple, but the accuracy is normal.
(Caetano et al. 2019a)	Using a skeleton image for 3D behavior recognition based on tree structure and reference joint.	Training efficiency is not high.
(Caetano et al. 2019b)	The time dynamics are encoded by calculating the motion amplitude and direction values of skeleton joints. Using different time scales to calculate the motion values of joints to filter the noise.	It can effectively filter the motion noise in the data.
(Li et al. 2019)	Re-encoding the skeleton joint information by set algebra.	Fast speed but low accuracy.

TABLE 8.4 Some Recent Techniques Using LSTM

Techniques	Feature and Description	Relative Merits
(Liu et al. 2017a)	Adding a multi-mode feature fusion strategy to the trust gate of spatio-temporal LSTM.	Recognition accuracy is improved but training efficiency is reduced.
(Liu et al. 2017b)	Using a global context-aware attention LSTM network (GCA-LSTM) that is mainly composed of two layers of LSTM. The first layer generates global background information, and the second layer adds attention mechanism to better focus on the key joint points of each frame.	It can better focus the key joint points in each frame.
(Liu et al. 2018)	Extending GCA-LSTM, adding coarse-grained and fine-grained attention mechanisms.	Recognition accuracy is improved but training efficiency is reduced.
(Zheng et al. 2019)	A dual flow attention-cycle LSTM network is proposed. The cycle relation network learns the spatial features in a single skeleton and the multi-layer LSTM learns the temporal features in the skeleton sequence.	Make full use of joint information to improve recognition accuracy.

8.7.1.3 Using GCN as Backbone

The set of human skeleton joints can be regarded as a topological graph. A topological graph is a kind of data with non-Euclidean structure, in which the number of adjacent vertices of each node may be different, so it is difficult to calculate the convolution with a fixed-size convolution kernel, hence impossible to process it directly with CNN. The graph convolution network (GCN) can directly process the topological graph. Here, it is only needed to represent the joint point data as a topological graph, in which the vertices in the spatial domain are connected by the spatial edge line, the corresponding joints between adjacent frames in the time domain are connected by the temporal edge line, and the spatial coordinate vector is used as the attribute feature of each joint point.

Some recent techniques taking GCN as network structure for behavior recognition are briefly described in Table 8.5.

8.7.1.4 Using Hybrid Network as Backbone

The research on behavior recognition based on joints can also use a hybrid network that makes full use of the feature extraction capabilities of CNN and graph convolutional networks in the spatial domain and the advantages of RNN in time-series classification. In this case, the original joint point data should be represented by the corresponding data formats according to the needs of different hybrid networks.

Some recent techniques using hybrid networks are briefly described in Table 8.6.

8.7.2 Detection of Video Anomalous Events

Video anomaly detection (VAD, also called video anomaly detection and localization) is designed to detect anomalous events that occur in videos and locate where they occur in the video.

Video anomaly event detection can be divided into two parts: the extraction of video features and the establishment of the anomaly event detection model. Commonly used

TABLE 8.5 Some Recent Techniques Using GCN

Techniques	Feature and description	Relative merits
(Li et al. 2019a)	Designing an encoder decoder method to capture the implicit joint correlation and obtain the physical structure link between joints using the high-order polynomial of the adjacency matrix.	High model complexity.
(Peng et al. 2019)	Using the neural architecture search to construct the graph convolution network, in which the cross-entropy evolution strategy is combined with the importance of hybrid methods to improve sampling efficiency and storage efficiency.	High sampling and storage efficiency.
(Wu et al. 2019)	Introducing the spatial residual layer and dense connection block enhancement into the spatio-temporal graph convolution network to improve the processing efficiency of spatio-temporal information.	Easy to combine with mainstream spatio-temporal graph convolution method.
(Shi et al. 2019)	Improving the two-stream adaptive graph convolution network by representing the skeleton data as a directed acyclic graph based on the motion dependence between natural human joints and bones.	High recognition accuracy.
(Li et al. 2019b)	Proposing a novel symbiotic graph convolution network to include not only the functional module of behavior recognition, but also the action prediction module.	The two modules promote each other in improving the accuracy of behavior recognition and action prediction.
(Yang et al. 2020)	Using a pseudo graph convolution network with time and channel attention mechanism, not only can the key frames be extracted, but also the input frames containing more features can be screened.	Key frames can be extracted, but some key information may be omitted.

video features are hand-designed features and features extracted by deep models. Video anomaly event detection models can be based on traditional probability inference or on deep learning.

8.7.2.1 Detection with Convolutional Auto-Encoder Block Learning

For complete representation and description of video events, multiple features may be required. The fusion of multiple features has stronger expressive power than a single feature. In the method based on convolutional auto-encoder and block learning (Li et al. 2021), both appearance feature and motion feature are used in combination.

The flow chart of video anomaly event detection with convolutional auto-encoder is shown in Figure 8.13. The video frames are first divided into non-overlapping small blocks. Then, the motion features (optical flow) representing the motion state and the appearance (histogram of gradient, HOG) features representing the existence of object are extracted.

For each optical flow feature and HOG feature of a certain block, an anomaly detection convolutional auto-encoder (AD-ConvAE) is set for training and testing, respectively. AD-ConvAE on a location block only pays attention to the crowd movement in this video location region, and the local features can be learned more effectively by using the block learning method. In the training process, the video only contains normal samples, and

TABLE 8.6 Some Recent Techniques Using Hybrid Networks

Techniques	Feature and Description	Relative Merits
CNN+LSTM (Zhang et al. 2019)	Designing a view-adaptive scheme that includes two view-adaptive neural networks. The view-adaptive recurrent network is composed of the main LSTM and a view-adaptive sub-network, and the joint point representation under the new viewpoint is sent to the main LSTM network to determine the behavior recognition. The view-adaptive convolutional network is composed of the main CNN and a view-adaptive sub-network, and the joint point representations under the new observation viewpoint are sent to the main CNN to determine the behavior category. Finally, the classification scores of the two parts of the network are fused.	Combined advantages of CNN for extracting behavioral features in the spatial domain, and of RNN for extracting behavioral features in the time domain. Less influence of different perspectives on the recognition results.
CNN+GCN (Hu et al. 2019)	Combining CNN and graph convolutional network not only considers the extraction of behavioral features in spatial and temporal domains, but also learns frequency patterns with the help of the residual frequency attention method.	Added learning of frequency.
GCN+LSTM (Si et al. 2019)	Using attention-enhanced graph convolutional LSTM network (AGC-LSTM), not only extract the behavioral features of the spatial and temporal domains, but also increase the time of the top AGC-LSTM layer by increasing the time receptive fields to enhance the ability for learning advanced features, thereby reducing computational cost.	Enhanced ability to learn advanced features and reduced computational cost.
GCN+LSTM (Gao et al. 2019)	Using a bidirectional attention graph convolutional network to learn spatio-temporal context information from human joint point data with focusing and diffusion mechanisms.	High recognition accuracy.
GCN+CNN (Zhang et al. 2020)	The semantics of the joints (frame index and joint type) are fed as part of the network input together with the position and velocity of the joints into the semantic-aware graph convolutional layer and the semantic-aware convolutional layer.	Reduced model complexity and improved recognition accuracy.

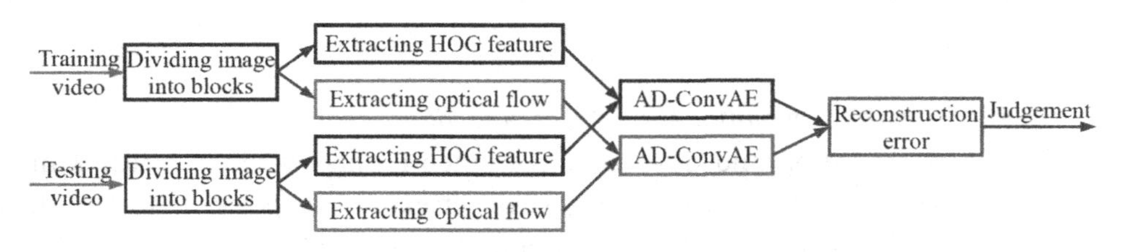

FIGURE 8.13 The flow chart of video anomaly event detection with convolutional auto-encoder.

AD-ConvAE learns the normal motion pattern of a certain region through the optical flow and HOG features of video frame blocks. During the testing process, the optical flow and HOG features of the block in the test video are put into AD-ConvAE for reconstruction, and the weighted reconstruction error is calculated according to the reconstruction error of the optical flow and the reconstruction error of the HOG feature. If the reconstruction error is large enough, it means that there is an abnormal event in the block. In this way, in addition to detecting anomalous events, the location of anomalous events is also completed.

The network structure of AD-ConvAE includes two parts: encoding and decoding. In the encoding part, multiple pairs of convolutional layers and pooling layers are used to obtain deep features. In the decoding part, multiple pairs of convolution operations and up-sampling operations are used to reconstruct the depth representation of the feature, and an image of the same size as the input image is output.

8.7.2.2 Detection Using One-Class Neural Network

One-class neural network (ONN) is an extension of one-class classifiers under the framework of deep learning. One-class SVM (OC-SVM) is a widely used unsupervised anomaly detection method. In fact, it is a special form of SVM that can learn a hyperplane that separates all data points in the kernel Hilbert space from the origin, and maximizes the distance from the hyperplane to the origin. In the OC-SVM model, all data except the origin are labeled as positive samples, and the origin is labeled as a negative sample.

ONN can be regarded as a neural network structure designed using the equivalent loss function of OC-SVM. In ONN, the data representation in the hidden layer is directly driven by the ONN, so it can be designed for the task of anomaly detection, whereby the two stages of feature extraction and anomaly detection are combined for joint optimization. ONN combines the layer-by-layer data representation ability of the auto-encoder with the one-class classification ability, which can distinguish all normal samples from abnormal samples.

The method of video anomaly event detection (Jiang and Li 2021) trains ONN separately on the local region blocks of the same size of video frame and optical flow graph to detect appearance anomaly and motion anomaly, and fuses the two to determine the final detection result. The flow chart of video anomaly event detection with ONN is shown in Figure 8.14. In the training phase, two auto-encoding networks are learned with the help of the RGB images and optical flow images of the training samples, respectively, and the encoder layer of the pre-trained auto-encoder and the ONN network are combined to jointly optimize the parameters and learn the anomaly detection model; in the testing phase, the RGB image and optical flow image of the given test region are input into the appearance anomaly detection model and the motion anomaly detection model, respectively, the output scores are fused, and the detection threshold is set to determine whether the region is abnormal.

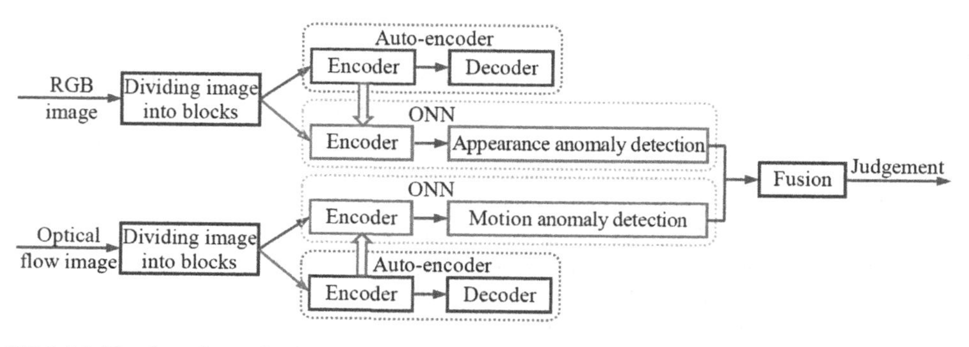

FIGURE 8.14 The flow chart of video anomaly event detection with ONN.

REFERENCES

Blank B, Gorelick L, Shechtman E, et al. 2005. Actions as space-time shapes. *Proceedings of the ICCV*, 2: 1395–1402.

Caetano, C., F. Brémond and W.R. Schwartz. 2019a. Skeleton image representation for 3D action recognition based on tree structure and reference joints. *SIBGRAPI Conference on Graphics, Patterns and Images*, 16–23.

Caetano, C., J. Sena, F. Brémond, et al. 2019b. SkeleMotion: A new representation of skeleton joint sequences based on motion information for 3D action recognition. *IEEE International Conference on Advanced Video and Signal Based Surveillance*, 1–8.

Gao, J.L., T. He, X. Zhou, et al. 2019. Focusing and diffusion: Bidirectional attentive graph convolutional networks for skeleton-based action recognition. *arXiv preprint* arXiv: 1912.11521.

Hu, G.Y., B. Cui and S.Yu. 2019. Skeleton-based action recognition with synchronous local and non-local spatiotemporal learning and frequency attention. *IEEE International Conference on Multimedia and Expo*, 1216–1221.

Ji, X.F., Qin, L.L. and Y.Y. Wang. 2019. Human interaction recognition based on RGB and skeleton data fusion model. *Journal of Computer Applications*, 39(11): 3349–3354.

Jiang, W.X. and G. Li. 2021. One-class neural network for video anomaly detection and localization. *Journal of Electronic Measurement and Instrumentation*, 35(7): 60–65.

Jia, H.X. and Y.-J. Zhang. 2007. A survey of computer vision based pedestrian detection for driver assistance systems. *Acta Automatica Sinica*, 33(1): 84–90.

Jia, H.X. and Y.-J. Zhang. 2009. Automatic people counting based on machine learning in intelligent video surveillance. *Video Engineering*, (4): 78–81.

Laptev, I. 2005. On space-time interest points. *International Journal of Computer Vision*, 64(2/3): 107–123.

Li, M.S., S.H. Chen, X. Chen, et al. 2019a. Actional-structural graph convolutional networks for skeleton-based action recognition. *IEEE Conference on Computer Vision and Pattern Recognition*, 3595–3603.

Li, M.S., S.H. Chen, X. Chen, et al. 2019b. Symbiotic graph neural networks for 3D skeleton-based human action recognition and motion prediction. *arXiv preprint* arXiv: 1910.02212.

Li, X.L., G.L. Ji and B. Zhao. 2021. Convolutional auto-encoder patch learning based video anomaly event detection and localization. *Journal of Data Acquisition and Processing*, 36(3): 489–497.

Li, Y.S., R.J. Xia, X. Liu, et al. 2019. Learning shape motion representations from geometric algebra spatiotemporal model for skeleton-based action recognition. *IEEE International Conference on Multimedia and Expo*, 1066–1071.

Liu, J., A. Shahroudy, D. Xu, et al. 2017a. Skeleton-based action recognition using spatio-temporal LSTM network with trust gates. *IEEE Transactions on Pattern Analysis and Machine Intelligence*, 40(12): 3007–3021.

Liu, J., G. Wang, P. Hu, et al. 2017b. Global context-ware attention LSTM networks for 3D action recognition. *IEEE Conference on Computer Vision and Pattern Recognition*, 1647–1656.

Liu, J., G. Wang, L.Y. Duan, et al. 2018. Skeleton-based human action recognition with global context-aware attention LSTM networks. *IEEE Transactions on Image Processing*, 27(4): 1586–1599.

Liu, Y., P.P. Xue and H. Li. 2021. A review of action recognition using joints based on deep learning. *Journal of Electronics & Information Technology*, 43(6): 1789–1802.

Moeslund, T.B., A. Hilton and V. Krüger. 2006. A survey of advances in vision-based human motion capture and analysis. *Computer Vision and Image Understanding*, 104: 90–126.

Morris, B.T. and M.M. Trivedi. 2008. A survey of vision-based trajectory learning and analysis for surveillance. *IEEE Transactions on Circuits and Systems for Video Technology*, 18(8): 1114–1127.

Makris, D. and T. Ellis. 2005. Learning semantic scene models from observing activity in visual surveillance, *IEEE Transactions on Systems Man & Cybernetics Part B Cybernetics*, 35(3): 397–408.

Peng, W., X.P. Hong, H.Y. Chen, et al. 2019. Learning graph convolutional network for skeleton-based human action recognition by neural searching. *arXiv preprint* arXiv: 1911.04131.

Poppe, R. 2010. A survey on vision-based human action recognition. *Image and Vision Computing*, 28: 976–990

Shi, L., Y.F. Zhang, J. Cheng, et al. 2019. Skeleton-based action recognition with directed graph neural networks. *IEEE Conference on Computer Vision and Pattern Recognition*, 7912–7921.

Si, C.Y., W.T. Chen, W. Wang, et al. 2019. An attention enhanced graph convolutional LSTM network for skeleton-based action recognition. *IEEE Conference on Computer Vision and Pattern Recognition*, 1227–1236.

Turaga, P., R. Chellappa, V.S. Subrahmanian, et al. 2008. Machine recognition of human activities: A survey. *IEEE Transactions on Circuits and Systems for Video Technology*, 18(11): 1473–1488.

Wu, C., X.J. Wu and J. Kittler. 2019. Spatial residual layer and dense connection block enhanced spatial temporal graph convolutional network for skeleton-based action recognition. *IEEE International Conference on Computer Vision Workshop*, 1–5.

Xu, C.L., S.H. Hsieh, C.M. Xiong, et al. 2015. Can humans fly? Action understanding with multiple classes of actors. *Proceedings of the CVPR*, 2264–2273.

Yan, A., Y.L. Wang, Z.F. Li, et al. 2019. PA3D: Pose-action 3D machine for video recognition. *IEEE Conference on Computer Vision and Pattern Recognition*, 7922–7931.

Yang, H.Y., Y.Z. Gu, J.C. Zhu, et al. 2020. PGCNTCA: Pseudo graph convolutional network with temporal and channel-wise attention for skeleton-based action recognition. *IEEE Access*, 8: 10040–10047.

Zhang, P.F., C.L. Lan, J.L. Xing, et al. 2019. View adaptive neural networks for high performance skeleton-based human action recognition. *IEEE Transactions on Pattern Analysis and Machine Intelligence*, 41(8): 1963–1978.

Zhang, P.F., C.L. Lan, W.J. Zeng, et al. 2020. Semantics-guided neural networks for efficient skeleton-based human action recognition. *IEEE Conference on Computer Vision and Pattern Recognition*, 1109–1118.

Zhang, Y.-J. 2018. The understanding of spatial-temporal behaviors. *Encyclopedia of Information Science and Technology, 4th Ed.*, Chapter 115 (1344-1354).

Zheng, W., L. Li, Z.X. Zhang, et al. 2019. Relational network for skeleton-based action recognition. *IEEE International Conference on Multimedia and Expo*, 826–831.

Index

M

N

O

P

Q

R

S

T

U

V

W

Z